Mathematical Logic
An Introduction to Model Theory

MATHEMATICAL CONCEPTS AND METHODS IN SCIENCE AND ENGINEERING

Series Editor: **Angelo Miele**
Mechanical Engineering and Mathematical Sciences
Rice University

Volume 1	**INTRODUCTION TO VECTORS AND TENSORS** Volume 1: Linear and Multilinear Algebra *Ray M. Bowen and C.-C. Wang*
Volume 2	**INTRODUCTION TO VECTORS AND TENSORS** Volume 2: Vector and Tensor Analysis *Ray M. Bowen and C.-C. Wang*
Volume 3	**MULTICRITERIA DECISION MAKING AND DIFFERENTIAL GAMES** *Edited by George Leitmann*
Volume 4	**ANALYTICAL DYNAMICS OF DISCRETE SYSTEMS** *Reinhardt M. Rosenberg*
Volume 5	**TOPOLOGY AND MAPS** *Taqdir Husain*
Volume 6	**REAL AND FUNCTIONAL ANALYSIS** *A. Mukherjea and K. Pothoven*
Volume 7	**PRINCIPLES OF OPTIMAL CONTROL THEORY** *R. V. Gamkrelidze*
Volume 8	**INTRODUCTION TO THE LAPLACE TRANSFORM** *Peter K. F. Kuhfittig*
Volume 9	**MATHEMATICAL LOGIC** An Introduction to Model Theory *A. H. Lightstone*
Volume 10	**SINGULAR OPTIMAL CONTROLS** *R. Gabasov and F. M. Kirillova*
Volume 11	**INTEGRAL TRANSFORMS IN SCIENCE AND ENGINEERING** *Kurt Bernardo Wolf*
Volume 12	**APPLIED MATHEMATICS** An Intellectual Orientation *Francis J. Murray*

Mathematical Logic

An Introduction to Model Theory

A.H. Lightstone

Queen's University

Kingston, Ontario, Canada

Edited by

H.B. Enderton

University of California

Los Angeles, California

PLENUM PRESS · NEW YORK AND LONDON

Library of Congress Cataloging in Publication Data

Lightstone, A H
Mathematical logic.

(Mathematical concepts and methods in science and engineering; 9)
Bibliography: p.
Includes index.
1. Logic, Symbolic and mathematical. 2. Model theory. I. Title.
QA9.L54 511'.3 77-17838
ISBN 0-306-30894-0

A Division of Plenum Publishing Corporation
227 West 17th Street, New York, N.Y. 10011

Printed in the United States of America

To the memory of Abraham Robinson

Foreword

Before his death in March, 1976, A. H. Lightstone delivered the manuscript for this book to Plenum Press. Because he died before the editorial work on the manuscript was completed, I agreed (in the fall of 1976) to serve as a surrogate author and to see the project through to completion.

I have changed the manuscript as little as possible, altering certain passages to correct oversights. But the alterations are minor; this is Lightstone's book.

H. B. Enderton

Preface

This is a treatment of the predicate calculus in a form that serves as a foundation for nonstandard analysis. Classically, the predicates and variables of the predicate calculus are kept distinct, inasmuch as no variable is also a predicate; moreover, each predicate is assigned an *order*, a unique natural number that indicates the length of each tuple to which the predicate can be prefixed. These restrictions are dropped here, in order to develop a flexible, expressive language capable of exploiting the potential of nonstandard analysis.

To assist the reader in grasping the basic ideas of logic, we begin in Part I by presenting the *propositional calculus* and *statement systems*. This provides a relatively simple setting in which to grapple with the sometimes foreign ideas of mathematical logic. These ideas are repeated in Part II, where the *predicate calculus* and *semantical systems* are studied.

Finally, in Part III, we present some applications. There is a substantial discussion of *nonstandard analysis*, a treatment of the *Löwenheim–Skolem Theorem*, a discussion of *axiomatic set theory* that utilizes semantical systems, and an account of *complete theories*. The presentation of complete theories includes Vaught's test, but is mainly devoted to an exposition of Robinson's notion of *model completeness* and its connection with *completeness*. Chapter 16 is taken from the author's *The Axiomatic Method: An Introduction to Mathematical Logic*,* with only a few minor changes.

This book contains many ideas due to Abraham Robinson, the father of nonstandard analysis. The author is also indebted to Prof. Ernest Heighton for several stimulating conversations and many valuable suggestions.

A. H. Lightstone

* Prentice-Hall, Englewood Cliffs (1964).

Contents

Introduction

A theory of deduction utilizes various ideas of logic that may appear strange, even foreign, to mathematics students with little background in logic. The main concern of this book is to develop the important theory of deduction known as the *predicate calculus*. In an effort to overcome the strangeness of the logical ideas and methods involved, we shall first present the theory of deduction based on the connectives $\neg$ (not) and $\vee$ (or). This theory, known as the *propositional calculus*, characterizes the conclusions, or consequences, of a given set of assumptions, and so provides us with the formal side of arguments. The question of the validity of a given argument of this sort is easy to solve by the truth-table method, and so is really trivial. Therefore, in studying the accompanying theory of deduction we are able to concentrate on the formal apparatus and methods of a theory of deduction, without the complications owing to the subject matter under investigation. In short, the propositional calculus is a convenient device for making clear the nature of a theory of deduction.

In rough outline, the steps in setting up a theory of deduction are as follows. First, the propositions (statements) of a language are characterized. This is achieved by actually creating a specific formal language possessing its own alphabet and rules of grammar; in fact, the sentences of the formal language are effectively spelled out by suitably chosen rules of grammar. Finally, the notion of *truth* within this specialized and highly artificial language is characterized in terms of the concept of a "proof."

To explain in a little more detail, a theory of deduction is based on an alphabet, which consists of symbols of several sorts. There are the *connectives* and *parentheses*, which are used to construct compound propositions from given propositions of the language; and there are symbols that yield the basic, atomic propositions of the language. As a first step toward characterizing the propositions of the language, it is convenient to introduce the notion of an *expression* of the language. An expression is any finite string of symbols of the language, with repetitions allowed. Certain sequences of propositions are recognized as "proofs," and the last proposition of each proof is said to be *provable*. Using the notion of a provable proposition, we can characterize in a purely formal manner the consequences of a set of propositions.

As we have suggested, we shall present two theories of deduction in this book. The first of these, the propositional calculus, involves a language whose connectives are "not" and the "inclusive or." We shall motivate this theory of deduction, which is highly abstract, by considering *statement systems*, which are simple and concrete. In this setting, we can utilize the truth-table approach of symbolic logic.

The second theory of deduction that we develop in this book is a version of the classical predicate calculus. This involves a language whose connectives are "not," the "inclusive or," and "for each" (the universal quantifier). We shall motivate this theory of deduction by considering *semantical systems*, a much simpler notion.

PART I
Statement Systems and Propositional Calculus

I

Statement Systems

1.1. Statement Systems

A statement system consists of a given set of statements, each of which is assigned a truth-value *true* or *false*. Each statement system has its own language, which is built up from its initial statements by connecting them with the logical connectives *not* (denoted by $\neg$) and the *inclusive or* (denoted by $\vee$). Applying the truth-table definitions of these connectives, we easily compute a unique truth-value for each of the compound statements in the language of the statement system. We shall go into this in Section 1.2.

Since a statement system involves a set of objects, called statements, and since each of these objects has a unique truth-value, we shall identify a statement system with a map whose range is included in {true, false}. The domain of this map is the set of initial statements of the statement system. By a statement system, then, we mean any map $\sum$ with a nonempty domain and with range included in {true, false}. We regard the members of dom $\sum$ as *statements* (so we use this term in a generalized sense); each of these objects is assigned a truth-value by the map $\sum$. Thus, a statement system $\sum$ consists of objects, called statements, each of which is assigned a truth-value by the map $\sum$.

Of course, a statement system may involve actual statements, in the usual sense of the term.

Example 1. Let $\sum$ be the map with range {true, false} and domain

{grass is green, oil is cheap, logic is easy, Washington is the capital of the United States} such that $\sum$ associates *true* with "grass is green" and "Washington is the capital of the United States," and $\sum$ associates *false* with "oil is cheap" and "logic is easy." Then $\sum$ is a statement system.

Note that we obtain different statement systems from a given set of statements by assigning truth-values in different ways to the statements of the system. To illustrate, let us change the truth-values assigned to the statements of the statement system of Example 1.

Example 2. Let $\sum$ be the map with domain

{grass is green, oil is cheap, logic is easy, Washington is the capital of the United States}

such that $\sum$ associates *false* with each statement in its domain. Then $\sum$ is a statement system.

We point out that the statement system of Example 2 is different from the statement system of Example 1. By a statement system we mean a map whose range is included in {true, false} and whose domain is non-empty; the maps of the two examples are certainly different.

The members of the domain of a statement system (i.e., its statements) need not be actual statements. Here is an example.

Example 3. Let $\sum$ be the map of $\{S, T, U\}$ into {true, false} that associates *true* with T and associates *false* with S and U. Then $\sum$ is a statement system.

1.2. Language of a Statement System

Each statement system $\sum$ has its own language, which is built up from the statements in dom $\sum$ by means of the logical connectives *not* ($\neg$) and the *inclusive or* ($\vee$). Each grammatical expression of this language is said to be a *statement well-formed formula*, or *swff* for short. Thus, each swff of $\sum$ consists of a finite number of objects in dom $\sum$ linked by the connectives $\neg$ and $\vee$ (and parentheses).

More formally, we say that the expression (S) is an *atomic* swff of $\sum$ for each $S \in \text{dom} \sum$. The remaining swffs of $\sum$ are defined as follows. Let A and B be any swffs of $\sum$; then we say that $(\neg A)$ and $(A \vee B)$ are swffs of $\sum$. This is subject to the requirement that each swff of $\sum$ involves only a finite number of instances of connectives.

We obtain the truth-value of each atomic swff of $\sum$ directly from the map $\sum$ itself. The truth-value of each compound swff of $\sum$ is obtained by considering the significance of the connectives *not* and the *inclusive or*. Bearing this in mind, we formulate the following definition.

Definition. (i) An atomic swff (S) is *true* for $\sum$ if $\sum$ associates "true" with S; (S) is *false* for $\sum$ if $\sum$ associates "false" with S. Here, $S \in \text{dom} \sum$.

(ii) $(\neg A)$ is *true* for $\sum$ if A is false for $\sum$; $(\neg A)$ is *false* for $\sum$ if A is true for $\sum$. Here, A is any swff of $\sum$.

(iii) $(A \vee B)$ is *true* for $\sum$ if A is true for $\sum$, B is true for $\sum$, or both A and B are true for $\sum$; $(A \vee B)$ is *false* for $\sum$ if A is false for $\sum$ and B is false for $\sum$. Here, A and B are any swffs of $\sum$.

We point out that this definition assigns a unique truth-value to each swff of $\sum$.

To illustrate, let $\sum$ be the statement system of Example 1, Section 1.1. Let

g = grass is green
o = oil is cheap
l = logic is easy
W = Washington is the capital of the United States

Then the following swffs of $\sum$ are each true for $\sum$:

$$(g), \quad (\neg(o)), \quad ((g) \vee (o)), \quad (\neg((\neg(W)) \vee (o)))$$

The following swffs of $\sum$ are each false for $\sum$:

$$(o), \quad (\neg(g)), \quad ((o) \vee (\neg(g)))$$

1.3. Names for Swffs

The purpose of the parentheses that appear in each swff is to avoid ambiguous statements. For example, "$\neg S \vee T$" could represent either $(\neg S) \vee T$ or $\neg(S \vee T)$. Moreover, parentheses impose a certain structure on swffs which we shall find very useful (see Section 2.2).

On the other hand, it is difficult to read a given swff if it involves many parentheses. We can obtain the best of both worlds by introducing conventions for omitting parentheses. This is achieved by introducing *names* for swffs; of course, we must avoid ambiguity, i.e., each of our names must name a unique swff.

We shall usually omit parentheses whenever this does not produce an ambiguous expression. For example, the outermost pair of parentheses of any swff and the pair of parentheses involved in each atomic swff can usually be suppressed without harm. For example, let S, $T \in \operatorname{dom} \sum$ and consider the swff

$$A = ((\neg(S)) \vee ((T) \vee (S)))$$

Under our agreement, "$(\neg S) \vee (T \vee S)$" is a name for A.

Short names for certain swffs are obtained by introducing the logical connectives $\wedge$ (*and*), $\rightarrow$ (*if . . . then*), and $\leftrightarrow$ (*if and only if*). Our agreement is that for any swffs A and B,

$$\begin{array}{lll} (A \wedge B) & \text{is a name for} & (\neg((\neg A) \vee (\neg B))) \\ (A \rightarrow B) & \text{is a name for} & ((\neg A) \vee B) \\ (A \leftrightarrow B) & \text{is a name for} & ((A \rightarrow B) \wedge (B \rightarrow A)) \end{array}$$

We emphasize that $\neg$ and $\vee$ are the basic connectives of our language, whereas the connectives $\wedge$, $\rightarrow$, and $\leftrightarrow$ are defined in terms of $\neg$ and $\vee$. We shall freely drop the outermost pair of parentheses of a *name* for a swff. For example, "$(S \wedge T) \rightarrow T$" is a name for the swff named by

$$(\neg(\neg((\neg S) \vee (\neg T)))) \vee T$$

i.e., the swff $((\neg(\neg((\neg(S)) \vee (\neg(T))))) \vee (T))$.

Here is a very useful convention for omitting parentheses. We shall attribute a built-in bracketing power, or reach, to the connectives in the following order: $\neg$, $\vee$, $\wedge$, $\rightarrow$, $\leftrightarrow$, where the connectives weakest in reach are written first in this list. Under this convention, $\rightarrow$ is the principal connective of the swff $S \vee T \rightarrow S$, since the bracketing power of $\rightarrow$ is stronger than the bracketing power of $\vee$. In other words, the reach of $\vee$ is blocked by the stronger connective $\rightarrow$ in the name "$S \vee T \rightarrow S$." So, this is an abbreviation for "$(S \vee T) \rightarrow S$," which itself is a name for a swff.

It is sometimes possible to make a dot do the work of several pairs of parentheses. The idea is to strengthen the bracketing power of a connective by placing a dot above it. This means that the bracketing power of the connectives is as follows: $\neg$, $\vee$, $\wedge$, $\rightarrow$, $\leftrightarrow$, $\dot{\neg}$, $\dot{\vee}$, $\dot{\wedge}$, $\dot{\rightarrow}$, $\dot{\leftrightarrow}$, $\ddot{\neg}$, $\ddot{\vee}$, $\ddot{\wedge}$, $\ddot{\rightarrow}$, $\ddot{\leftrightarrow}$, and so on.

Under this "dot" convention, the swff

$$(S \rightarrow T) \rightarrow ((U \vee S) \rightarrow (T \vee U))$$

can be written as

$$S \rightarrow T \mathrel{\dot{\rightarrow}} U \vee S \rightarrow T \vee U$$

without ambiguity.

The intention of the above conventions is not to eliminate all parentheses, but merely to reduce the number of parentheses that appear in a name for a given swff so that the eye will not be lost in a maze of parentheses. One or two pairs of parentheses may not be objectionable in a name for a swff and may even be desirable. Our goal is that names for swffs should be readable. For this reason, we prefer to write

$$S \vee \neg T \rightarrow \neg(S \vee T)$$

rather than $S \vee \neg T \mathbin{\dot{\rightarrow}} \dot{\neg} S \vee T$.

In this section we have followed the usual custom of mentioning a swff A by writing down a name for A. For example, we introduced "$A \wedge B$" as a name for a certain swff C by writing down a name for C rather than C itself.

Exercises

1. Let Σ be the statement system such that dom $\Sigma = \{S_1, \ldots, S_9\}$, where S_i = "i is prime," for $i = 1, \ldots, 9$, and let Σ assign "true" to S_i when i is prime; i.e., S_2, S_3, S_5, and S_7 are true for Σ, whereas S_1, S_4, S_6, S_8, and S_9 are false for Σ. Compute the truth-value of each of the following swffs of Σ:

   (a) $S_1 \vee S_3$.
   (b) $S_1 \vee \neg S_2$.
   (c) $S_2 \wedge S_3 \rightarrow S_4$.
   (d) $S_3 \wedge S_7 \leftrightarrow \neg S_9$.
   (e) $S_5 \vee \neg S_6 \rightarrow \neg S_5 \vee S_6$.

2. Prove that for every statement system Σ, each swff of Σ has a unique truth-value. *Hint:* If there is a swff of Σ that does not have a unique truth-value, then there is a shortest swff of Σ that does not have a unique truth-value.

3. Let A, B, and C be swffs of Σ, a statement system. Show that each of the following swffs is true for Σ:

   (a) $A \vee A \rightarrow A$.
   (b) $A \rightarrow A \vee B$.
   (c) $A \rightarrow B \mathbin{\dot{\rightarrow}} C \vee A \rightarrow B \vee C$.

4. Let A and $A \rightarrow B$ be true swffs of a statement system Σ. Prove that B is true for Σ.

5. Let A, B, and C be swffs of a statement system Σ. Prove that:

   (a) $\neg(A \vee B)$ is true for Σ iff $\neg A \wedge \neg B$ is true for Σ.
   (b) $\neg(A \wedge B)$ is true for Σ iff $\neg A \vee \neg B$ is true for Σ.
   (c) $A \wedge (B \vee C)$ is true for Σ iff $(A \wedge B) \vee (A \wedge C)$ is true for Σ.
   (d) $A \vee (B \wedge C)$ is true for Σ iff $(A \vee B) \wedge (A \vee C)$ is true for Σ.
   (e) $A \mathbin{\dot{\rightarrow}} B \rightarrow C$ is true for Σ iff $A \wedge B \rightarrow C$ is true for Σ.

2

Propositional Calculus

2.1. Well-Formed Formulas

Ordinarily the symbols appearing in a mathematical investigation constitute names for the mathematical objects involved in that particular branch of mathematics. On the other hand, in the study of a *formal system* the objects under discussion are certain expressions built up from given symbols that in themselves possess no denotation in the usual sense. In mathematics, a symbol is used to denote a mathematical object (e.g., a numeral denotes a number); here, on the other hand, when a symbol of a formal system is written, that symbol is the object we wish to denote. In short, each symbol of a formal system denotes itself!

Beyond question, this approach, in which it is the symbol as an object in itself that interests us, is highly sophisticated and requires some getting used to. Furthermore, we shall consider as an object in itself an array of symbols obtained by writing down several symbols one after another.

One purpose of the propositional calculus is to formalize and study those tautologies and syllogisms that can be built up without using quantifiers as connectives. So, the language of the propositional calculus must be capable of representing and analyzing statements that do not include quantifiers among their connectives. The first step in introducing a new language is to announce its alphabet—the basic symbols used in constructing the language. Thus, we require a stock of symbols to denote the basic, atomic statements of the language. Also, we need two connectives *not* and *or* to help us form compound statements (one connective, e.g., Sheffer's stroke, will do, but this results in an extremely cumbersome language). Finally, we shall use pairs of parentheses for punctuation. Thus, the alphabet of the propositional calculus consists of three kinds of symbols, as follows.

Propositions: $X, Y, Z, \ldots$. (This list is infinite and excludes the symbols below.)

Connectives: $\neg, \vee$.

Parentheses: (,).

Remember that the basic symbols of a language possess no intrinsic meaning; however, as a guide in following the discussion, we have indicated the *intended* meaning or function of the above symbols. We emphasize again that each of these symbols denotes itself only.

As a first step toward defining the sentences of our language, which are called *well-formed formulas* or *wffs* for short, we present the notion of an *expression* of the propositional calculus.

Definition of Expression. Each finite string of symbols of the propositional calculus is called an *expression.*

Here repetitions are allowed, i.e., the same symbol may occur several times in an expression; but the total number of instances of symbols must be finite. We regard zero as a finite number, so the empty string is also an expression. Here are some examples:

$$),\quad \neg\vee,\quad Y \vee (\neg Z),\quad XXX\neg\vee\vee\vee\vee$$

On the other hand, none of the following is an expression:

$$\begin{matrix}X\\X'\end{matrix},\quad \begin{matrix}(\\)\end{matrix},\quad \bar{X},\quad [X],\quad XXX\cdots X\cdots \text{ (infinitely many } X\text{'s)},\quad ((}$$

Having defined expressions, we now select certain expressions, which we shall call *well-formed formulas* or *wffs* for short.

Definition of Wff. First we say that (P) is a wff whenever P is a proposition. Next, we say that $(\neg A)$ is a wff whenever A is a wff. Finally, we say that $(A \vee B)$ is a wff whenever A and B are wffs. Each wff contains only a finite number of instances of connectives.

For example, Y is a proposition, so (Y) is a wff; a wff of this sort is called an *atomic* wff. Thus, by an atomic wff we mean (P) provided that P is a proposition. Each wff of the propositional calculus is built up from a finite number of atomic wffs by connecting them with parentheses and connectives, as given by the above definition. Since (Y) is a wff, so is $(\neg(Y))$. Similarly, $((\neg(Y)) \vee (X))$ is a wff.

Notice our use of capital letters at the beginning of the alphabet (e.g., A, B, and C) as placeholders for wffs, and our use of "P" as a placeholder for propositions. Placeholders allow us to talk about wffs in a general way, and to communicate facts about wffs in general (as opposed to a specific wff).

Since each wff is an expression and each expression is finite, it follows that each wff is finite. We mean that each wff is a string of symbols of finite length; i.e., the number of instances of symbols in the string is finite.

A wff is atomic if neither $\neg$ nor $\vee$ occurs in the wff. We say that a wff is *composite* if it is not atomic. Thus a composite wff involves at least one instance of a connective $\neg$ or $\vee$. By the *length* of a wff we shall mean the number of instances of these connectives in the wff. The length of each atomic wff is zero; the length of each composite wff is a positive integer. For example, the length of the wff $(((X) \vee (Y)) \vee (\neg(X)))$ is 3.

We now present a general method of proving that each wff possesses a stated property. For concreteness we study the property involved by considering the set of all wffs that have the property. Essentially, this means that the notion of *property* is reduced to membership in a specified set of wffs. Here is our method.

Fundamental Theorem about Wffs. A set of wffs, say S, is the set of all wffs iff:

1. $A \in S$ whenever A is atomic.
2. $(\neg A) \in S$ whenever $A \in S$.
3. $(A \vee B) \in S$ whenever $A \in S$ and $B \in S$.

Demonstration. First note that if S is the set of all wffs, then certainly statements 1, 2, and 3 are correct. Next, let S be a set of wffs for which the statements 1, 2, and 3 are correct; we shall show that S is the set of all wffs. Assume that S is not the set of all wffs. Then, by 1, some composite wff is not a member of S. Now, each wff has a length, some nonnegative integer. Consider the lengths of the wffs that are not members of S; this yields a set of positive integers. But each nonempty set of positive integers has a smallest member; let k be the smallest member of the set of positive integers that we have just described. This means that there is a composite wff of length k that is not a member of S, and that each composite wff whose length is less than k is a member of S. Let A be a wff of length k that is not a member of S. In view of our definition of a wff, there are just two possibilities:

(a) There is a wff B such that $A = (\neg B)$. The length of B is $k - 1$; so $B \in S$. Thus, by 2, $A \in S$. But $A \notin S$. This contradiction shows that this case is impossible.

(b) There are wffs C and D such that $A = (C \vee D)$. Clearly the length of C and the length of D are both less than the length of A; so $C \in S$ and $D \in S$. Thus, by 3, $A \in S$. But $A \notin S$. This contradiction shows that this case is impossible.

We are forced to conclude that the assumption of our argument is false; i.e., "S is not the set of all wffs" is false. In other words, S is the set of all wffs. This establishes the Fundamental Theorem about Wffs.

Let P be a property which is meaningful for wffs, i.e., it is true or false that a specific wff has the property. By forming the set of all wffs that have property P and applying the Fundamental Theorem about Wffs, we easily derive the following algorithm.

Algorithm to Prove That Each Wff Has Property P:

1. Prove that each atomic wff has property P.
2. Prove that $(\neg A)$ has property P whenever A is a wff with property P.
3. Prove that $(A \vee B)$ has property P whenever A and B are wffs with property P.

As a first application of the Fundamental Theorem about Wffs, we establish the following obvious fact.

Fact 1. Each wff begins with a left-hand parenthesis and ends with a right-hand parenthesis.

Dem. Let $S = \{A \mid A$ is a wff whose first symbol is "(" and whose last symbol is ")"$\}$. The three conditions of the Fundamental Theorem about Wffs are true for S. Thus, S is the set of all wffs; i.e., each wff begins with "(" and ends with ")."

It is convenient to denote the set of all wffs of the propositional calculus by "W"; i.e., $\mathsf{W} = \{A \mid A$ is a wff$\}$.

Exercises

1. Use the Fundamental Theorem about Wffs to prove that at least one proposition (not necessarily the same one) occurs in each wff. *Note:* A proposition

P *occurs* in a wff A provided that there are expressions ϕ and θ such that $A = \phi P \theta$.

2. Use the Fundamental Theorem about Wffs to prove that at least one atomic wff (not necessarily the same one) occurs in each wff. *Note:* An atomic wff (P) *occurs* in a wff A provided that there are expressions ϕ and θ such that $A = \phi(P)\theta$.

3. Use the Fundamental Theorem about Wffs to prove that the number of occurrences of left-hand parentheses in each composite wff exceeds one.

4. Let S be a set of composite wffs such that:

   (1) $(\neg A) \in S$ if $A \in S$ and if A is composite.
   (2) $(A \vee B) \in S$ if $A, B \in S$ and if A and B are both composite.

   Is it true that S is the set of all composite wffs? Justify your answer.

5. Let l be a map of W, the set of all wffs, into R, the set of all real numbers, such that:

   (1) $l[A] = 0$ for each atomic wff A.
   (2) $l[(\neg B)] = 1 + l[B]$ for each wff B.
   (3) $l[(C \vee D)] = 1 + l[C] + l[D]$ for all wffs C and D.

   (a) Describe l in colloquial language.
   (b) Use the Fundamental Theorem about Wffs to prove that l has the property specified in your answer to part (a).
   (c) Use part (b) to prove that there is just one map of W into R that meets the three conditions listed above.

6. Let m be a map of W into R such that:

   (1) $m[A] = 1$ for each atomic wff A.
   (2) $m[(\neg B)] = m[B]$ for each wff B.
   (3) $m[(C \vee D)] = m[C] + m[D]$ for all wffs C and D.

   (a) Describe m in colloquial language.
   (b) Use the Fundamental Theorem about Wffs to prove that m has the property specified in your answer to part (a).
   (c) Use part (b) to prove that there is just one map of W into R that meets the three conditions listed above.

7. Let n be a map of W into R such that:

   (1) $n[A] = 0$ for each atomic wff A.
   (2) $n[(\neg B)] = 1 + n[B]$ for each wff B.
   (3) $n[(C \vee D)] = n[C] + n[D]$ for all wffs C and D.

   (a) Describe n in colloquial language.
   (b) Use the Fundamental Theorem about Wffs to prove that n has the property specified in your answer to part (a).
   (c) Use part (b) to prove that there is just one map of W into R that meets the three conditions listed above.

8. Let o be a map of **W** into R such that:
   (1) $o[A] = 0$ for each atomic wff A.
   (2) $o[(\neg B)] = o[B]$ for each wff B.
   (3) $o[(C \vee D)] = 1 + o[C] + o[D]$ for all wffs C and D.

   (a) Describe o in colloquial language.
   (b) Use the Fundamental Theorem about Wffs to prove that o has the property specified in your answer to part (a).
   (c) Use part (b) to prove that there is just one map of **W** into R that meets the three conditions listed above.

9. Let p be a map of **W** into R such that:
   (1) $p[A] = 2$ for each atomic wff A.
   (2) $p[(\neg B)] = 2 + p[B]$ for each wff B.
   (3) $p[(C \vee D)] = 2 + p[C] + p[D]$ for all wffs C and D.

   (a) Describe p in colloquial language.
   (b) Use the Fundamental Theorem about Wffs to prove that p has the property specified in your answer to part (a).
   (c) Use part (b) to prove that there is just one map of **W** into R that meets the three conditions listed above.

10. Let q be a map of **W** into R such that:
   (1) $q[A] = 3$ for each atomic wff A.
   (2) $q[(\neg B)] = 3 + q[B]$ for each wff B.
   (3) $q[(C \vee D)] = 3 + q[C] + q[D]$ for all wffs C and D.

   (a) Describe q in colloquial language.
   (b) Use the Fundamental Theorem about Wffs to prove that q has the property specified in your answer to part (a).
   (c) Use part (b) to prove that there is just one map of **W** into R that meets the three conditions listed above.

11. Let r be a map of **W** into R such that:
   (1) $r[(X)] = 1$.
   (2) $r[A] = 0$ for each atomic wff A such that $A \neq (X)$.
   (3) $r[(\neg B)] = r[B]$ for each wff B.
   (4) $r[(C \vee D)] = r[C] + r[D]$ for all wffs C and D.

   (a) Describe r in colloquial language.
   (b) Use the Fundamental Theorem about Wffs to prove that r has the property specified in your answer to part (a).
   (c) Use part (b) to prove that there is just one map of **W** into R that meets the four conditions listed above.

2.2. Parentheses

The purpose of introducing a pair of parentheses at each construction step in our definition of wffs is to impose a certain structure on wffs.

In order to study this structure effectively, we now present the notion of the *mate* of a parenthesis in a wff.

Definition of Mate. The *mate* of a left-hand (LH) parenthesis occurring in a wff is the first right-hand (RH) parenthesis to its right such that an equal number of LH parentheses and RH parentheses occur in between. The *mate* of a RH parenthesis occurring in a wff is the first LH parenthesis to its left such that equal numbers of RH parentheses and LH parentheses occur in between.

For example, the mate of each parenthesis of the following wff is indicated by placing the same letter below a parenthesis and its mate:

$$\underset{abc}{(((}X\underset{c}{)} \vee \underset{d}{(}\neg\underset{e}{(}Y\underset{edb}{)))} \vee \underset{f}{(}Z\underset{fa}{))}$$

From our definition, a particular parenthesis occurring in a wff has at most one mate; it is conceivable, of course, that some parenthesis occurring in a wff has no mate. We shall prove that this does not happen (see Lemma 4); moreover, we shall prove that the mate of the leftmost LH parenthesis of any wff is the rightmost RH parenthesis of the wff (see Corollary 1).

Lemma 1. The number of LH parentheses in a wff is the same as the number of its RH parentheses.

Dem. Let $S = \{A \in \mathsf{W} \mid$ the number of LH parentheses in A is the same as the number of RH parentheses in $A\}$. We shall prove that S meets the three requirements of the Fundamental Theorem about Wffs.

1. Let A be any atomic wff; then $A = (P)$, where P is a proposition. Clearly, A has one LH parenthesis and one RH parenthesis; thus $A \in S$. This proves that each atomic wff is in S.

2. Let $B \in S$; we shall show that $(\neg B) \in S$. Let n be the number of LH parentheses in B; by assumption, there are n RH parentheses in B. Clearly, $(\neg B)$ has exactly one more LH parenthesis than B, and $(\neg B)$ has one more RH parenthesis than B. So $(\neg B)$ has exactly $1 + n$ LH parentheses and exactly $1 + n$ RH parentheses. Therefore, $(\neg B) \in S$.

3. Let $C, D \in S$; we shall show that $(C \vee D) \in S$. Let n be the number of LH parentheses in C and let m be the number of LH parentheses in D. By assumption, C has exactly n RH parentheses and D has exactly m RH parentheses. Thus $(C \vee D)$ has exactly $1 + n + m$ LH parentheses and exactly $n + m + 1$ RH parentheses; therefore $(C \vee D) \in S$.

Since S satisfies the three requirements of the Fundamental Theorem about Wffs, we conclude that $S = \mathsf{W}$. This establishes Lemma 1.

Lemma 2. Counting from left to right, the number of instances of LH parentheses in a wff is greater than the number of instances of RH parentheses until the rightmost RH parenthesis is reached.

Dem. Let S be the set of all wffs that satisfy this lemma; apply the Fundamental Theorem about Wffs to prove that $S = \mathsf{W}$.

Corollary 1. The rightmost RH parenthesis of a wff is the mate of its leftmost LH parenthesis.

Dem. Consider Lemmas 1 and 2.

Corollary 2. The leftmost LH parenthesis of a wff is the mate of its rightmost RH parenthesis.

Dem. The result of interchanging each instance of "left" and "right" throughout a statement concerning parentheses is called the *dual* of the statement. We point out that the dual of Lemma 2 is true. Using Lemma 1 and the dual of Lemma 2, we easily establish Corollary 2.

Lemma 3. Each LH parenthesis of a wff has a mate.

Dem. Let $S = \{A \in \mathsf{W} \mid \text{each LH parenthesis of } A \text{ has a mate}\}$; apply the Fundamental Theorem about Wffs to prove that $S = \mathsf{W}$.

Lemma 4. Each parenthesis of a wff has a mate.

Dem. First, we must establish the dual of Lemma 3. Let $S = \{A \in \mathsf{W} \mid \text{each RH parenthesis of } A \text{ has a mate}\}$. Applying the Fundamental Theorem about Wffs, we easily prove that $S = \mathsf{W}$; so each RH parenthesis of a wff has a mate. By Lemma 3, each LH parenthesis of a wff has a mate; thus each parenthesis of a wff has a mate.

Lemma 5. No two LH parentheses of a wff have the same mate.

Dem. Let $S = \{A \in \mathsf{W} \mid \text{no two LH parentheses of } A \text{ have the same mate}\}$; applying the Fundamental Theorem about Wffs, we can prove that $S = \mathsf{W}$.

Lemma 6. No two parentheses of a wff have the same mate.

Dem. It is easy to prove the dual of Lemma 5, i.e., no two RH parentheses of a wff have the same mate.

Before continuing, we present a second method of establishing Lemma 5, which penetrates a little further into the concept. Let A be a wff such that two LH parentheses of A, say a and b, have the same mate, say c. Then c is to the right of both a and b. We may assume that b is to the right of a. By assumption, (i) there are equal numbers of LH parentheses and RH parentheses between a and c, and (ii) there are equal numbers of LH parentheses and RH parentheses between b and c. Therefore, there is exactly one more RH parenthesis between a and b than there are LH parentheses. It follows that there is a RH parenthesis between a and b, say e, such that there are equal numbers of LH parentheses and RH parentheses between a and e. Therefore, c is *not* the mate of a. This contradiction proves that A does not exist.

Since a pair of parentheses is introduced into a wff A at each construction step involved in building up A from its atomic wff, we are tempted to conjecture that each RH parenthesis of a wff A is the mate of a LH parenthesis of A iff the LH parenthesis is the mate of the RH parenthesis. The following definition brings out the basic idea more clearly.

Definition. A parenthesis of a wff is said to be *mated* if it is the mate of its mate.

For example, each parenthesis of an atomic wff is mated.

Lemma 7. The leftmost LH parenthesis of each wff is mated; the rightmost RH parenthesis of each wff is mated.

Dem. Apply Corollaries 1 and 2.

It is easy, now, to prove that each parenthesis of each wff is mated.

Theorem 1. Each parenthesis of each wff is mated.

Dem. Let $S = \{A \in \mathsf{W} \mid \text{each parenthesis of } A \text{ is mated}\}$; we shall apply the Fundamental Theorem about Wffs to prove that $S = \mathsf{W}$.

1. Let A be any atomic wff, say (P). Clearly, each parenthesis of (P) is mated. Thus $(P) \in S$.

2. Let $B \in S$; we shall show that $(\neg B) \in S$. By Lemma 7, the leftmost LH parenthesis and the rightmost RH parenthesis of $(\neg B)$ are each mated. By assumption, the remaining parentheses of $(\neg B)$ are mated. Thus $(\neg B) \in S$.

3. Let $B, C \in S$; we shall show that $(B \vee C) \in S$. By Lemma 7, the leftmost LH parenthesis and the rightmost RH parenthesis of $(B \vee C)$ are each mated. By assumption, the remaining parentheses of $(B \vee C)$ are mated. Thus $(B \vee C) \in S$. We conclude that $S = \mathsf{W}$. This establishes Theorem 1.

Exercises

1. Prove Lemma 2.
2. Prove the dual of Lemma 2: Counting from right to left, the number of instances of RH parentheses in a wff is greater than the number of instances of LH parentheses until the leftmost LH parenthesis is reached.
3. Prove Lemma 3.
4. Prove the dual of Lemma 3.
5. Prove Lemma 5.
6. Prove the dual of Lemma 5.
7. For the purpose of this problem, we say that a LH parenthesis and a RH parenthesis of a wff are *paired* if and only if there are equal numbers of LH parentheses and RH parentheses in between. Exhibit a wff for which a LH parenthesis is paired with two RH parentheses.
8. Prove Lemma 7.
9. Find the mate of the third symbol of the following wff:

$$((((\neg(Z)) \vee (Y)) \vee (\neg(Y))) \vee (\neg(Y)))$$

Verify that these parentheses are mated.

2.3. Main Connective of Wffs

As we shall soon see, our discussion of the parentheses of a wff has an impact on the notion of a *main* connective of a composite wff, which we now present.

Definition of Main Connective. Let A be any composite wff. An occurrence of a connective in A is said to be a *main* connective of A if:

1. $A = (\neg B)$, where B is a wff. In this case, the displayed instance of the connective $\neg$ is said to be a *main* connective of A.
2. $A = (C \vee D)$, where C and D are wffs. In this case, the displayed instance of the connective $\vee$ is said to be a *main* connective of A.

For example, the second symbol of $(\neg((Y) \vee (X)))$ is a main connective of this wff; the eighth symbol of $((\neg(Y)) \vee (X))$ is a main connective of this wff.

Of course, it is conceivable that a specific wff A has several main connectives. This will be the case if there are wffs B, C, and D such that $A = (\neg B)$ and $A = (C \vee D)$; or if there are wffs B, C, D, and E such that $A = (B \vee C)$ and $A = (D \vee E)$, where $B \neq D$ and $C \neq E$. In the former case, both $\neg$ and $\vee$ are main connectives of A; in the latter case, two instances of $\vee$ in A are main connectives of A. We shall prove that this cannot happen. Indeed, we shall prove that each composite wff has exactly one main connective. Of course, each composite wff has at least one main connective. We must show that no composite wff has more than one main connective.

Theorem 1. No composite wff has two main connectives.

Dem. Assume that $\neg$ is a main connective of a composite wff A; then $A = (\neg B)$, where B is a wff. Thus "$\neg$" is the second symbol of A. Now, if $\vee$ is also a main connective of A, then $A = (C \vee D)$, where C and D are wffs; notice that the first symbol of C is a LH parenthesis, so the second symbol of A is a LH parenthesis. But the second symbol of A is "$\neg$"; this contradiction proves that $\vee$ is not a main connective of A. Of course, no other instance of $\neg$ in A (except the second symbol of A) is a main connective of A. The situation is a little different in case an instance of $\vee$ is a main connective of a wff A; at first sight, it is conceivable that another occurrence of $\vee$ in this wff is also a main connective of A. This means that there are wffs B, C, D, and E such that

$$A = (B \vee C) \qquad \text{and} \qquad A = (D \vee E)$$

where $B \neq D$ and $C \neq E$. Now, the mate of the second symbol of A is the last symbol of B, since B is a wff. Moreover, D is a wff; so the mate of the second symbol of A is the last symbol of D. But each parenthesis of A has a unique mate; therefore, the last symbol of D is also the last symbol of B, and it follows that $B = D$. Thus, the two displayed occurrences of $\vee$ in A are actually the same occurrence of $\vee$. We conclude that each composite wff has at most one main connective.

Corollary 1. Each composite wff has a unique main connective.

We now consider the problem of locating the main connective of a given composite wff. Each wff is a finite string of symbols, some of which may be connectives. We need an algorithm that will mechanically present the main connective of a composite wff. Again, it is merely a matter of counting parentheses, forming two totals, one for LH parentheses and the other for RH parentheses. Here is our algorithm.

Algorithm for Main Connective. Counting from left to right, the first connective reached for which the number of LH parentheses is one greater than the number of RH parentheses is the main connective of the wff.

Dem. Let $S = \{A \in \mathbf{W} \mid A \text{ is atomic or the algorithm is true for } A\}$. Apply the Fundamental Theorem about Wffs to prove that $S = \mathbf{W}$.

To illustrate this algorithm, consider the wff

$$((\neg((\neg(Y)) \vee (X))) \vee (Y))$$

The 17th symbol of this wff, counting from left to right, is the first connective that is preceded by one more LH parenthesis than RH parentheses. We conclude that the second instance of $\vee$ in this wff is its main connective; of course, this means that the given wff has the form $(A \vee B)$.

In operational terms this algorithm can be expressed as follows.

Operational Form of Algorithm for Main Connective. Add 1 for a LH parenthesis and -1 for a RH parenthesis, and take subtotals starting at the leftmost symbol of A. Then the connective reached when the subtotal is 1 is the main connective of A.

This result provides us with an easy method of determining whether a given expression is a wff. Of course, we can decide whether an expression consisting of three symbols is a wff: It must have the form (P), where P is a proposition, to be a wff. The problem is to decide whether a longer expression is a wff.

Consider an expression consisting of more than three symbols. Whether or not it is a wff, the *algorithm for main connective* either (i) fails, or (ii) yields a unique connective of the expression. In the first case, we conclude that the given expression is *not* a wff. In the second case, our problem is reduced to considering one or two expressions each shorter than the given expression. Repeating this program sufficiently often, we either prove that the given expression is not a wff, or else we obtain the atomic wffs from which the expression was constructed. In the latter case, the expression is a wff.

To illustrate, consider the expression:

$$((\neg((\neg(Y)) \vee (X))) \vee ((\neg((Z) \vee (Y))) \vee ((X) \vee (Z)))) \qquad (1)$$

Our algorithm points out the second "$\vee$" of (1); so we split (1) into two expressions, deleting a pair of parentheses and the "$\vee$" in question, namely

$$(\neg((\neg(Y)) \vee (X))) \qquad (2)$$

and

$$((\neg((Z) \vee (Y))) \vee ((X) \vee (Z))) \qquad (3)$$

Our algorithm points out the first "$\neg$" of (2); so (2) yields

$$((\neg(Y)) \vee (X)) \qquad (2')$$

which, in turn, yields the expressions $(\neg(Y))$ and (X), which are wffs. Turning to (3), we find that our algorithm points out the second "$\vee$"; so (3) yields

$$(\neg((Z) \vee (Y))) \qquad (4)$$

and

$$((X) \vee (Z)) \qquad (5)$$

From (4) we obtain $((Z) \vee (Y))$, which yields (Z) and (Y). From (5) we obtain (X) and (Z). We conclude that (1) is a wff.

Exercises

1. Apply the Definition of Main Connective to show that the 17th symbol of the wff

$$((\neg((Y) \vee (\neg(X)))) \vee ((X) \vee (Y)))$$

is a main connective of this wff. Does it have another main connective?

2. Apply the Definition of Main Connective to show that the 23rd symbol of the wff

$$((((\neg(Z)) \vee (Y)) \vee (\neg(Y))) \vee (\neg(X)))$$

is a main connective of this wff. Does it have another main connective?

3. Prove the Algorithm for Main Connective.

4. Apply the Algorithm for Main Connective to find the main connective of the wff of Exercise 1.

5. Apply the Algorithm for Main Connective to find the main connective of the wff of Exercise 2.

Determine which of the following expressions are wffs.

6. $((\neg((\neg(Y)) \vee (X))) \vee ((\neg((Z) \vee (Y))) \vee ((X) \vee (Z))))$.

7. $((\neg((Y) \vee (\neg(X)))) \vee ((Z) \vee (\neg(Y)))$.

8. $(((\neg(X)) \vee (Y)) \vee ((\neg(Y)) \vee (\neg(X))))$.

9. $((\neg((\neg(X)) \vee (Y))) \vee ((\neg(\neg(Y))) \vee (\neg(X)))$.

2.4. Names for Wffs; Principal Connective of Name

As we have demonstrated, parentheses impose a certain structure on wffs that we find extremely useful. On the other hand, it is difficult to read a specific wff when it involves many parentheses. We can obtain the best of both worlds by introducing parentheses-omitting conventions, just as for swffs. This is achieved by introducing *names* for wffs; of course, we must avoid ambiguity, i.e., each of our names must be a name for a unique wff.

Let A be any wff; there is no ambiguity in dropping the outermost pair of parentheses of A; moreover, we can safely suppress the pair of parentheses involved in each atomic wff (P) occurring in A. For example, let

$$A = ((\neg(Y)) \vee ((X) \vee (Y)))$$

Under our agreement, "$(\neg Y) \vee (X \vee Y)$" is a name for A.

Just as for swffs, short names for certain wffs are obtained by introducing the connectives $\wedge$ (*and*), $\rightarrow$ (*if . . . then*), $\leftrightarrow$ (*if and only if*). Our agreement is that for any wff A and B,

$$\begin{array}{lll} (A \wedge B) & \text{is a name for} & (\neg((\neg A) \vee (\neg B))) \\ (A \rightarrow B) & \text{is a name for} & ((\neg A) \vee B) \\ (A \leftrightarrow B) & \text{is a name for} & ((A \rightarrow B) \wedge (B \rightarrow A)) \end{array}$$

It is important to distinguish between the *basic* connectives $\neg$ and $\vee$ of a propositional calculus and the *defined* connectives $\wedge$, $\rightarrow$, and $\leftrightarrow$ just introduced.

Again, we shall freely drop the outermost pair of parentheses of a *name* for a wff. For example, "$(X \wedge Y) \rightarrow Y$" is a name for the wff named by "$\neg(\neg((\neg X) \vee (\neg Y))) \vee Y$," i.e., the wff

$$((\neg(\neg((\neg(X)) \vee (\neg(Y))))) \vee (Y))$$

As for swffs, we obtain short names for wffs by assigning a bracketing power, or reach, to the connectives $\neg$, $\vee$, $\wedge$, $\rightarrow$, $\leftrightarrow$ in that order. We shall use the same *dot* convention as for swffs to assist us in naming wffs (see page 8).

Our purpose in introducing the defined connectives $\wedge$, $\rightarrow$, and $\leftrightarrow$ is to provide short, easy-to-read names for wffs; moreover, the connective $\wedge$ is of fundamental importance in our discussion of duality (see Chapter 5).

We want to extend the notion of the main connective of a composite wff by taking into account the defined connectives $\wedge$, $\rightarrow$, and $\leftrightarrow$; here we shall be dealing with *names* for wffs, rather than wffs themselves. Accordingly, we now present the notion of the *principal* connective of a name for a wff. We emphasize that we are involved with names for wffs, whereas in the case of a *main* connective we deal with wffs only, not their names.

Definition of Principal Connective. Let N be a name for a wff, and let c be one of $\vee$, $\wedge$, $\rightarrow$, or $\leftrightarrow$. Then:

1. c is the principal connective of N if $N = (N_1)\, c\, (N_2)$, where N_1 and N_2 are names for wffs.
2. $\neg$ is the principal connective of N if $N = \neg(N_1)$, where N_1 is a name for a wff.

For example, $\wedge$ is the principal connective of the name $(\neg X) \wedge Y$; $\neg$ is the principal connective of the name $\neg((\neg(\neg X)) \vee (\neg Y))$.

Exercises

Write down a name, as short as possible, for each of the following wffs:

1. $((Z) \vee (\neg(X)))$.
2. $((\neg(X)) \vee (Y))$.
3. $((\neg(\neg((\neg(Y)) \vee (\neg(X))))) \vee (X))$.

Write down the wff, in full, whose name is:

4. $X \vee Y \to \neg Y$.

5. $X \leftrightarrow \neg Y$.

6. $X \to Y \dot{\to} Z \vee X \to Y \vee Z$.

Use one dot to simplify each of the following:

7. $(X \wedge Y) \vee (Y \wedge X)$.

8. $(\neg X \to Y) \wedge (X \vee \neg Y)$.

9. $(X \to Y) \to (Z \to W)$.

10. $(X \leftrightarrow \neg Y) \vee (X \wedge Y)$.

Indicate the principal connective of each of the following by inserting one or two pairs of parentheses:

11. $\neg X \to Y \dot{\wedge} X \vee Y$.

12. $X \to Y \leftrightarrow \neg Y \to \neg X$.

13. $W \wedge \neg Y$.

14. $X \to X \vee X$.

15. $X \wedge Y \dot{\vee} Z \vee W$.

16. $X \wedge Y \to Z$.

17. $X \dot{\wedge} Y \to Z$.

18. $X \dot{\to} Y \leftrightarrow Z$.

19. $X \to Y \leftrightarrow Z$.

2.5. Valuations

We point out that the procedure for constructing wffs of our propositional calculus is precisely the method that yields the swff of a statement system. Whereas each swff of a statement system possesses a unique truth-value, wffs do not have truth-values.

It is important that we inject truth-values into our propositional calculus in the following sense. First, we introduce symbols t and f, which we regard as standing for *true* and *false*, respectively. Let σ be a map whose domain is the set of all atomic wffs of our propositional calculus, and whose range is included in $\{t, f\}$; then σ is said to be an *assignment* for this propositional calculus. The idea is that σ assigns a truth-value to each atomic wff. The purpose of an assignment is to reflect the basic feature of a statement system, namely that each of its atomic statements has a unique truth-value.

In the same way that compound swffs receive unique truth-values, we use an assignment σ to provide each composite wff with a truth-value. In fact, we extend σ to a map ν of W into $\{\mathsf{t}, \mathsf{f}\}$, which is defined as follows:

1. $\nu[A] = \sigma[A]$ for each atomic wff A.
2. $\nu[\neg B] = \begin{cases} \mathsf{t} & \text{if } \nu[B] = \mathsf{f} \\ \mathsf{f} & \text{if } \nu[B] = \mathsf{t}. \end{cases}$
3. $\nu[C \vee D] = \begin{cases} \mathsf{f} & \text{if } \nu[C] = \nu[D] = \mathsf{f} \\ \mathsf{t} & \text{otherwise.} \end{cases}$

The map ν is called a *valuation*; for each wff A, $\nu[A]$ (also written "νA") is said to be the truth-value of A under the assignment σ.

We must prove that our definition of a valuation does indeed associate a unique truth-value with each wff. For that purpose we use the next theorem.

Construction Theorem for Wffs. In order to define an operation on all wffs, it suffices to give the following:

(a) the definition of the value of the operation at each atomic wff;

(b) the definition of the value of the operation at $(\neg B)$, possibly using its value at B;

(c) the definition of the value of the operation at $(C \vee D)$, possibly using its values at C and D.

Dem. The thing to prove is that this way of defining an operation is noncircular, and that (by virtue of the results in Section 2.3) items (a)–(c) do not conflict with one another.

Assume that we are given the definitions demanded by (a)–(c). Let W_n be the set of all wffs of length n or less. (Recall that the length of a wff is the number of instances of connectives in it.) For example, W_0 is the set of atomic wffs. Call an integer n *reachable* if there is a function with domain W_n that obeys the conditions given in (a)–(c). Such a function will be said to *reach n*. We now proceed as follows:

1. Suppose that f reaches m and g reaches n and that $m \leq n$. We claim that g extends f, i.e., the two functions agree on the common domain W_m. This claim is verified by induction on the lengths of the formulas in W_m. The functions f and g must agree on the atomic formulas, since both must satisfy (a). And if they agree on formulas of length k (where $k < m$), then they must agree also on formulas of length $k + 1$, by (b) and (c).

2. It follows from the preceding paragraph that for each reachable number n, there is only one function that reaches n. Call this unique function f_n.

3. Next we claim that every nonnegative integer is reachable. This claim is verified by induction. Item (a) yields a function that reaches 0, so 0 is reachable. Now suppose we have a function f_k that reaches k. We need to extend it [in accordance with (b) and (c)] to a function reaching $k + 1$. Consider any wff C of length $k + 1$. Then either C has the form $(\neg A)$ or it has the form $(A \vee B)$, *but not both.* Furthermore, the components wffs A and B *are uniquely determined.* The italicized phrases in these last two sentences follow from Section 2.3; they are the keys to the proof. Thus we may use (b) and (c), without conflict, to define unambiguously the correct function value at the formula C. We thereby obtain a function reaching $k + 1$. So by induction, every nonnegative integer is reachable.

4. We can now define the desired operation F at any wff A. We take $F(A) = f_n(A)$, where n is any integer at least as large as the length of A. Then it is easy to see that F obeys the conditions (a)–(c). Thus there is at least one operation meeting the conditions. And as in step 1, there can be at most one. Thus we have a well-defined operation.

Lemma 1. Each valuation is a map of W into $\{\mathsf{t}, \mathsf{f}\}$.

Dem. Apply the Construction Theorem for Wffs.

Here are some facts about valuations. Throughout ν is any valuation and A and B are any wffs.

Fact 1.

$$\nu[A \wedge B] = \begin{cases} \mathsf{t} & \text{if } \nu A = \nu B = \mathsf{t} \\ \mathsf{f} & \text{otherwise} \end{cases}$$

Dem. We point out that $A \wedge B = \neg(\neg A \vee \neg B)$.

Fact 2.

$$\nu[A \rightarrow B] = \begin{cases} \mathsf{t} & \text{if } \nu A = \mathsf{f} \text{ or } \nu B = \mathsf{t} \\ \mathsf{f} & \text{if } \nu A = \mathsf{t} \text{ and } \nu B = \mathsf{f} \end{cases}$$

Dem. We point out that $A \rightarrow B = \neg A \vee B$.

Fact 3.

$$\nu[A \leftrightarrow B] = \begin{cases} \mathsf{t} & \text{if} \quad \nu A = \nu B \\ \mathsf{f} & \text{if} \quad \nu A \neq \nu B \end{cases}$$

Dem. We point out that $A \leftrightarrow B = (A \rightarrow B) \wedge (B \rightarrow A)$.

We shall be specially interested in wffs A such that $\nu A = \mathsf{t}$ for every valuation ν. Here are some examples: $X \vee \neg X$, $X \vee Y \rightarrow Y \vee X$, $X \vee X \rightarrow X$, $X \rightarrow X \vee Y$, $X \rightarrow Y \dot{\rightarrow} Z \vee X \rightarrow Y \vee Z$.

Exercises

1. (a) Find a valuation ν such that $\nu[X \vee \neg Y] = \mathsf{t}$.
   (b) Find a valuation ν such that $\nu[X \vee \neg Y] = \mathsf{f}$.

2. (a) Exhibit a wff A such that $\nu A = \mathsf{t}$ for every valuation ν.
   (b) Exhibit a wff B such that $\nu B = \mathsf{f}$ for every valuation ν.
   (c) For the wffs A and B of parts (a) and (b), compute $\nu[\neg(A \vee B)]$.

3. Prove Fact 1.

4. Prove Fact 2.

5. Prove Fact 3.

6. Let ν be the valuation such that $\nu A = \mathsf{t}$ for each atomic wff A. Compute:

   (a) $\nu[X \rightarrow \neg Y \dot{\rightarrow} X \rightarrow Y]$.
   (b) $\nu[X \rightarrow \neg Y \leftrightarrow X \rightarrow Y]$.

7. Let ν be the valuation such that $\nu A = \mathsf{f}$ for each atomic wff A. Compute:

   (a) $\nu[X \rightarrow \neg Y \dot{\rightarrow} X \rightarrow Y]$.
   (b) $\nu[X \rightarrow \neg Y \leftrightarrow X \rightarrow Y]$.

3

Provable Wffs

3.1. $\sum$-Interpreters

The language of each statement system is closely linked to the propositional calculus; indeed, we may regard the propositional calculus as the blueprint for the language of any statement system. The idea is to interpret each of the propositions $X, Y, Z, \ldots$ of the propositional calculus as a basic statement of the given statement system, i.e., a member of its domain.

To be specific, let $\sum$ be a statement system with basic statements $P_1, P_2, P_3, \ldots$; i.e., dom $\sum = \{P_1, P_2, P_3, \ldots\}$. Let μ be a map of $\{X, Y, Z, \ldots\}$ into dom $\sum$; so μ associates some P_i with each proposition of the propositional calculus. In this way μ does the interpreting mentioned in the preceding paragraph. Moreover, for each wff A of the propositional calculus, let "$\mu[A]$" (also written "μA") denote the swff of $\sum$ obtained by replacing each proposition occurring in A by its image under μ, and regarding the symbols "$\neg$," "$\vee$," "(," and ")" as the corresponding symbols in the language of $\sum$. In this way μ translates each wff of the propositional calculus into a swff of $\sum$. The map μ is said to be a $\sum$-interpreter. In general, there are many $\sum$-interpreters.

Example 1. Let $\sum$ be the statement system whose domain is

{the moon is round, the earth is flat, the sun is hot}

and let $\sum$[the moon is round] = f, $\sum$[the earth is flat] = t, and

$\sum$[the sun is hot] = f. Let μ be the $\sum$-interpreter such that

$$\begin{aligned} \mu[X] &= \text{the moon is round,} \\ \mu[Y] &= \text{the earth is flat,} \\ \mu[P] &= \text{the sun is hot,} \end{aligned}$$

for any other proposition P of the propositional calculus. Then

$$\mu[X \vee \neg Y] = (\text{the moon is round}) \vee \neg(\text{the earth is flat})$$

This is a swff of $\sum$ that is false for $\sum$ since both of its disjuncts are false for $\sum$.

Our first lemma incorporates a method of constructing a valuation from a given statement system and interpreter.

Lemma 1. Let $\sum$ be any statement system and let μ be any $\sum$-interpreter. Let ν be the valuation such that for each atomic wff (P)

$$\nu[(P)] = \begin{cases} \mathsf{t} & \text{if } \mu(P) \text{ is true for } \sum \\ \mathsf{f} & \text{if } \mu(P) \text{ is false for } \sum \end{cases}$$

Then for each wff A

$$\nu[A] = \mathsf{t} \quad \text{iff } \mu A \text{ is true for } \sum \tag{1}$$

Dem. Let $S = \{A \in \mathsf{W} \mid \nu[A] = \mathsf{t} \text{ iff } \mu A \text{ is true for } \sum\}$. We can show that S meets the three conditions of the Fundamental Theorem about Wffs; thus $S = \mathsf{W}$. The details are left as an exercise.

We now introduce a procedure for constructing a statement system from the propositional calculus and a given valuation ν. First, we define the domain of the map $\sum$; next, we announce which member of {true, false} $\sum$ associates with a given member of its domain.

Definition of $\sum$:

1. dom $\sum = \{P \mid P \text{ is a proposition of the propositional calculus}\}$.
2. $\sum$ associates "true" with a member of its domain, say P, if $\nu[(P)] = \mathsf{t}$; $\sum$ associates "false" with P if $\nu[(P)] = \mathsf{f}$.

By construction, the swffs of $\sum$ are precisely the wffs of the propositional calculus. To avoid confusion we shall use the identity interpreter ι to indicate a swff of $\sum$; i.e., for each wff A, the corresponding swff of $\sum$ is denoted by "ιA."

Utilizing these conventions, we now formulate the following important relation between the given valuation ν and truth-values in Σ.

Lemma 2. For each wff A

$$\iota A \text{ is true for } \Sigma \text{ iff } \nu[A] = \mathbf{t} \tag{2}$$

Dem. Let $S = \{A \in \mathsf{W} \mid \iota A \text{ is true for } \Sigma \text{ iff } \nu[A] = \mathbf{t}\}$. We can show that S meets the three conditions of the Fundamental Theorem about Wffs; thus $S = \mathsf{W}$. The details are left as an exercise.

The importance of the lemmas of this section is brought out in our demonstration of the Criterion for True Wffs (see page 34).

Exercises

1. Usually, each statement system Σ possesses many Σ-interpreters. Let Σ be a statement system that has just one Σ-interpreter. Describe dom Σ.

2. Let Σ be the statement system of Example 3, page 6, and let μ be the Σ-interpreter for which $\mu P = U$ for each proposition P of the propositional calculus.

   (a) Exhibit a wff A such that μA is true for Σ.
   (b) Exhibit a wff of the form $B \vee C$ such that $\mu(B \vee C)$ is false for Σ.
   (c) Is there an atomic wff A such that μA is true for Σ? Explain.

3. Characterize the valuation ν constructed from the statement system Σ and Σ-interpreter μ of Exercise 2 by applying the procedure in Lemma 1. Verify in this case that

$$\nu[A] = \mathbf{t} \quad \text{iff } \mu A \text{ is true for } \Sigma$$

   for each wff A.

4. Complete the proof of Lemma 1.

5. Let ν be the valuation such that $\nu[A] = \mathbf{t}$ for each atomic wff A. Exhibit the statement system Σ constructed from ν by applying the definition preceding Lemma 2. Verify in this case that

$$\iota A \text{ is true for } \Sigma \text{ iff } \nu[A] = \mathbf{t}$$

   for each wff A.

6. Complete the proof of Lemma 2.

3.2. True Wffs

One goal of the propositional calculus is to analyze logical arguments of the sort studied in symbolic logic. Now, an argument with assumptions $A_1, \ldots, A_n$ and conclusion C is valid iff the statement

$$A_1 \wedge \cdots \wedge A_n \to C$$

is a tautology, i.e., takes the truth-value "true" in each truth-value case. The purpose of this chapter is to characterize tautologies without using truth-values; we shall do this in Section 3.3.

In this section we present some basic facts about tautologies, which will motivate our formal definition of Section 3.3. Although our wffs represent mathematical statements, some of which are tautologies, it is important to maintain a clear distinction between wffs and mathematical statements. Indeed, we want to discuss wffs, not mathematical statements. Motivated by the notion of a tautologous statement, we now present the corresponding concept for wffs; a wff that represents a tautology is said to be ***true*** (the word "valid" is used by some authors in place of "true").

Definition of True Wff. A wff A is said to be ***true*** if μA is true for $\sum$ whenever $\sum$ is a statement system and μ is a $\sum$-interpreter.

For example, each of the following is a true wff:

$$X \vee \neg X, \quad X \vee Y \to Y \vee X, \quad X \vee (Y \vee Z) \leftrightarrow (X \vee Y) \vee Z,$$
$$X \wedge Y \to Z \leftrightarrow X \to (Y \to Z).$$

More generally, let A, B, and C be any wffs; then each of the following is a true wff:

$$A \vee A \to A$$
$$A \to A \vee B$$
$$A \to B \dot{\to} C \vee A \to B \vee C$$

Moreover, the following rule of inference is easy to establish.

Modus Ponens. B is a true wff if both A and $A \to B$ are true wffs.

We now present a criterion for true wffs that involves valuations.

Criterion for True Wffs. Let B be any wff; then B is true iff $\nu B = \mathrm{t}$ for each valuation ν.

Dem. 1. Assume that B is a true wff. Let ν be any valuation and let $\sum$ be the statement system obtained from ν as in Lemma 2, page 33, i.e.,

$$\text{dom} \sum = \{P \mid P \text{ is a proposition of the propositional calculus}\}$$

and for each $P \in \text{dom} \sum$, $\sum P = \text{true}$ iff $\nu(P) = \mathtt{t}$. By Lemma 2, page 33, for each wff A,

$$\nu A = \mathtt{t} \qquad \text{iff } \iota A \text{ is true for } \sum$$

where ι is the identity interpreter. By assumption, ιB is true for $\sum$; so $\nu B = \mathtt{t}$. We conclude that for each valuation ν, $\nu B = \mathtt{t}$.

2. Assume that $\nu B = \mathtt{t}$ for each valuation ν. We shall show that B is a true wff. Let $\sum$ be any statement system and let μ be any $\sum$-interpreter. Let ν be the valuation of Lemma 1, page 32, i.e., for each atomic wff (P),

$$\nu(P) = \mathtt{t} \qquad \text{iff } \mu(P) \text{ is true for } \sum$$

By Lemma 1, page 32, for each wff A,

$$\nu A = \mathtt{t} \qquad \text{iff } \mu A \text{ is true for } \sum$$

By assumption, $\nu B = \mathtt{t}$; so μB is true for $\sum$. We conclude that B is a true wff.

Here is an important fact.

Lemma 1. $A \wedge B$ is a true wff iff both A and B are true wffs.

Dem. We shall apply our Criterion for True Wffs.

1. Assume that $A \wedge B$ is a true wff. Then $\nu(A \wedge B) = \mathtt{t}$ for each valuation ν. It follows that $\nu A = \mathtt{t}$ and $\nu B = \mathtt{t}$ for each valuation ν. Thus, both A and B are true wffs.

2. Assume that both A and B are true wffs. Then $\nu A = \mathtt{t}$ and $\nu B = \mathtt{t}$ for each valuation ν. Thus, $\nu(A \wedge B) = \mathtt{t}$ for each valuation ν. Applying our Criterion for True Wffs, we conclude that $A \wedge B$ is true.

We now present the important notion of a *model* of a set of wffs. A statement system $\sum$ is said to be a model of K, a set of wffs, provided that there is a $\sum$-interpreter μ such that μA is true for $\sum$ for each $A \in \mathsf{K}$. In this case, we also say that $\sum$ is a model of K under μ.

The statement system $\sum$ of Example 1, page 31, is a model of $\mathsf{K} = \{\neg X, Y, \neg Z\}$ under the $\sum$-interpreter μ of that example. Note that $\mu(\neg X)$, μY, and $\mu(\neg Z)$ are true swffs of $\sum$.

Clearly, each statement system is a model of K if each member of K is a true wff. In particular, each statement system is a model of the empty set.

The ideas of this section center around the truth-values t and f. These symbols are not part of the alphabet of the propositional calculus; we introduce truth-values and the notion of a *true* wff in an effort to motivate the work of Section 3.3, where we shall characterize the notion of a true wff within our formal language.

Exercises

1. Let A, B, and C be any wffs; prove that each of the following wffs is true:
   (a) $A \vee A \rightarrow A$.
   (b) $A \rightarrow A \vee B$.
   (c) $A \rightarrow B \dot{\rightarrow} C \vee A \rightarrow B \vee C$.
2. Show that B is a true wff if both A and $A \rightarrow B$ are true wffs.
3. Show that $A \rightarrow C$ is a true wff given that both $A \rightarrow B$ and $B \rightarrow C$ are true wffs.
4. Show that $\neg B \rightarrow \neg A$ is true iff $A \rightarrow B$ is true.
5. Let Σ be any statement system and let μ be any Σ-interpreter.
   (a) Show that $\mu[A \wedge B]$ is true for Σ iff μA and μB are both true for Σ.
   (b) Show that $\mu[A \rightarrow B]$ is true for Σ if μA is false for Σ or μB is true for Σ.
   (c) Show that $\mu[A \rightarrow B]$ is false for Σ if μA is true for Σ and μB is false for Σ.
   (d) Show that $\mu[A \leftrightarrow B]$ is true for Σ iff μA and μB have the same truth-value in Σ.
6. Let A and B be wffs such that $\nu A = \nu B$ for each valuation ν. Prove that both $A \rightarrow B$ and $B \rightarrow A$ are true wffs.
7. Let A and B be wffs such that both $A \rightarrow B$ and $B \rightarrow A$ are true. Prove that $\nu A = \nu B$ for each valuation ν.
8. Given that $A \rightarrow C$ and $B \rightarrow C$ are true wffs, show that $A \vee B \rightarrow C$ is a true wff.

Use the Criterion for True Wffs to prove that none of the following wffs is true:

9. (Y).
10. $X \vee Y$.
11. $X \vee \neg Y$.
12. $X \rightarrow Y$.
13. $X \rightarrow Y \dot{\rightarrow} X \vee Y$.
14. $X \rightarrow Y \dot{\rightarrow} X \vee \neg Y$.

15. Justify the statement that each statement system is a model of the empty set.

16. Prove that each statement system is a model of $\{X \vee \neg X\}$.

17. Let K be a set of true wffs; prove that each statement system is a model of K.

Use the Criterion for True Wffs to prove that each of the following wffs is true, where A, B, and C are any wffs:

18. $(A \rightarrow B) \wedge (B \rightarrow C) \dot{\rightarrow} A \rightarrow C$.

19. $A \rightarrow B \dot{\rightarrow} C \vee \neg B \rightarrow C \vee \neg A$.

20. $A \rightarrow B \dot{\rightarrow} C \wedge A \rightarrow C \wedge B$.

21. $(A \rightarrow C) \wedge (B \rightarrow C) \dot{\rightarrow} A \vee B \rightarrow C$.

22. $(A \rightarrow B) \wedge (A \rightarrow C) \dot{\rightarrow} A \rightarrow B \wedge C$.

23. $A \vee (B \vee C) \leftrightarrow (A \vee B) \vee C$.

24. $A \wedge (B \wedge C) \leftrightarrow (A \wedge B) \wedge C$.

25. $A \vee (B \wedge C) \leftrightarrow (A \vee B) \wedge (A \vee C)$.

26. $A \wedge (B \vee C) \leftrightarrow (A \wedge B) \vee (A \wedge C)$.

3.3. Proofs and Provable Wffs

The word *proof* is used in a highly technical sense in a theory of deduction; a proof is a finite sequence of wffs satisfying certain formal requirements. In order to understand the formal requirements that appear in the definition of *proof*, we recall that each wff in the following three sets of wffs is true:

AS 1. $\{A \vee A \rightarrow A \mid A \text{ is a wff}\}$.
AS 2. $\{A \rightarrow A \vee B \mid A \text{ and } B \text{ are wffs}\}$.
AS 3. $\{A \rightarrow B \dot{\rightarrow} C \vee A \rightarrow B \vee C \mid A, B, \text{ and } C \text{ are wffs}\}$.

Moreover, B is a true wff if both A and $A \rightarrow B$ are true wffs.

Now, our goal is to characterize the set of all true wffs without referring to truth-values. We achieve this in two steps. First, we shall define the notion of a *proof*; next, we shall define the notion of a *provable* wff. Later we shall see that the set of all provable wffs is the set of all true wffs.

Here is our definition of *proof*.

Definition of Proof. A finite sequence of wffs is called a *proof* if each of its terms, say E, satisfies at least one of the following conditions:

(a) E is a member of one of the three sets of wffs displayed above.
(b) There is a wff D such that both D and $D \rightarrow E$ precede E in the sequence.

The three sets of wffs involved in our definition, i.e., 1, 2, and 3 above, are known as *axiom schemes*. The second part of our definition involves Modus Ponens, a rule of inference.

Here is a proof.

Example 1.

$$X \vee X \rightarrow X \mathbin{\dot{\rightarrow}} \neg X \vee (X \vee X) \rightarrow X \vee \neg X,\ X \vee X \rightarrow X,$$
$$\neg X \vee (X \vee X) \rightarrow X \vee \neg X,\ X \rightarrow X \vee X,\ X \vee \neg X.$$

We regard a proof as being an entity in itself, a mathematical object worthy of study in its own right and possessing interesting and useful properties. To illustrate this point of view, consider the following facts.

Lemma 1. Let π_1 and π_2 be proofs; then the sequence π_1, π_2 obtained by adjoining the sequence π_2 to the sequence π_1 is a proof.

Dem. It is easy to show that each term of the sequence π_1, π_2 is either a member of an axiom scheme or is obtained from preceding terms of the sequence by applying Modus Ponens. The details are left as an exercise. Clearly π_1, π_2 is a finite sequence of wffs.

Lemma 2. Let π_1 be a proof whose last term is A, and let π_2 be a proof whose last term is $A \rightarrow B$; then the sequence π_1, π_2, B is a proof.

Dem. The details are left as an exercise.

Lemma 3. Let π be a proof, let (P) be an atomic wff, and let B be any wff. Let π' be the sequence obtained from π by replacing each instance of (P) in each wff of π by B. Then π' is a proof.

Dem. If π' is not a proof, there is a term of π' that violates the requirements for a proof; indeed, there is a first term of π' with this property, say E' (i.e., each term of π' that precedes E' meets the requirements of a proof). Consider the corresponding term of π, which we denote by E. Then E is not a member of an axiom scheme [replacing (P) by B throughout an axiom yields a member of the same axiom scheme]. Therefore E is preceded in π by wffs of the form $D \rightarrow E$ and D. So, in π',

E' is preceded by wffs of the form $D' \to E'$ and D'. Thus, the presence of E' in the sequence π' is justified by Modus Ponens. This contradiction proves that π' is a proof.

To illustrate this result, let B be any wff; then the sequence obtained from the proof of Example 1 by replacing each instance of (X) by B is a proof.

Next, we introduce the notion of a *provable* wff.

Definition. A wff, say C, is said to be *provable* if there is a proof whose last term is C.

We shall abbreviate the statement "C is provable" by writing "$\vdash C$." The last term of the proof of Example 1 is $X \vee \neg X$, so $\vdash X \vee \neg X$, i.e., the wff $X \vee \neg X$ is provable.

Let C be a wff such that $\vdash C$; then there exists a proof π whose last term is C. In this case we say that π is a *proof of* C. So, the proof of Example 1 is a proof of $X \vee \neg X$.

Analogous to the Fundamental Theorem about Wffs is the useful Fundamental Theorem about Provable Wffs, which we now present.

Fundamental Theorem about Provable Wffs. A set of provable wffs, say S, is the set of all provable wffs provided that:

1. $A \in S$ if A is a member of an axiom scheme.
2. $B \in S$ if there is a wff A such that both $A \to B$ and A are in S.

Dem. Let S be a set of provable wffs that meets the conditions of our theorem, but does not contain all provable wffs. Then there is a provable wff, say A, which is not in S. Let π be a proof of A; clearly there is a term of π, say B, such that $B \notin S$ but each term of π that precedes B is in S. By (1), B is not an axiom. But π is a proof; therefore, B is preceded in π by wffs $C \to B$ and C (for some wff C). From (2) we conclude that $B \in S$. This contradiction demonstrates that the assumption of this argument is false; we conclude that S contains all provable wffs.

As we have mentioned, we wish to show that the set of provable wffs *is* the set of true wffs; indeed, our purpose in introducing the notion of a provable wff is to characterize the concept of a true wff within our formal language. We are now in a position to prove that the set of all provable wffs is a subset of the set of all true wffs.

Lemma 4. Each provable wff is true.

Dem. Let $S = \{A \in \mathsf{W} \mid A \text{ is true and } \vdash A\}$. We have seen earlier that each axiom is true; so $A \in S$ if A is an axiom. We have also pointed out that Modus Ponens applies to true wffs, i.e., E is true if both D and $D \rightarrow E$ are true. Therefore, $B \in S$ if there is a wff A such that both $A \rightarrow B$ and A are in S. Applying the Fundamental Theorem about Provable Wffs, we conclude that each provable wff is in S; i.e., each provable wff is true.

Later we shall prove that each true wff is provable.

Corollary 1. No atomic wff is provable.

Dem. No atomic wff is true.

We can say a little more in this direction.

Corollary 2. The length of each provable wff is at least two.

Dem. No wff of length one is true; so, by Lemma 4, no wff of length one is provable.

Since $\vdash X \vee \neg X$, and this wff has length two, Corollary 2 is the best we can do in this direction.

The axiomatic definition of *proof* given here is based on the approach of Hilbert and Ackermann (1950) (the original German edition is dated 1928), which involved four axiom schemes. Their approach was, in turn, based on *Principia Mathematica* by Whitehead and Russell (1910). The reduction to three axioms schemes was begun by Götlind (1947) and completed by Rasiowa (1949).

Exercises

1. Show that the following sequence is a proof: $Y \vee X \rightarrow (Y \vee X) \vee Y \rightarrow \neg Y \vee (Y \vee X) \rightarrow ((Y \vee X) \vee Y) \vee \neg Y$, $Y \vee X \rightarrow (Y \vee X) \vee Y$, $\neg Y \vee (Y \vee X) \rightarrow ((Y \vee X) \vee Y) \vee \neg Y$, $Y \rightarrow Y \vee X$, $((Y \vee X) \vee Y) \vee \neg Y$.

2. Let A be any wff; use Example 1 to construct a proof whose last term is $A \vee \neg A$.

3. Show that $\vdash A \vee \neg A$ provided that A is a wff.

4. Prove Lemma 1.

5. Prove Lemma 2.

6. Let A be any provable wff, let π be any proof of A, and let B be any wff. Show that the sequence $\pi, A \rightarrow A \vee B, A \vee B$ is a proof.

7. Prove that $\vdash A \vee B$, given that $\vdash A$ and that B is a wff.

8. Suppose that $\vdash A \rightarrow B$, π is a proof of $A \rightarrow B$, and C is a wff. Show that the sequence $\pi, A \rightarrow B \mathrel{\dot{\rightarrow}} C \vee A \rightarrow B \vee C, C \vee A \rightarrow B \vee C$ is a proof.

9. Prove that $\vdash C \vee A \rightarrow B \vee C$, given that $\vdash A \rightarrow B$ and given that C is a wff.

10. Show that $\vdash A$ if $\vdash A \vee A$.

Use Lemma 4 to prove that none of the following wffs is provable.

11. (Y).

12. $X \vee Y$.

13. $X \vee \neg Y$.

14. $X \rightarrow Y$.

15. $X \rightarrow Y \mathrel{\dot{\rightarrow}} X \vee Y$.

16. $X \rightarrow Y \mathrel{\dot{\rightarrow}} X \vee \neg Y$.

3.4. Rules of Inference

By a *rule of inference* we mean a statement that asserts that a wff is provable if certain other wffs are provable. Here are some examples.

Modus Ponens. $\vdash B$ if $\vdash A$ and $\vdash A \rightarrow B$.

Dem. See Lemma 2, page 38.

Lemma 1. $\vdash C \vee A \rightarrow B \vee C$ if $\vdash A \rightarrow B$.

Dem. Let π be a proof of $A \rightarrow B$. Then $\pi, A \rightarrow B \mathrel{\dot{\rightarrow}} C \vee A \rightarrow B \vee C, C \vee A \rightarrow B \vee C$ is a proof. So, $\vdash C \vee A \rightarrow B \vee C$.

Lemma 2. $\vdash C \vee \neg A$ if $\vdash A \rightarrow B$ and $\vdash B \rightarrow C$.

Dem. By Lemma 1, $\vdash \neg A \vee B \rightarrow C \vee \neg A$; so $\vdash C \vee \neg A$ by Modus Ponens.

Lemma 3. $\vdash \neg B \rightarrow \neg A$ iff $\vdash A \rightarrow B$.

Dem. 1. Assume that $\vdash \neg B \rightarrow \neg A$. By Lemma 1, $\vdash B \vee \neg B \rightarrow \neg A \vee B$. But $\vdash B \vee \neg B$ for each wff B (see Example 1, page 38, and the comment following Lemma 3, page 38). So $\vdash A \rightarrow B$ by Modus Ponens.

2. Assume that $\vdash A \rightarrow B$. By Example 1, page 38, $\vdash B \rightarrow \neg(\neg B)$. By Lemma 1, $\vdash \neg A \vee B \rightarrow \neg(\neg B) \vee \neg A$; i.e., $\vdash A \rightarrow B \dot{\rightarrow} \neg B \rightarrow \neg A$. So, by Modus Ponens, $\vdash \neg B \rightarrow \neg A$.

Corollary 1. $\vdash \neg(\neg A) \rightarrow A$ for each wff A.

Dem. By Example 1, page 38, $\vdash \neg A \rightarrow \neg(\neg(\neg A))$. Thus, by Lemma 3, $\vdash \neg(\neg A) \rightarrow A$.

Of course, Corollary 1 is not a rule of inference. Our next lemma is a rule of inference.

Lemma 4. $\vdash A \rightarrow \neg(\neg C)$ if $\vdash A \rightarrow B$ and $\vdash B \rightarrow C$.

Dem. By Lemma 3, $\vdash \neg C \rightarrow \neg B$ and $\vdash \neg B \rightarrow \neg A$. So, by Lemma 2, $\vdash \neg A \vee \neg(\neg C)$; i.e., $\vdash A \rightarrow \neg(\neg C)$.

Our next three results are not rules of inference; they assert that each wff possessing a certain form is provable.

Lemma 5. $\vdash A \vee \neg(\neg A) \rightarrow \neg(\neg A)$ for each wff A.

Dem. By Corollary 1, $\vdash \neg(\neg A) \rightarrow A$; so by Lemma 1, $\vdash A \vee \neg(\neg A) \rightarrow A \vee A$. But $\vdash A \vee A \rightarrow A$; thus, by Lemma 4, $\vdash A \vee \neg(\neg A) \rightarrow \neg(\neg A)$.

Lemma 6. $\vdash A \rightarrow A$ for each wff A.

Dem. By AS 2, $\vdash A \rightarrow A \vee \neg(\neg A)$; by Lemma 5, $\vdash A \vee \neg(\neg A) \rightarrow \neg(\neg A)$. Thus, by Lemma 2, $\vdash \neg(\neg A) \vee \neg A$. Application of Lemma 3 yields our result.

Lemma 7. $\vdash A \vee B \rightarrow B \vee A$ for any wff A and B.

Dem. By Lemma 6, $\vdash B \rightarrow B$; thus, by Lemma 1, $\vdash A \vee B \rightarrow B \vee A$.

Here is a rule of inference.

Corollary 2. $\vdash B \vee A$ if $\vdash A \vee B$.

Dem. Apply Modus Ponens to $\vdash A \vee B$ and $\vdash A \vee B \to B \vee A$.

We can now prove that $\to$ is transitive; this is a rule of inference.

Theorem 1. $\vdash A \to C$ if $\vdash A \to B$ and $\vdash B \to C$.

Dem. By Lemma 2, $\vdash C \vee \neg A$; so, by Corollary 2, $\vdash A \to C$.

In our next lemma we show that a certain variation of the form involved in AS 3 also yields provable wffs.

Lemma 8. $\vdash A \to B \mathrel{\dot{\to}} C \vee A \to C \vee B$ for any wffs A, B, and C.

Dem. By AS 3, $\vdash A \to B \mathrel{\dot{\to}} C \vee A \to B \vee C$. In Exercise 5, we show that $\vdash C \vee A \to B \vee C \mathrel{\dot{\to}} C \vee A \to C \vee B$. Thus, by Theorem 1, $\vdash A \to B \mathrel{\dot{\to}} C \vee A \to C \vee B$.

Here is a rule of inference that we shall need later.

Lemma 9. $\vdash A \vee C \to B \vee D$ provided that $\vdash A \to B$ and $\vdash C \to D$.

Dem. We are given that $\vdash A \to B$; so by Lemma 1, $\vdash C \vee A \to B \vee C$. By Lemma 7, $\vdash A \vee C \to C \vee A$. So, by Theorem 1,

$$\vdash A \vee C \to B \vee C \tag{1}$$

Also, $\vdash C \to D$ is given; so $\vdash B \vee C \to D \vee B$ by Lemma 1. By Lemma 7, $\vdash D \vee B \to B \vee D$. So, by Theorem 1,

$$\vdash B \vee C \to B \vee D \tag{2}$$

Application of Theorem 1 to (1) and (2) yields $\vdash A \vee C \to B \vee D$.

The following corollary is also useful.

Corollary 3. $\vdash A \wedge C \to B \wedge D$ provided that $\vdash A \to B$ and $\vdash C \to D$.

Dem. By Lemma 3, $\vdash \neg B \to \neg A$ and $\vdash \neg D \to \neg C$; thus, by Lemma 9,

$$\vdash \neg B \vee \neg D \to \neg A \vee \neg C$$

Applying Lemma 3 to the preceding provable wff, we obtain

$$\vdash \neg(\neg A \vee \neg C) \to \neg(\neg B \vee \neg D)$$

i.e.,

$$\vdash A \wedge C \rightarrow B \wedge D$$

Our next lemma is needed to establish Lemma 11.

Lemma 10. $\vdash A \rightarrow A \wedge A$ for each wff A.

Dem. By AS 1, $\vdash \neg A \vee \neg A \rightarrow \neg A$; thus, by Lemma 3,

$$\vdash \neg(\neg A) \rightarrow \neg(\neg A \vee \neg A)$$

i.e.,

$$\vdash \neg(\neg A) \rightarrow A \wedge A$$

This suffices to prove the lemma.

Here is Lemma 11; we shall use it to establish our Criterion for Deducibility, page 90, and Fact 8, page 88.

Lemma 11. $\vdash A \rightarrow C$ if $\vdash A \rightarrow B$ and $\vdash A \wedge B \rightarrow C$.

Dem. $\vdash A \rightarrow A$ and $\vdash A \rightarrow B$; by Corollary 3, $\vdash A \wedge A \rightarrow A \wedge B$. By Theorem 1, $\vdash A \wedge A \rightarrow C$, and by Lemma 10, $\vdash A \rightarrow A \wedge A$. Thus, by Theorem 1, $\vdash A \rightarrow C$.

Exercises

1. Let A be any wff; show that the sequence: $A \vee A \rightarrow A \dot{\rightarrow} \neg A \vee (A \vee A) \rightarrow A \vee \neg A$, $A \vee A \rightarrow A$, $A \rightarrow (A \vee A) \dot{\rightarrow} A \vee \neg A$, $A \rightarrow A \vee A$, $A \vee \neg A$ is a proof.
2. Prove that $\vdash A \vee \neg A$ for each wff A.
3. Show that $\vdash \neg A \vee \neg(\neg A)$ for each wff A.
4. Use Lemmas 1 and 7 of this section to show that

   $$\vdash C \vee A \rightarrow B \vee C \dot{\rightarrow} (C \vee B) \vee \neg(C \vee A)$$

   for any wffs A, B, and C.
5. Apply Lemma 7 and Theorem 1 of this section to prove that

   $$\vdash C \vee A \rightarrow B \vee C \dot{\rightarrow} C \vee A \rightarrow C \vee B$$

   for any wffs A, B, and C.
6. Prove that $\vdash C \vee A \rightarrow C \vee B$ if $\vdash A \rightarrow B$ and C is a wff.

7. Prove that $\vdash A \vee C \rightarrow B \vee C$ if $\vdash A \rightarrow B$ and C is a wff.
8. Prove that $\vdash A \rightarrow B \mathrel{\dot{\rightarrow}} (C \rightarrow A) \rightarrow (C \rightarrow B)$ for any wffs A, B, and C.
9. Let $\vdash A \rightarrow B$ and let C be any wff; prove that $\vdash C \wedge A \rightarrow C \wedge B$.

3.5. Equivalent Wffs

We now introduce an important binary relation on the set of all wffs.

Definition. We shall say that wffs A and B are *equivalent* (in symbols $A \equiv B$) if $\vdash A \rightarrow B$ and $\vdash B \rightarrow A$.

For example, $A \equiv A$ since $\vdash A \rightarrow A$ whenever A is a wff. It is easy to verify that for any wffs A and B, $\neg(\neg A) \equiv A$ and $A \vee B \equiv B \vee A$.

Lemma 1. $\equiv$ is an equivalence relation on the set of all wffs.

Dem. We have already pointed out that $\equiv$ is reflexive; it is easy to verify that $\equiv$ is symmetric and transitive.

Lemma 2. $A \equiv B$ iff $\neg A \equiv \neg B$.

Dem. 1. Assume that $A \equiv B$. Then $\vdash A \rightarrow B$ and $\vdash B \rightarrow A$. By Lemma 3, page 41, $\vdash \neg B \rightarrow \neg A$ and $\vdash \neg A \rightarrow \neg B$; so $\neg A \equiv \neg B$.

2. Assume that $\neg A \equiv \neg B$. Then $\vdash \neg A \rightarrow \neg B$ and $\vdash \neg B \rightarrow \neg A$; thus $\vdash B \rightarrow A$ and $\vdash A \rightarrow B$ (by Lemma 3, page 41). So $A \equiv B$.

Lemma 3. If $A \equiv B$ and C is any wff, then $A \vee C \equiv B \vee C$ and $C \vee A \equiv C \vee B$.

Dem. 1. By Lemma 8, page 43, $\vdash A \rightarrow B \mathrel{\dot{\rightarrow}} C \vee A \rightarrow C \vee B$. Thus, by Modus Ponens, $\vdash C \vee A \rightarrow C \vee B$. Similarly, $\vdash B \rightarrow A \mathrel{\dot{\rightarrow}} C \vee B \rightarrow C \vee A$; so $\vdash C \vee B \rightarrow C \vee A$; thus $C \vee A \equiv C \vee B$.

2. We have already observed that $A \vee C \equiv C \vee A$ and $C \vee B \equiv B \vee C$; in the first part of this demonstration we have shown that $C \vee A \equiv C \vee B$. Thus

$$A \vee C \equiv C \vee A \equiv C \vee B \equiv B \vee C$$

so $A \vee C \equiv B \vee C$ since $\equiv$ is transitive.

We now present some connections between equivalent wffs, valuations, truth-values of swffs of a statement system, and true wffs.

Lemma 4. If $A \equiv B$, then $\nu A = \nu B$ for each valuation ν.

Dem. Let ν be a valuation such that $\nu A = \mathsf{t}$ and $\nu B = \mathsf{f}$. Now, each provable wff is true, so $A \to B$ is true; thus $\nu[A \to B] = \mathsf{t}$. But $\nu[A \to B] = \mathsf{f}$ since $\nu A = \mathsf{t}$ and $\nu B = \mathsf{f}$. We conclude that $\nu A = \nu B$.

Corollary 1. Let $A \equiv B$; then both A and B are true, or neither is true.

Although the converse of Lemma 4 is correct, we are not yet in a position to prove so. To this purpose we require the following lemma and the fact that each true wff is provable, which we shall soon establish.

Lemma 5. If $\nu A = \nu B$ for each valuation ν, then $A \to B$ and $B \to A$ are both true wffs.

Dem. We must show that $\nu[A \to B] = \nu[B \to A] = \mathsf{t}$ for each valuation ν. But $\nu A = \nu B$; so $\nu[A \to B] = \mathsf{t}$ and $\nu[B \to A] = \mathsf{t}$.

We turn now to statement systems.

Lemma 6. Let $\sum$ be any statement system, let μ be any $\sum$-interpreter, and let $A \equiv B$; then the swffs μA and μB have the same truth-value.

Dem. Each provable wff is true; so $A \to B$ is true. It follows that the swff $\mu(A \to B)$ is true for $\sum$, i.e.,

$$\mu A \to \mu B \tag{1}$$

is true for $\sum$. Considering the true wff $B \to A$, we find that the swff

$$\mu B \to \mu A \tag{2}$$

is true for $\sum$. Assume that μA is true for $\sum$; from (1), μB is true for $\sum$. Assume that μA is false for $\sum$; from (2), μB is false for $\sum$. This establishes Lemma 6.

Here is a related result.

Lemma 7. If $A \equiv B$, then $A \leftrightarrow B$ is a true wff.

Dem. By assumption $\vdash A \to B$ and $\vdash B \to A$. Each provable wff is true; thus $A \rightarrow B$ and $B \rightarrow A$ are true wffs. In view of Lemma 1, page 35,

$$(A \rightarrow B) \wedge (B \rightarrow A)$$

is a true wff; i.e., $A \leftrightarrow B$ is true.

We mention that Lemma 6 follows from Lemma 7.

Our next result is important.

Lemma 8. $A \rightarrow B \equiv \neg B \rightarrow \neg A$.

Dem.

$$\begin{aligned} \neg B \rightarrow \neg A &= \neg(\neg B) \vee \neg A \\ &\equiv \neg A \vee \neg(\neg B) && \text{(since } \vee \text{ is commutative)} \\ &\equiv \neg A \vee B && \text{(by Lemma 3)} \end{aligned}$$

It follows that $\neg A \vee B \equiv \neg B \rightarrow \neg A$.

By the definition of equivalent wff, Lemma 8 yields the following fact.

Corollary 2. $\vdash A \rightarrow B \mathrel{\dot{\rightarrow}} \neg B \rightarrow \neg A$ and $\vdash \neg B \rightarrow \neg A \mathrel{\dot{\rightarrow}} A \rightarrow B$.

We can now establish the following rule of inference.

Lemma 9. If $\vdash C \mathrel{\dot{\rightarrow}} A \rightarrow B$, then $\vdash C \mathrel{\dot{\rightarrow}} \neg B \rightarrow \neg A$.

Dem. $\vdash C \mathrel{\dot{\rightarrow}} A \rightarrow B$ is given, and $\vdash A \rightarrow B \mathrel{\dot{\rightarrow}} \neg B \rightarrow \neg A$ by Corollary 2; so $\vdash C \mathrel{\dot{\rightarrow}} \neg B \rightarrow \neg A$ by Theorem 1, page 43.

Exercises

1. Prove Lemma 1.

2. Show that $A \equiv B$ if $\vdash A$ and $\vdash B$ (i.e., provable wffs are equivalent).

3. Let A and B be wffs such that $A \vee C \equiv B \vee C$ for some wff C. Is it necessarily the case that $A \equiv B$? Justify your answer.

4. Let A and B be wffs such that $A \vee C \equiv B \vee C$ for each wff C. Is it necessarily the case that $A \equiv B$? Justify your answer.

5. Let A and B be wffs such that $A \wedge C \equiv B \wedge C$ for some wff C. Is it necessarily the case that $A \equiv B$? Justify your answer.

6. Prove that $A \wedge B \equiv B \wedge A$ for any wffs A and B.

7. Let A and B be wffs such that $A \wedge C \equiv B \wedge C$ for each wff C. Is it necessarily the case that $A \equiv B$? Justify your answer.

8. Let $A \equiv B$ and let C be any wff; prove that $C \wedge A \equiv C \wedge B$.

4

Substitution Theorems

4.1. Subwffs, Components, and Wff-Builders

Here is the notion of a *subwff* of a wff.

Definition of Subwff. We say that B is a *subwff* of a wff A provided that B is a wff and there are expressions ϕ and θ such that $A = \phi B \theta$.

So, a subwff of a wff A is a block of symbols contained in A that is itself a wff. For example, each of (X), (Y), $(\neg(X))$, $((Y) \vee (\neg(X)))$ is a subwff of $((Y) \vee (\neg(X)))$.

In order to utilize our Fundamental Theorem about Wffs in this connection, we need to formulate the notion of a subwff in a manner that apes the definition of the notion of wff itself. Accordingly, we now introduce the concept of a *component* of a wff.

Definition of Component. Each atomic wff, say A, has exactly one component, namely A itself; the components of $\neg B$ are $\neg B$ and each component of B; the components of $C \vee D$ are $C \vee D$, each component of C, and each component of D.

For example, the components of $((Y) \vee (\neg(X)))$ are the following wffs: $((Y) \vee (\neg(X)))$, (Y), $(\neg(X))$, (X).

We must show that our definition of *component* is proper.

Lemma 1. The definition of *component* associates a unique set of wffs with each wff.

Dem. Apply the Construction Theorem for Wffs.

We want to prove that our notions of *subwff* and *component* are the same; i.e., for any wff A and B, B is a subwff of A iff B is a component of A. We shall do this in two steps.

Lemma 2. Each component of a wff A is a subwff of A.

Dem. Let $S = \{A \in \mathsf{W} \mid \text{each component of } A \text{ is a subwff of } A\}$. We now apply the Fundamental Theorem about Wffs to prove that $S = \mathsf{W}$.

1. Let A be any atomic wff. But the only component of an atomic wff A is A itself; by definition, A is a subwff of A. Thus $A \in S$.

2. Let $B \in S$; we shall prove that $\neg B \in S$. The components of $\neg B$ are $\neg B$ and the components of B. Clearly, $\neg B$ is a subwff of $\neg B$; moreover, each component of B is a subwff of B by assumption, but each subwff of B is a subwff of $\neg B$, so each component of B is a subwff of $\neg B$. Thus, each component of $\neg B$ is a subwff of $\neg B$. Therefore, $\neg B \in S$.

3. Let $C, D \in S$; we shall prove that $C \vee D \in S$. The components of $C \vee D$ are $C \vee D$, each component of C, and each component of D. Each of these wffs is a subwff of $C \vee D$. Therefore, $C \vee D \in S$.

Applying the Fundamental Theorem about Wffs, we conclude that $S = \mathsf{W}$; this establishes Lemma 2.

Lemma 3. Each subwff of a wff A is a component of A.

Dem. Let $S = \{A \in \mathsf{W} \mid \text{each subwff of } A \text{ is a component of } A\}$. We now apply the Fundamental Theorem about Wffs to prove that $S = \mathsf{W}$.

1. Let A be any atomic wff. The only subwff of an atomic wff A is A itself; by definition, A is a component of A. Thus $A \in S$.

2. Let $C \in S$; we shall prove that $\neg C \in S$. Let B be a subwff of $\neg C$; then there are expressions ϕ and θ such that $(\neg C) = \phi B \theta$. Now, the leftmost LH parenthesis of B has a mate in B, namely the rightmost RH parenthesis of B. There are two possibilities: (i) The leftmost LH parenthesis of B is the leftmost symbol of $(\neg C)$. In this case, the rightmost symbol of B is the rightmost symbol of $(\neg C)$; so $B = (\neg C)$, hence, B is a component of C. (ii) The leftmost LH parenthesis of B is a symbol of C. But the mate of this LH parenthesis of C is a RH parenthesis of C, so B is a subwff of C. Since $C \in S$, it follows that B is a component of C. Thus, by definition, B is a component of $\neg C$.

3. Assume that $D, E \in S$; we shall show that $D \vee E \in S$. Let B be any subwff of $D \vee E$. Considering the leftmost LH parenthesis of B, we can show, as in the preceding case, that B is a component of $D \vee E$. This completes our proof.

Corollary 1. Let A and B be any wffs; then B is a component of A iff B is a subwff of A.

We now turn our attention from a subwff B of a wff $\phi B \theta$ to the accompanying expressions ϕ and θ.

Definition of Wff-Builder. We shall say that $[\phi, \theta]$ is a *wff-builder* if there is a wff A such that $\phi A \theta$ is a wff.

Note. Here, A is a wff rather than a *name* for a wff.

For example, $[(\neg,)]$ is a wff-builder since $(\neg(X))$ is a wff. Also, $[((Y) \vee ,)]$ is a wff-builder since $((Y) \vee (Z))$ is a wff. Notice that $[\ ,\ ]$ is a wff-builder (here, both ϕ and θ are the empty expression). On the other hand, $[(,)]$ is *not* a wff-builder.

Let ϕ and θ be expressions such that for *some* wff A, $\phi A \theta$ is a wff; we shall now prove that $\phi B \theta$ is a wff for *each* wff B. This fact about wff-builders can be formulated as follows.

Lemma 4. Let C be any wff and let C' be any expression obtained from C by replacing a subwff of C by any wff. Then C' is a wff.

Dem. We rely on the Fundamental Theorem about Wffs. Let $S = \{C \in \mathsf{W} \mid$ each expression obtained from C by replacing any subwff of C by any wff is a wff$\}$.

1. Let C be any atomic wff. Then C has just one subwff, itself. Substituting any wff B for C yields $C' = B$; so C' is a wff. Thus $C \in S$.
2. Let $D \in S$; we shall show that $\neg D \in S$. The subwffs of $\neg D$ are $\neg D$ and each subwff of D. Substituting any wff, say B, for $\neg D$ yields B, which is a wff. Let A be any subwff of D; so $D = \phi A \theta$. Substituting any wff B for A yields $D' = \phi B \theta$; this is a wff since $D \in S$. Therefore, $[(\neg D)]' = (\neg \phi B \theta)$ is a wff.
3. Let $E, F \in S$; we shall show that $E \vee F \in S$. The subwffs of $E \vee F$ are: $E \vee F$, each subwff of E, and each subwff of F. Substituting any wff B for $E \vee F$ yields B, a wff. Let A be any subwff of E; so $E = \phi A \theta$. Substituting B for A yields $E' = \phi B \theta$; so $[(E \vee F)]' = (\phi B \theta \vee F)$, which is a wff. Similarly, substituting a wff B for a subwff of F yields a wff. Thus $E \vee F \in S$.

We conclude that S is the set of all wffs; this establishes Lemma 4.

We are now ready to prove the comment that precedes Lemma 4.

Corollary 2. Let $[\phi, \theta]$ be any wff-builder and let B be any wff; then $\phi B\theta$ is a wff.

Dem. By assumption, there is a wff A such that $\phi A\theta$ is a wff. Clearly A is a subwff of $\phi A\theta$. Substituting B, where B is any wff, for this subwff of $\phi A\theta$ yields $\phi B\theta$. By Lemma 4, $\phi B\theta$ is a wff.

Exercises

List the subwffs of each of the following wffs.

1. $X \rightarrow Y \dot{\rightarrow} \neg Z \vee X \rightarrow Y \vee \neg Z.$
2. $X \rightarrow Y \dot{\rightarrow} \neg Y \rightarrow \neg X.$
3. $X \wedge Y \leftrightarrow X \vee \neg Y.$
4. Criticize the following argument: "[(,)] is a wff-builder because X is a wff and (X) is a wff."

4.2. Substitution Theorem for Wffs

We are now ready to consider an important fact.

Substitution Theorem for Wffs. Let $A \equiv B$ and let $[\phi, \theta]$ be any wff-builder; then $\phi A\theta \equiv \phi B\theta$.

To establish this theorem, it is convenient to formulate it in terms of the following many-valued operation on W. Let $C \in \mathsf{W}$; then C' denotes any wff obtained from C by replacing a subwff of C by an *equivalent* wff.

Substitution Theorem for Wffs. For each wff C and for each corresponding wff C', $C \equiv C'$.

Dem. Let $S = \{C \in \mathsf{W} \mid C \equiv C' \text{ for each } C'\}$. We shall apply the Fundamental Theorem about Wffs to prove that $S = \mathsf{W}$.

1. Consider an atomic C. Let $B \equiv C$; substituting B for C yields B, i.e., $C' = B$. So $C \equiv C'$.

2. Let $D \in S$; we shall show that $\neg D \in S$. Let A be any subwff of $\neg D$ and let $B \equiv A$. If $A = \neg D$, then $[\neg D]' = B$; so $\neg D \equiv [\neg D]'$. If A is a subwff of D, then $D = \phi A\theta$; thus $D \equiv \phi B\theta$ by assumption. Therefore, by Lemma 2, page 45, $\neg D \equiv \neg\phi B\theta$. So, $\neg D \equiv [\neg D]'$. Thus $\neg D \in S$.

3. Let $E, F \in S$; we shall show that $E \vee F \in S$. Let A be any subwff of $E \vee F$ and let $B \equiv A$. If $A = E \vee F$, then $[E \vee F]' = B$; so $E \vee F \equiv [E \vee F]'$. If A is a subwff of E, then $E = \phi A\theta$; so $E \equiv \phi B\theta$ by assumption. Thus, by Lemma 3, page 45, $E \vee F \equiv \phi B\theta \vee F$, i.e., $E \vee F \equiv [E \vee F]'$. Similarly, if A is a subwff of F, we see that $E \vee F \equiv [E \vee F]'$. Thus $E \vee F \in S$. Applying the Fundamental Theorem about Wffs, we conclude that $S = \mathsf{W}$. This establishes our Substitution Theorem for Wffs.

Our result has the following useful corollary.

Substitution Theorem for Provable Wffs. $\vdash \phi B\theta$ if $\vdash \phi A\theta$ and if $A \equiv B$.

Dem. Since $A \equiv B$, it follows that $\phi A\theta \equiv \phi B\theta$; so $\vdash \phi A\theta \rightarrow \phi B\theta$. But $\vdash \phi A\theta$; by Modus Ponens, $\vdash \phi B\theta$.

We now illustrate these substitution theorems.

Example 1. Show that $A \vee C \equiv B \vee D$ if $A \equiv B$ and $C \equiv D$.

Solution. We shall apply the Substitution Theorem for Wffs twice. First, we apply the wff-builder $[(, \vee C)]$ to $A \equiv B$; this yields $A \vee C \equiv B \vee C$. Next, we apply the wff-builder $[(B \vee,)]$ to $C \equiv D$; this yields $B \vee C \equiv B \vee D$. But $\equiv$ is transitive, so these two equivalences yield $A \vee C \equiv B \vee D$. Alternatively, we can use subwffs instead of wff-builders; thus $A \vee C \equiv B \vee C$ since $A \equiv B$; and $B \vee C \equiv B \vee D$ since $C \equiv D$. Thus $A \vee C \equiv B \vee D$.

Example 2. Show that $\vdash \phi B\theta$ if $\vdash \phi(\neg(\neg B))\theta$.

Solution. Now, $\neg(\neg B) \equiv B$; so, by the Substitution Theorem for Provable Wffs, $\vdash \phi B\theta$.

Example 3. Show that $\vdash \phi(B \vee A)\theta$ if $\vdash \phi(A \vee B)\theta$ and $[\phi, \theta]$ is a wff-builder.

Solution. Recall that $A \vee B \equiv B \vee A$ and apply the Substitution Theorem for Provable Wffs.

Example 4. Show that $A \to B \equiv \neg(A \wedge \neg B)$ for any wffs A and B.

Solution. Now, $A \wedge \neg B = \neg(\neg A \vee \neg(\neg B))$; so

$$
\begin{aligned}
\neg(A \wedge \neg B) &= \neg(\neg(\neg A \vee \neg(\neg B))) \\
&\equiv \neg(\neg(\neg A \vee B)) && \text{[substitute } B \text{ for } \neg(\neg B)] \\
&\equiv \neg A \vee B && \text{[since } \neg(\neg C) \equiv C \text{ for any wff } C] \\
&= A \to B
\end{aligned}
$$

Thus $\quad A \to B \equiv \neg(A \wedge \neg B)$.

Let A and B be subwffs of a wff C; we shall say that A and B are *disjoint* subwffs of C provided that

$$C = \phi_1 A \phi_2 B \phi_3 \qquad \text{or} \qquad C = \phi_1 B \phi_2 A \phi_3$$

where ϕ_1, ϕ_2, and ϕ_3 are expressions. For example, (X) and (Y) are disjoint subwffs of $((\neg(X)) \vee (Y))$.

More generally, let $A_1, \ldots, A_n$ be subwffs of a wff C. We shall say that $A_1, \ldots, A_n$ are *disjoint* subwffs of C provided that

$$C = \phi_1 A_{i_1} \phi_2 A_{i_2} \phi_3 \cdots \phi_n A_{i_n} \phi_{n+1}$$

where $\{i_1, \ldots, i_n\} = \{1, \ldots, n\}$ and $\phi_1, \ldots, \phi_n$ are expressions.

We sometimes wish to apply the Substitution Theorem for Wffs (or the Substitution Theorem for Provable Wffs) several times, operating on disjoint subwffs of some wff, i.e., replacing disjoint subwffs by equivalent wffs. We can regard this as a single application of a generalized substitution theorem.

Generalized Substitution Theorem for Wffs. Let $A_1, \ldots, A_n$ be disjoint subwffs of a wff C, let $A_i \equiv B_i$ for $i = 1, \ldots, n$, and let D be the wff obtained from C by substituting B_i for A_i, $i = 1, \ldots, n$. Then $C \equiv D$.

Dem. Apply the Substitution Theorem for Wffs n times.

Here are some examples.

Example 5. Show that $\neg(A \vee B) \equiv \neg A \wedge \neg B$ for each wff A and B.

Solution. Now,

$$\begin{aligned} \neg A \wedge \neg B &= \neg(\neg(\neg A) \vee \neg(\neg B)) && \text{(by definition)} \\ &\equiv \neg(A \vee B) && \text{(by the Generalized Substitution Theorem for Wffs)} \end{aligned}$$

Next, we rework Example 1 from the viewpoint of disjoint subwffs.

Example 6. Show that $A \vee C \equiv B \vee D$ if $A \equiv B$ and $C \equiv D$.

Solution. Notice that A and C are disjoint subwffs of $A \vee C$; thus, by the Generalized Substitution Theorem for Wffs,

$$A \vee C \equiv B \vee D$$

Example 7. Show that $\vdash A \rightarrow B \dot{\rightarrow} C \wedge A \rightarrow C \wedge B$ for any wffs A, B, and C.

Solution. By Lemma 8, page 43,

$$\vdash \neg B \rightarrow \neg A \dot{\rightarrow} \neg C \vee \neg B \rightarrow \neg C \vee \neg A \tag{1}$$

By Lemma 8, page 47,

$$\text{LHS} \equiv A \rightarrow B$$

and

$$\begin{aligned} \text{RHS} &\equiv \neg(\neg C \vee \neg A) \rightarrow \neg(\neg C \vee \neg B) \\ &\equiv \neg(\neg C) \wedge \neg(\neg A) \rightarrow \neg(\neg C) \wedge \neg(\neg B) && \text{(by Example 5)} \\ &\equiv C \wedge A \rightarrow C \wedge B && \text{(by the Generalized Substitution Theorem for Wffs)} \end{aligned}$$

Substituting in (1), we obtain

$$\vdash A \rightarrow B \dot{\rightarrow} C \wedge A \rightarrow C \wedge B$$

by the Substitution Theorem for Provable Wffs.

Exercises

1. Prove that $\neg(\neg A \land \neg B) \equiv A \lor B$ for any wffs A and B.
2. Using the fact that $\vdash \neg X \lor \neg(\neg X)$, show that $\vdash \neg X \lor X$.
3. Given that $\vdash A \rightarrow B$, prove that $\vdash B \lor A \rightarrow B$.
4. Generalize the Substitution Theorem for Provable Wffs following the procedure used in generalizing the Substitution Theorem for Wffs.

5

Duality

5.1. Normal Form

In Section 2.4 we introduced the connective $\wedge$ (and) as part of a parentheses-omitting program. Here and in Section 5.4 we shall bring out an important aspect of this connective. Indeed, we shall establish an algorithm that yields a wff equivalent to the negation of a given wff, say A. The first step is to put A into a certain standard form, which we call *normal* form. This involves finding a wff B equivalent to A, such that B has a name expressed in terms of $\neg$, $\vee$, and $\wedge$, in which each instance of $\neg$ is prefixed to an atomic wff.

We begin the task of putting A into normal form by expressing all occurrences of $\rightarrow$ and $\leftrightarrow$ in terms of $\neg$, $\vee$, and $\wedge$. Next, we must recognize each $\wedge$-like (read "and-like") subwff of A. A wff is said to be $\wedge$-like if it has the form "$\neg(C \vee D)$," where C and D are wffs. To ***recognize*** an $\wedge$-like subwff of A, say $\neg(C \vee D)$, we replace this subwff by $\neg C \wedge \neg D$, an equivalent wff.

Recall the following facts:

(a) $\neg(\neg D) \equiv D$ for any wff D.
(b) $\neg(D \vee E) \equiv \neg D \wedge \neg E$ for any wffs D and E.
(c) $\neg(D \wedge E) \equiv \neg D \vee \neg E$ for any wffs D and E.

These equivalences and the Substitution Theorem for Wffs allow us to obtain a wff equivalent to a given wff; moreover, each instance of $\neg$ in the resulting *name* of a wff is prefixed to an atomic wff. The idea is to apply (a), (b), and (c) repeatedly until each occurrence of $\neg$ is prefixed to an atomic wff; we call this operation *bringing* $\neg$ *inside.*

We now illustrate this procedure for putting a wff into normal form.

Example 1. Put $E = \neg(X \vee \neg Y) \leftrightarrow Z$ into normal form.

Solution. First, we eliminate $\rightarrow$ and $\leftrightarrow$:

$$\begin{aligned} E &= \neg(X \vee \neg Y) \rightarrow Z \mathbin{\dot\wedge} Z \rightarrow \neg(X \vee \neg Y) \\ &= \neg\neg(X \vee \neg Y) \vee Z \mathbin{\dot\wedge} \neg Z \vee \neg(X \vee \neg Y) \end{aligned}$$

Next, we recognize each $\wedge$-like subwff:

$$E \equiv \neg(\neg X \wedge \neg\neg Y) \vee Z \mathbin{\dot\wedge} \neg Z \vee (\neg X \wedge \neg\neg Y)$$

Finally, we bring $\neg$ inside:

$$\begin{aligned} E &\equiv (\neg\neg X \vee \neg\neg\neg Y) \vee Z \mathbin{\dot\wedge} \neg Z \vee (\neg X \wedge Y) \\ &\equiv (X \vee \neg Y) \vee Z \mathbin{\dot\wedge} \neg Z \vee (\neg X \wedge Y) \end{aligned}$$

We have put E into normal form.

Here is our algorithm for obtaining a wff equivalent to the negation of a given wff.

Algorithm. Let A be any wff; then $\neg A \equiv C$, where C is obtained from A in three steps, as follows.

Step 1. Put A into normal form, say B.

Step 2. Interchange $\vee$ and $\wedge$ throughout B.

Step 3. If an atomic subwff of B is prefixed by $\neg$, delete this $\neg$; if an atomic subwff of B is not prefixed by $\neg$, then insert $\neg$.

We shall justify this algorithm in Section 5.4 (see Lemma 3, page 69). Meanwhile, we illustrate our algorithm.

Example 2. Find a wff equivalent to $\neg E$, where E is the wff of Example 1.

Solution. From Example 1, Step 1 of the algorithm yields a wff B, where

$$B = (X \vee \neg Y) \vee Z \mathbin{\dot\wedge} \neg Z \vee (\neg X \wedge Y)$$

Step 2, applied to B, yields

$$(X \wedge \neg Y) \wedge Z \mathbin{\dot\vee} \neg Z \wedge (\neg X \vee Y) \tag{1}$$

Step 3, applied to (1), yields

$$(\neg X \wedge Y) \wedge \neg Z \mathbin{\dot\vee} Z \wedge (X \vee \neg Y)$$

By our algorithm, we conclude that this wff is equivalent to $\neg E$.

The notion of *normal* form discussed here is highly intuitive; this discussion is intended to motivate the transform N defined on page 62.

Exercises

Put each of the following wffs into normal form; in each case, show that your answer is equivalent to the given wff.

1. $X \vee \neg(Y \vee \neg X)$.
2. $Y \wedge \neg(X \vee \neg X) \dot{\vee} \neg Y$.
3. $\neg(Y \vee (X \rightarrow \neg Y))$.
4. $X \rightarrow Y \dot{\rightarrow} \neg Z \vee X \rightarrow \neg Z \vee Y$.
5. $\neg X \rightarrow Y \leftrightarrow \neg Y \rightarrow X$.
6. For each of the five preceding wffs, find a wff equivalent to its negation.

5.2. Syntactical Transforms

In order to formalize our notion of *normal form* introduced in Section 5.1, it is helpful to consider the general concept of a syntactical transform. By a *syntactical transform* we mean any map of W, the set of all wffs, into W. Following the usual code, "T[A]" denotes the wff that a syntactical transform T associates with a wff A; usually we shall suppress the brackets, writing "TA," but only if no ambiguity results.

We point out that a map is a syntactical transform provided that its domain is W and its range is a subset of W. For example, each of the following maps is a syntactical transform:

$$\{(A, A) \mid A \in \mathsf{W}\}, \quad \{(A, Y) \mid A \in \mathsf{W}\}, \quad \{(A, \neg A) \mid A \in \mathsf{W}\},$$
$$\{(A, A \vee A) \mid A \in \mathsf{W}\}$$

In each of these examples we characterize a syntactical transform by providing a specific rule for determining the image of A, where A is any wff.

We now present an efficient method of characterizing or exhibiting specific syntactical transforms. Of course, we have characterized a syntactical transform T iff we can compute TA for each wff A. In particular, we must be able to determine TA for each atomic wff A; we must be able to determine T[$\neg B$] in terms of B and the T-images of a finite number of wffs, each shorter than $\neg B$; we must be able to compute

T[$C \vee D$] in terms of C, D, and the T-images of a finite number of wffs, each shorter than $C \vee D$. It turns out that this information completely characterizes T.

First, we shall establish the following basic result.

Lemma 1. A map T of S into W is a syntactical transform if S is a set of wffs such that:

1. $A \in S$ for each atomic wff A.
2. $\neg B \in S$ provided that $B \in S$.
3. $C \vee D \in S$ provided that $C, D \in S$.

Dem. We must show that $S =$ W, i.e., T is a map of W into W. But S satisfies the conditions of the Fundamental Theorem about Wffs; thus $S =$ W.

We come now to our main result. Notice that we define T[$\neg B$] and T[$C \vee D$] in terms of certain T-images. In a sense, then, we define T in terms of T. Of course, this is generally unacceptable; here, our procedure is sound because of the requirement that T[$\neg B$] be defined in terms of the T-images of wffs *shorter* than $\neg B$ and T[$C \vee D$] be defined in terms of the T-images of wffs *shorter* than $C \vee D$.

Theorem 1. Let T be any syntactical transform; then T is characterized by providing the following information.

(a) Announce TA for each atomic wff A. The definition may involve A but may not involve a T-image.
(b) Announce T[$\neg B$] in terms of B and the T-images of a finite number of wffs, each shorter than $\neg B$; here B is any wff.
(c) Announce T[$C \vee D$] in terms of C, D, and the T-images of a finite number of wffs each shorter than $C \vee D$; here C and D are any wffs.

Dem. This is similar to the Construction Theorem for Wffs. A modification of the proof given there will be applicable here. The details are left to Exercise 6.

In view of this result we can characterize a specific syntactical transform, say T, by announcing the T-image of each atomic wff, the T-image of $\neg B$ in terms of B and the T-images of wffs shorter than $\neg B$, and the T-image of $C \vee D$ in terms of C, D, and the T-images of wffs shorter than $C \vee D$. In Section 5.3 we shall define the syntactical transforms N, M, R, and D by following this prescription.

Example 1. We can now clarify the construction involved in Lemma 3, page 38. There, we replaced each occurrence of a given atomic wff (P) throughout a proof π by a specific wff B. Let U be a map of a subset of W into W, such that:

1. $\mathsf{U}A = \begin{cases} B & \text{if} \quad A = (P) \\ A & \text{if} \quad A \neq (P) \text{ and if } A \text{ is atomic} \end{cases}$
2. $\mathsf{U}[\neg A] = \neg\mathsf{U}A$ for each wff A.
3. $\mathsf{U}[A \vee C] = \mathsf{U}A \vee \mathsf{U}C$ for each wff $A \vee C$.

Here, think of P as a specific proposition; actually, we are defining a family of syntactical transforms, one for each proposition P. By Lemma 1, U is a syntactical transform. By Theorem 1, there is just one syntactical transform that meets conditions 1, 2, and 3. So, these three conditions specify or characterize a unique syntactical transform. In colloquial language, U replaces each occurrence of (P) in a given wff by B; i.e., for each wff A, $\mathsf{U}A$ is the wff obtained by substituting B for (P) throughout A. Of course, $\mathsf{U}A = A$ in case (P) does not occur in A.

In Lemma 3, page 38, the sequence π' is obtained from the given proof π by replacing each term of π by its U-image. In other words, $\pi' = \mathsf{U}\pi$. In Lemma 3, we proved that $\mathsf{U}\pi$ is a proof if π is a proof.

Exercises

1. Let N be a map whose domain is W and whose range is a subset of W such that:

   (1) $\mathsf{N}A = A$ for each atomic wff A.

   (2) $\mathsf{N}[\neg A] = \begin{cases} \neg A & \text{if} \quad A \text{ is atomic} \\ \mathsf{N}B & \text{if} \quad A = \neg B \\ \mathsf{N}[\neg C] \wedge \mathsf{N}[\neg D] & \text{if} \quad A = C \vee D. \end{cases}$

   (3) $\mathsf{N}[C \vee D] = \mathsf{N}C \vee \mathsf{N}D$ for any wffs C and D.

   Prove directly, i.e., without using Theorem 1, that there is just one syntactical transform that satisfies conditions 1–3.

2. Let N be the syntactical transform of Exercise 1.

   (a) Use the Fundamental Theorem about Wffs to prove that $\mathsf{N}[\neg A] \equiv \neg\mathsf{N}[A]$ for each wff A.

   (b) Use part (a) and the Fundamental Theorem about Wffs to prove that $\mathsf{N}A \equiv A$ for each wff A.

3. Let M be a map whose domain is W and whose range is a subset of W such that:

(1) $\mathsf{M}A = \neg A$ for each atomic wff A.

(2) $\mathsf{M}[\neg A] = \begin{cases} A & \text{if } A \text{ is atomic} \\ \mathsf{M}B & \text{if } A = \neg B \\ \mathsf{M}[\neg C] \vee \mathsf{M}[\neg D] & \text{if } A = C \vee D. \end{cases}$

(3) $\mathsf{M}[C \vee D] = \mathsf{M}C \wedge \mathsf{M}D$ for any wffs C and D.

Prove directly, i.e., without using Theorem 1, that there is just one syntactical transform that satisfies conditions 1–3.

4. Let M be the syntactical transform of Exercise 3.

(a) Use the Fundamental Theorem about Wffs to prove that $\mathsf{M}[\neg A] \equiv \neg\mathsf{M}[A]$ for each wff A.

(b) Use part (a) and the Fundamental Theorem about Wffs to prove that $\mathsf{M}A \equiv \neg A$ for each wff A.

5. Let S be a map whose domain is a subset of W and whose range is a subset of W such that:

(1) $\mathsf{S}A = (X)$ for each atomic wff A.

(2) $\mathsf{S}[\neg A] = \neg\mathsf{S}A$ for each wff A.

(3) $\mathsf{S}[A \vee B] = \mathsf{S}A \vee \mathsf{S}B$ for each wff $A \vee B$.

(a) Compute $\mathsf{S}[Y \vee \neg Z \rightarrow X]$.

(b) Show that $\mathsf{S}[A \rightarrow B] = \mathsf{S}A \rightarrow \mathsf{S}B$ for any wffs A and B.

(c) Prove that $\vdash\mathsf{S}A$ if $\vdash A$.

(d) Show that $\mathsf{S}[\mathsf{S}A] = \mathsf{S}A$ for each wff A.

(e) Describe the map S in intuitive terms.

6. Prove Theorem 1.

5.3. Normal Transforms

In Section 5.1 we mentioned that a wff A is in normal form if each $\wedge$-component of A has been recognized and if $\neg$ is prefixed only to atomic wffs. Of course, this is merely an intuitive description of normal form. More precisely, A is in normal form iff there is a wff B such that $\mathsf{N}B = A$, where N is the following syntactical transform. Throughout this section we shall rely on Theorem 1, page 60, to characterize specific syntactical transforms. Here is our definition of N.

Definition of N:

1. $\mathsf{N}A = A$ for each atomic wff A.

2. $\mathsf{N}[\neg A] = \begin{cases} \neg A & \text{if } A \text{ is atomic} \\ \mathsf{N}B & \text{if } A = \neg B \\ \mathsf{N}[\neg C] \wedge \mathsf{N}[\neg D] & \text{if } A = C \vee D. \end{cases}$

3. $\mathsf{N}[C \vee D] = \mathsf{N}C \vee \mathsf{N}D$ for each wff $C \vee D$.

For example,

$$
\begin{aligned}
&\mathsf{N}[X \vee \neg Y \dot{\vee} \neg(\neg W \vee (Z \vee \neg X))] \\
&\quad = \mathsf{N}[X \vee \neg Y] \vee \mathsf{N}[\neg(\neg W \vee (Z \vee \neg X))] && \text{(by 3)} \\
&\quad = \mathsf{N}[X \vee \neg Y] \dot{\vee} \mathsf{N}W \wedge \mathsf{N}[\neg(Z \vee \neg X)] && \text{[by 2 (applied twice)]} \\
&\quad = \mathsf{N}X \vee \mathsf{N}[\neg Y] \dot{\vee} \mathsf{N}W \wedge (\mathsf{N}[\neg Z] \wedge \mathsf{N}X) && \text{(by 2 and 3)} \\
&\quad = X \vee \neg Y \dot{\vee} W \wedge (\neg Z \wedge X) && \text{(by 1 and 2)}
\end{aligned}
$$

Our goal in this chapter is to establish the Principle of Duality (see page 70). To this purpose we shall need four syntactical transforms N, M, R, and D. We have already defined N; we now present definitions of M, R, and D.

Definition of M:

1. $\mathsf{M}A = \neg A$ for each atomic wff A.

2. $\mathsf{M}[\neg A] = \begin{cases} A & \text{if } A \text{ is atomic} \\ \mathsf{M}B & \text{if } A = \neg B \\ \mathsf{M}[\neg C] \vee \mathsf{M}[\neg D] & \text{if } A = C \vee D. \end{cases}$

3. $\mathsf{M}[C \vee D] = \mathsf{M}C \wedge \mathsf{M}D$ for each wff $C \vee D$.

Definition of R:

1. $\mathsf{R}A = \neg A$ for each atomic wff A.

2. $\mathsf{R}[\neg A] = \begin{cases} A & \text{if } A \text{ is atomic} \\ \mathsf{R}B & \text{if } A = \neg B \\ \mathsf{R}[\neg C] \wedge \mathsf{R}[\neg D] & \text{if } A = C \vee D. \end{cases}$

3. $\mathsf{R}[C \vee D] = \mathsf{R}C \vee \mathsf{R}D$ for each wff $C \vee D$.

Definition of D:

1. $\mathsf{D}A = A$ for each atomic wff A.

2. $\mathsf{D}[\neg A] = \begin{cases} \neg A & \text{if } A \text{ is atomic} \\ \mathsf{D}B & \text{if } A = \neg B \\ \mathsf{D}[\neg C] \vee \mathsf{D}[\neg D] & \text{if } A = C \vee D. \end{cases}$

3. $\mathsf{D}[C \vee D] = \mathsf{D}C \wedge \mathsf{D}D$ for each wff $C \vee D$.

As we have mentioned, the syntactical transform N is designed to put each wff, say A, into normal form in the sense of Section 5.1. Notice that N has the following impact on A:

(a) N recognizes each $\wedge$-like subwff of A.
(b) N eliminates $\neg\neg$ wherever this appears in A.

To help understand these transforms, we now state their purpose. Now, M is designed first to transform a wff into normal form, then to interchange ∨ and ∧ throughout, and finally to attach ¬ to each atomic wff that is not prefixed by ¬, at the same time deleting ¬ from each atomic wff to which it is prefixed. This is achieved in three steps as follows:

(a) Write down the definition of N.
(b) Put ∧ for the displayed ∨ in the RHS of 3; put ∨ for the displayed ∧ in the RHS of the third line of 2.
(c) Attach ¬ to the RHS of 1; delete ¬ from the RHS of the first line of 2.

Moving on to R, we point out that this syntactical transform is designed first to put a wff into normal form and next to reverse the impact of M on atomic wff. Now, the operation of attaching ¬ or deleting ¬ is reversed by repeating the operation. So our definition of R is obtained in two steps:

(a) Write down the definition of N.
(b) Attach ¬ to the RHS of 1; delete ¬ from the RHS of the first line of 2.

Finally, we consider the transform D. This syntactical transform is designed first to put a wff into normal form, and then to interchange ∨ and ∧ throughout. Accordingly, our definition of D can be obtained in two steps:

(a) Write down the definition of N.
(b) Put ∧ for the displayed ∨ in the RHS of 3; put ∨ for the displayed ∧ in the RHS of the third line of 2.

It is important to observe that the definitions of N, M, R, and D follow a common pattern. Table 1 makes this pattern more evident.

Notice that for each of the transforms N, M, R, and D, say T, TA is either A or $\neg A$ if A is atomic, and T$[C \vee D]$ is either T$C \vee$ TD or

Table 1

	N	M	R	D
1. A atomic	A	$\neg A$	$\neg A$	A
2. (a) $\neg A$, A atomic	$\neg A$	A	A	$\neg A$
(b) $\neg\neg B$	NB	MB	RB	DB
(c) $\neg(C \vee D)$	N$[\neg C]$ ∧ N$[\neg D]$	M$[\neg C]$ ∨ M$[\neg D]$	R$[\neg C]$ ∧ R$[\neg D]$	D$[\neg C]$ ∨ D$[\neg D]$
3. $C \vee D$	NC ∨ ND	MC ∧ MD	RC ∨ RD	DC ∧ DD

$\mathsf{T}C \wedge \mathsf{T}D$. Moreover, the definition of $\mathsf{T}[\neg A]$ can be reconstructed from parts 1 and 3 of the definition. For example, if A is atomic, $\mathsf{T}[\neg A] = A$ in case $\mathsf{T}A = \neg A$, whereas $\mathsf{T}[\neg A] = \neg A$ in case $\mathsf{T}A = A$.

In summary, the definition of each of the transforms N, M, R, and D is characterized by two parameters, which we denote by n and d, where $\mathsf{T}A = nA$ for A atomic and $\mathsf{T}[C \vee D] = \mathsf{T}C\, d\, \mathsf{T}D$; so n is either $\neg$ or is blank, and d is either $\vee$ or $\wedge$.

In the spirit of this observation, we now present the notion of a normal transform.

Definition. A syntactical transform T is said to be *normal* if:

1. $\mathsf{T}A = nA$ whenever A is atomic.
2. $$\mathsf{T}[\neg A] = \begin{cases} mA & \text{if } A \text{ is atomic} \\ \mathsf{T}B & \text{if } A = \neg B \\ \mathsf{T}[\neg B]\, c\, \mathsf{T}[\neg C] & \text{if } A = B \vee C. \end{cases}$$
3. $\mathsf{T}[A \vee B] = \mathsf{T}A\, d\, \mathsf{T}B$ for each wff $A \vee B$.

where n is $\neg$ or is blank, m is $\neg$ or is blank, and $m \neq n$; and $\{d, c\} = \{\vee, \wedge\}$.

Clearly, there are exactly four normal transforms, namely N, M, R, and D, whose parameters are displayed in Table 2.

The following lemmas are easy to verify; throughout, T is a normal transform with parameters n and d.

Lemma 1. $\mathsf{T}[nA] = A$ for any atomic wff A.

Lemma 2. $\mathsf{T}[mA] = \neg A$ for any atomic wff A.

Lemma 3. $\mathsf{T}[A\, d\, B] = \mathsf{T}A \vee \mathsf{T}B$ for any wffs A and B.

Lemma 4. $\mathsf{T}[A\, c\, B] = \mathsf{T}A \wedge \mathsf{T}B$ for any wffs A and B.

Lemma 5. $\mathsf{T}[A \wedge B] = \mathsf{T}A\, c\, \mathsf{T}B$ for any wffs A and B.

Table 2

T	n	d
N		$\vee$
M	$\neg$	$\wedge$
R	$\neg$	$\vee$
D		$\wedge$

We want to show that normal transforms form a group under composition. Composition of maps is associative; so it is only necessary to verify the following:

1. The product of any two normal transforms is a normal transform.
2. N is the identity.
3. Each normal transform has an inverse, namely itself.

Proceeding intuitively, based on the motivating definition of the normal transforms N, M, R, and D, we obtain the multiplication table—Table 3—for our group operation. Examining this table, we can now verify 1–3. Moreover, we see from the table that the group operation is commutative.

Our results, which are highly intuitive, can be obtained rigorously by applying the following theorem, which expresses the parameters of a product $\mathsf{T}_1\mathsf{T}_2$ in terms of the parameters of T_1 and T_2. We shall express "n is blank" by writing "n = bl."

Theorem 1. Let T_1 be a normal transform with parameters n_1 and d_1, and let T_2 be a normal transform with parameters n_2 and d_2. Then $\mathsf{T}_1\mathsf{T}_2$ is a normal transform and its parameters are n and d, where

$$n = \begin{cases} n_1 & \text{if } n_2 = \text{bl} \\ m_1 & \text{if } n_2 = \neg \end{cases} \quad \text{and} \quad d = \begin{cases} d_1 & \text{if } d_2 = \vee \\ c_1 & \text{if } d_2 = \wedge \end{cases}$$

Dem. It is easy to verify that if A is atomic, then:

$$\text{(a)} \quad [\mathsf{T}_1\mathsf{T}_2]A = \begin{cases} n_1 A & \text{if } n_2 = \text{bl} \\ m_1 A & \text{if } n_2 = \neg \end{cases}$$

Similarly, it is clear that

$$\text{(b)} \quad [\mathsf{T}_1\mathsf{T}_2][B \vee C] = \begin{cases} [\mathsf{T}_1\mathsf{T}_2]B\ d_1\ [\mathsf{T}_1\mathsf{T}_2]C & \text{if } d_2 = \vee \\ [\mathsf{T}_1\mathsf{T}_2]B\ c_1\ [\mathsf{T}_1\mathsf{T}_2]C & \text{if } d_2 = \wedge \end{cases}$$

(c) Finally, we consider $[\mathsf{T}_1\mathsf{T}_2][\neg A]$. Here, there are three cases.

Table 3

	N	M	R	D
N	N	M	R	D
M	M	N	D	R
R	R	D	N	M
D	D	R	M	N

1. A atomic. Then

$$[\mathsf{T}_1\mathsf{T}_2][\neg A] = \mathsf{T}_1[m_2 A] = \begin{cases} m_1 A & \text{if} \quad m_2 = \neg \\ n_1 A & \text{if} \quad m_2 = \text{bl} \end{cases}$$

This, together with (a), verifies that $n = n_1$ if $n_2 = \text{bl}$ and that $n = m_1$ if $n_2 = \neg$.

2. $A = \neg B$. Then $[\mathsf{T}_1\mathsf{T}_2][\neg A] = [\mathsf{T}_1\mathsf{T}_2]B$.
3. $A = C \vee D$. Then

$$[\mathsf{T}_1\mathsf{T}_2][\neg A] = \mathsf{T}_1[\mathsf{T}_2[\neg C]\, c_2\, \mathsf{T}_2[\neg D]]$$

so

$$[\mathsf{T}_1\mathsf{T}_2][\neg A] = [\mathsf{T}_1\mathsf{T}_2][\neg C]\, d_1\, [\mathsf{T}_1\mathsf{T}_2][\neg D]$$

if $c_2 = \vee$; whereas by Lemma 5,

$$[\mathsf{T}_1\mathsf{T}_2][\neg A] = [\mathsf{T}_1\mathsf{T}_2][\neg C]\, c_1\, [\mathsf{T}_1\mathsf{T}_2][\neg D]$$

if $c_2 = \wedge$. This, together with (b), verifies that $d = d_1$ if $d_2 = \vee$ and that $d = c_1$ if $d_2 = \wedge$. We conclude that $\mathsf{T}_1\mathsf{T}_2$ is a normal transform provided T_1 and T_2 are both normal; moreover, the parameters of $\mathsf{T}_1\mathsf{T}_2$ are as stated in Theorem 1.

Using this result, it is easy to establish the following useful facts.

Lemma 6. $\mathsf{TT} = \mathsf{N}$ whenever T is normal.

Lemma 7. $\mathsf{TN} = \mathsf{T}$ whenever T is normal.

Theorem 2. $(\{\mathsf{N}, \mathsf{M}, \mathsf{R}, \mathsf{D}\}, \circ, \mathsf{N})$ is a group; here $\circ$ denotes composition.

Dem. Lemma 7 asserts that N is the group identity, and Lemma 6 asserts that each group element is its own inverse. But composition of mappings is associative; so, our algebraic system is a group.

Moreover, it is well known that any group that possesses the property of Lemma 6 is Abelian. This is easy to prove directly; assume that $xx = e$ whenever $x \in G$, where $(G, \cdot, e)$ is a group. Now $(xy)(xy) = e$, so $x(xy)(xy)y = xey$, i.e., $yx = xy$. So $(G, \cdot, e)$ is Abelian.

In the case of our group we obtain

Lemma 8. $\mathsf{T}_1\mathsf{T}_2 = \mathsf{T}_2\mathsf{T}_1$ whenever T_1 and T_2 are normal.

Thus the group of Theorem 2 is Abelian.

Next, we work out the parameters of MR. By Theorem 1 they are bl and $\wedge$; these are the parameters of D. So, we have verified the following lemma.

Lemma 9. MR = D.

Exercises

1. Show that $\mathsf{N}[A \wedge B] = \mathsf{N}A \wedge \mathsf{N}B$ for any wffs A and B.
2. Show that $\mathsf{N}[A \rightarrow B] = \mathsf{N}[\neg A] \vee \mathsf{N}B$ for any wffs A and B.
3. Show that $\mathsf{N}[A \leftrightarrow B] = (\mathsf{N}[\neg A] \vee \mathsf{N}B) \wedge (\mathsf{N}[\neg B] \vee \mathsf{N}A)$ for any wffs A and B.
4. Compute $\mathsf{N}[(X \rightarrow \neg Y) \leftrightarrow (X \wedge Y)]$.
5. Show that $\mathsf{M}[A \wedge B] = \mathsf{M}A \vee \mathsf{M}B$ for any wffs A and B.
6. Show that $\mathsf{M}[A \rightarrow B] = \mathsf{M}[\neg A] \wedge \mathsf{M}B$ for any wffs A and B.
7. Show that $\mathsf{M}[A \leftrightarrow B] = (\mathsf{M}[\neg A] \wedge \mathsf{M}B) \vee (\mathsf{M}[\neg B] \wedge \mathsf{M}A)$ for any wffs A and B.
8. Compute $\mathsf{M}[\neg X \wedge Y \rightarrow \neg Z]$.
9. Compute $\mathsf{R}[X \rightarrow Y \wedge \neg Z]$.
10. Compute $\mathsf{D}[\neg Y \leftrightarrow X \wedge \neg Y]$.
11. Prove Lemma 1.
12. Prove Lemma 2.
13. Prove Lemma 3.
14. Prove Lemma 4.
15. Prove Lemma 5.
16. Use Theorem 1 to prove Lemma 6.
17. Use Theorem 1 to prove Lemma 7.

5.4. Duality

We intend to prove that for each normal transform T, $\mathsf{T}A \equiv \mathsf{T}B$ iff $A \equiv B$. First, we need the following fact.

Lemma 1. Let T be any normal transform; then $\mathsf{T}[\neg A] \equiv \neg\mathsf{T}A$ for each wff A.

Dem. Let T be a normal transform with parameters n and d. We shall use the Fundamental Theorem about Wffs to establish Lemma 1. Accordingly, let $S = \{A \mid \mathsf{T}[\neg A] \equiv \neg\mathsf{T}A\}$.

1. A atomic. Now, $\mathsf{T}[\neg A] = mA$ and $\neg\mathsf{T}A = \neg nA \equiv mA = \mathsf{T}[\neg A]$; so $A \in S$.
2. Given that $B \in S$, show that $\neg B \in S$; i.e., $\mathsf{T}[\neg\neg B] \equiv \neg\mathsf{T}[\neg B]$. But

$$\mathsf{T}[\neg\neg B] = \mathsf{T}B \equiv \neg\mathsf{T}[\neg B]$$

So $\neg B \in S$.

3. Given that $C, D \in S$, show that $C \vee D \in S$; i.e., $\mathsf{T}[\neg(C \vee D)] \equiv \neg\mathsf{T}[C \vee D]$. Now,

$$\mathsf{T}[\neg(C \vee D)] = \mathsf{T}[\neg C] \, c \, \mathsf{T}[\neg D] \equiv \neg\mathsf{T}C \, c \, \neg\mathsf{T}D$$

and

$$\neg\mathsf{T}[C \vee D] = \neg(\mathsf{T}C \, d \, \mathsf{T}D) \equiv \neg\mathsf{T}C \, c \, \neg\mathsf{T}D$$

So $C \vee D \in S$. We conclude that S is the set of all wffs.

Next, we exhibit facts about N and M that will help us to verify that $\mathsf{T}A \equiv \mathsf{T}B$ if $A \equiv B$, where T is any normal transform.

Lemma 2. $\mathsf{N}A \equiv A$ for each wff A.

Dem. Apply the Fundamental Theorem about Wffs.

Lemma 3. $\mathsf{M}A \equiv \neg A$ for each wff A.

Dem. Apply the Fundamental Theorem about Wffs.

This result verifies our algorithm on page 58.
We can now establish the following result.

Lemma 4. $\mathsf{N}A \equiv \mathsf{N}B$ if $A \equiv B$.

Dem. By Lemma 2, $\mathsf{N}A \equiv A$ and $\mathsf{N}B \equiv B$; but $A \equiv B$ and $\equiv$ is transitive; so $\mathsf{N}A \equiv \mathsf{N}B$.

Lemma 5. $\mathsf{M}A \equiv \mathsf{M}B$ if $A \equiv B$.

Dem. Since $A \equiv B$, $\neg A \equiv \neg B$; by Lemma 3, $\mathsf{M}A \equiv \neg A$ and $\mathsf{M}B \equiv \neg B$. Thus, $\mathsf{M}A \equiv \mathsf{M}B$.

We need certain facts about R in order to prove that $\mathsf{R}A \equiv \mathsf{R}B$ whenever $A \equiv B$.

Lemma 6. $\mathsf{R}[A \rightarrow B] \equiv \mathsf{R}A \rightarrow \mathsf{R}B$ for any wffs A and B.

Dem. $\mathsf{R}[A \rightarrow B] = \mathsf{R}[\neg A \vee B] = \mathsf{R}[\neg A] \vee \mathsf{R}B \equiv \neg \mathsf{R}A \vee \mathsf{R}B$ by Lemma 1 and the Substitution Theorem for Wffs.

Lemma 7. $\vdash \mathsf{R}A$ if $\vdash A$.

Dem. We shall use the Fundamental Theorem about Provable Wffs. Let

$$S = \{A \in \mathsf{W} \mid \vdash A \text{ and } \vdash \mathsf{R}A\}$$

It is easy to verify that $\vdash \mathsf{R}A$ if A is an axiom; indeed $\mathsf{R}A$ is equivalent to a member of the same axiom scheme as A. Thus, each axiom is in S. Next, we must show that $B \in S$ if $A \in S$ and if $A \rightarrow B \in S$. By assumption, $\vdash A$ and $\vdash A \rightarrow B$; so $\vdash B$. Moreover, $\vdash \mathsf{R}A$ and $\vdash \mathsf{R}[A \rightarrow B]$. It follows that $\vdash \mathsf{R}A \rightarrow \mathsf{R}B$. So, by Modus Ponens, $\vdash \mathsf{R}B$. Therefore, $B \in S$. We conclude that S is the set of all provable wffs. This establishes Lemma 7.

Corollary 1. $\vdash \mathsf{R}A$ iff $\vdash A$.

Dem. If $\vdash \mathsf{R}A$, then by Lemma 7, $\vdash \mathsf{R}[\mathsf{R}A]$; i.e., $\vdash \mathsf{N}A$, so $\vdash A$. If $\vdash A$, then $\vdash \mathsf{R}A$ by Lemma 7.

Lemma 8. $\mathsf{R}A \equiv \mathsf{R}B$ if $A \equiv B$.

Dem. We are given that $\vdash A \rightarrow B$ and $\vdash B \rightarrow A$. Applying Lemma 7 to the first provable wff yields $\vdash \mathsf{R}[A \rightarrow B]$; thus $\vdash \mathsf{R}A \rightarrow \mathsf{R}B$. In the same way, $\vdash B \rightarrow A$ yields $\vdash \mathsf{R}B \rightarrow \mathsf{R}A$. Thus $\mathsf{R}A \equiv \mathsf{R}B$ if $A \equiv B$.

It is a simple matter, now, to establish the following result.

Principle of Duality. $\mathsf{D}A \equiv \mathsf{D}B$ if $A \equiv B$.

Dem. Let $A \equiv B$; by Lemma 8, $\mathsf{R}A \equiv \mathsf{R}B$, so by Lemma 5, $\mathsf{M}[\mathsf{R}A] \equiv \mathsf{M}[\mathsf{R}B]$; i.e., $\mathsf{D}A \equiv \mathsf{D}B$.

We can establish the converse of Lemmas 4, 5, and 8 and the Principle of Duality by applying Lemma 6, page 67, Lemma 2, and the fact that $\equiv$ is transitive.

Corollary 2. Let T be any normal transform; then $\mathsf{T}A \equiv \mathsf{T}B$ iff $A \equiv B$.

Dem. Assume that $\mathsf{T}A \equiv \mathsf{T}B$; then $\mathsf{T}[\mathsf{T}A] \equiv \mathsf{T}[\mathsf{T}B]$ as we have just established; i.e., $\mathsf{N}A \equiv \mathsf{N}B$ by Lemma 6, page 67; by Lemma 2, $\mathsf{N}A \equiv A$ and $\mathsf{N}B \equiv B$, so $A \equiv B$.

Here is a well-known property of D that follows from our results.

Lemma 9. $\vdash A$ iff $\vdash \neg\mathsf{D}A$.

Dem. Observe that $\neg\mathsf{D}A \equiv \mathsf{M}[\mathsf{D}A] = \mathsf{R}A$, since $\mathsf{MD} = \mathsf{R}$. But $\vdash A$ iff $\vdash \mathsf{R}A$; it follows that $\vdash A$ iff $\vdash \neg\mathsf{D}A$.

Corollary 3. $\vdash \mathsf{D}[\neg A]$ iff $\vdash A$.

Dem. By Lemma 9, $\vdash A$ iff $\vdash \neg\mathsf{D}A$; but $\neg\mathsf{D}A \equiv \mathsf{D}[\neg A]$, so $\vdash A$ iff $\vdash \mathsf{D}[\neg A]$.

Corollary 4. $\vdash \mathsf{D}A$ iff $\vdash \neg A$.

Dem. By Corollary 3, $\vdash \mathsf{D}[\neg\neg A]$ iff $\vdash \neg A$; i.e., $\vdash \mathsf{D}A$ iff $\vdash \neg A$.

We can now establish the following result.

Lemma 10. $\vdash A \rightarrow B$ iff $\vdash \mathsf{D}B \rightarrow \mathsf{D}A$.

Dem. By Corollary 3, $\vdash A \rightarrow B$ iff $\vdash \mathsf{D}[\neg(A \rightarrow B)]$; now,

$$\begin{aligned} \mathsf{D}[\neg(A \rightarrow B)] &= \mathsf{D}[\neg(\neg A \vee B)] \\ &= \mathsf{D}[\neg\neg A] \vee \mathsf{D}[\neg B] \equiv \mathsf{D}A \vee \neg\mathsf{D}B \end{aligned}$$

by Lemma 1 and the Substitution Theorem for Wffs. Of course, $\mathsf{D}A \vee \neg\mathsf{D}B \equiv \neg\mathsf{D}B \vee \mathsf{D}A$; so $\vdash \mathsf{D}B \rightarrow \mathsf{D}A$ iff $\vdash A \rightarrow B$.

We now illustrate the Principle of Duality.

Example 1. Show that $A \wedge B \equiv B \wedge A$ if A and B are wffs.

Solution. We have established that $A \vee B \equiv B \vee A$ if A and B are any wffs; therefore, for any wffs A and B, $\mathsf{D}A \vee \mathsf{D}B \equiv \mathsf{D}B \vee \mathsf{D}A$.

Thus, by the Principle of Duality, $\mathsf{D}[\mathsf{D}A \vee \mathsf{D}B] \equiv \mathsf{D}[\mathsf{D}B \vee \mathsf{D}A]$; i.e., $\mathsf{D}[\mathsf{D}A] \wedge \mathsf{D}[\mathsf{D}B] \equiv \mathsf{D}[\mathsf{D}B] \wedge \mathsf{D}[\mathsf{D}A]$. Applying the Substitution Theorem for Wffs and noting that $\mathsf{D}[\mathsf{D}A] = \mathsf{N}A \equiv A$, we obtain $A \wedge B \equiv B \wedge A$.

Example 2. Let $A \equiv B$ and let C be any wff; show that $A \wedge C \equiv B \wedge C$.

Solution. By the Principle of Duality, $\mathsf{D}A \equiv \mathsf{D}B$; thus, by Lemma 3, page 45,

$$\mathsf{D}A \vee \mathsf{D}C \equiv \mathsf{D}B \vee \mathsf{D}C$$

and it follows from the Principle of Duality that

$$\mathsf{D}[\mathsf{D}A] \wedge \mathsf{D}[\mathsf{D}C] \equiv \mathsf{D}[\mathsf{D}B] \wedge \mathsf{D}[\mathsf{D}C]$$

thus, by the Substitution Theorem for Wffs, $A \wedge C \equiv B \wedge C$.

Here is an example of the value of Lemma 10.

Example 3. Show that $\vdash A \wedge B \rightarrow A$ for any wffs A and B.

Solution. Let A and B be any wffs; then $\vdash \mathsf{D}A \rightarrow \mathsf{D}A \vee \mathsf{D}B$ since this wff is an axiom. By Lemma 10, $\vdash \mathsf{D}[\mathsf{D}A \vee \mathsf{D}B] \rightarrow \mathsf{D}[\mathsf{D}A]$, i.e., $\vdash \mathsf{D}[\mathsf{D}A] \wedge \mathsf{D}[\mathsf{D}B] \rightarrow \mathsf{D}[\mathsf{D}A]$; by the Substitution Theorem for Provable Wffs we conclude that $\vdash A \wedge B \rightarrow A$.

Note. More simply, we can show that $\vdash A \wedge B \rightarrow A$ as follows. By Axiom Scheme 2, $\vdash \neg A \rightarrow \neg A \vee \neg B$; thus, by Lemma 3, page 41,

$$\vdash \neg(\neg A \vee \neg B) \rightarrow \neg(\neg A)$$

i.e.,

$$\vdash A \wedge B \rightarrow \neg(\neg A)$$

so

$$\vdash A \wedge B \rightarrow A$$

Exercises

1. Prove Lemma 2.
2. Prove Lemma 3.

3. Show that $\vdash \mathsf{R}A$ for each axiom A.

4. Assuming that $A \vee (B \vee C) \equiv (A \vee B) \vee C$ for any wffs A, B, and C, prove that $A \wedge (B \wedge C) \equiv (A \wedge B) \wedge C$ for any wffs A, B, and C.

5. Assuming that $A \vee (B \wedge C) \equiv (A \vee B) \wedge (A \vee C)$ for any wffs A, B, and C, prove that $A \wedge (B \vee C) \equiv (A \wedge B) \vee (A \wedge C)$ for any wffs A, B, and C.

6. Show that $\vdash (A \wedge C) \wedge \neg(B \wedge C) \rightarrow A \wedge \neg B$ for any wffs A, B, and C. *Hint:* Note that $\vdash B \rightarrow A \mathrel{\dot{\rightarrow}} B \vee C \rightarrow A \vee C$ for any wffs A, B, and C. Apply Lemma 10.

7. Prove that $\vdash A \rightarrow B \mathrel{\dot{\rightarrow}} A \wedge C \rightarrow B \wedge C$ for any wffs A, B, and C. *Hint:* Apply Lemma 3, page 41, to the provable wff of Exercise 6.

6

Deducibility and Completeness

6.1. More Provable Wffs

In Section 6.3 we shall demonstrate that each true wff is provable. To this purpose we require more information about provable wffs; in particular, we must show that $\vee$ and $\wedge$ are associative, that $\vee$ distributes over $\wedge$, and that $\wedge$ distributes over $\vee$.

First, we shall demonstrate that $\vee$ is associative.

Lemma 1. $\vdash A \vee (C \vee B) \rightarrow ((B \vee A) \vee C) \vee A$ for any wffs A, B, and C.

Dem. Now,

$$\vdash B \rightarrow B \vee A \qquad \text{(by AS 2)}$$

Inserting C on both sides of the arrow yields

$$\vdash C \vee B \rightarrow (B \vee A) \vee C \qquad \text{(by Lemma 1, page 41)}$$

Inserting A on both sides of the arrow yields

$$\vdash A \vee (C \vee B) \rightarrow ((B \vee A) \vee C) \vee A$$

Lemma 2. $\vdash ((B \vee A) \vee C) \vee A \rightarrow (B \vee A) \vee C$ for any wffs A, B, and C.

Dem. Now,

$$\vdash A \to B \vee A \qquad (\text{since } \vdash A \to A \vee B \text{ and } A \vee B \equiv B \vee A)$$

and

$$\vdash B \vee A \to (B \vee A) \vee C \qquad (\text{by AS 2})$$

so

$$\vdash A \to (B \vee A) \vee C \qquad (\text{since} \to \text{is transitive})$$

Inserting $(B \vee A) \vee C$ on both sides of the arrow yields

$$\vdash ((B \vee A) \vee C) \vee A \to (B \vee A) \vee C$$

since $D \vee D \equiv D$ for any wff D (and the Substitution Theorem for Provable Wffs is available).

Lemma 3. $\vdash A \vee (C \vee B) \to (B \vee A) \vee C$ for any wffs A, B, and C.

Dem. Use Lemmas 1 and 2 and the fact that $\to$ is transitive.

Lemma 4. $\vdash A \vee (B \vee C) \to (A \vee B) \vee C$ for any wffs A, B, and C.

Dem. In Lemma 3, substitute $B \vee C$ for $C \vee B$ and substitute $A \vee B$ for $B \vee A$.

Lemma 5. $\vdash (C \vee B) \vee A \to C \vee (B \vee A)$ for any wffs A, B, and C.

Dem. In Lemma 3, substitute $(C \vee B) \vee A$ for the LHS and substitute $C \vee (B \vee A)$ for the RHS.

From Lemmas 4 and 5 we see that $\vee$ is associative, i.e., we have the following result.

Theorem 1. $A \vee (B \vee C) \equiv (A \vee B) \vee C$ for any wffs A, B, and C.

Applying the Principle of Duality, we conclude that $\wedge$ is associative, i.e., we have the result:

Corollary 1. $A \wedge (B \wedge C) \equiv (A \wedge B) \wedge C$ for any wffs A, B, and C.

In view of these results we can extend our parentheses-omitting conventions as follows. We shall denote both $A \vee (B \vee C)$ and $(A \vee B) \vee C$ by writing "$A \vee B \vee C$" and we shall denote both $A \wedge (B \wedge C)$ and $(A \wedge B) \wedge C$ by writing "$A \wedge B \wedge C$." Of course, we are entitled to use this convention only if we are working up to equivalence—for example, if we are constructing a chain of equivalent wffs or are working with a provable wff for which $A \vee B \vee C$ or $A \wedge B \wedge C$ is a subwff. This parentheses-omitting convention can be extended to a disjunction with any number of disjuncts and to a conjunction with any number of conjuncts. Here, we rely on the following lemma.

Lemma 6. Let n be any natural number and let $B_1, \ldots, B_n$ be any wffs; then any two wffs obtained by inserting parentheses in the expression "$B_1 \vee \cdots \vee B_n$" are equivalent.

Dem. Use mathematical induction.

In view of this result we shall denote any wff obtained by inserting parentheses in the expression $B_1 \vee \cdots \vee B_n$ by writing "$B_1 \vee \cdots \vee B_n$." Of course, the corresponding convention applies to $\wedge$.

We shall need the following fact.

Lemma 7. $A \rightarrow (B \rightarrow C) \equiv A \wedge B \rightarrow C$ for any wffs A, B, and C.

Dem. $A \rightarrow (B \rightarrow C) \equiv \neg A \vee \neg B \vee C \equiv \neg(A \wedge B) \vee C$.

Corollary 2. $\vdash A \dot{\rightarrow} B \rightarrow A \wedge B$ for any wffs A and B.

Dem. By Lemma 7, $A \dot{\rightarrow} B \rightarrow A \wedge B \equiv A \wedge B \dot{\rightarrow} A \wedge B$. The latter wff is provable, so the former wff is provable.

Corollary 3. $\vdash A \wedge B$ if $\vdash A$ and $\vdash B$.

Dem. By Corollary 2, $\vdash A \dot{\rightarrow} B \rightarrow A \wedge B$; thus, by Modus Ponens, $\vdash B \rightarrow A \wedge B$; again, by Modus Ponens, $\vdash A \wedge B$.

Theorem 2. $\vdash B_1 \wedge \cdots \wedge B_n$ if $\vdash B_1, \ldots,$ and $\vdash B_n$.

Dem. Mathematical induction.

We want to establish the converse of Theorem 2. First, consider the following facts.

Lemma 8. $\vdash A \wedge B \rightarrow A$ for any wffs A and B.

Dem. See Example 3, page 72.

Lemma 9. $\vdash A \wedge B \rightarrow B$ for any wffs A and B.

Dem. $A \wedge B \equiv B \wedge A$, so by Lemma 8 and the Substitution Theorem for Provable Wffs, $\vdash A \wedge B \rightarrow B$.

Theorem 3. If $\vdash B_1 \wedge \cdots \wedge B_n$, then $\vdash B_1, \ldots,$ and $\vdash B_n$.

Dem. We shall use mathematical induction on n. Clearly 1 has the property. Assume that $k \in N$ has the property (i.e., if the conjunction of k wffs is provable, then each of the k conjuncts is provable). We shall prove that $k + 1$ has the same property. Now, by Lemma 8,

$$\vdash (B_1 \wedge \cdots \wedge B_k) \wedge B_{k+1} \rightarrow B_1 \wedge \cdots \wedge B_k \tag{1}$$

We are assuming that $\vdash B_1 \wedge \cdots \wedge B_{k+1}$ and wish to show that each of its $k + 1$ conjuncts is provable. From (1) and Modus Ponens, we see that $\vdash B_1 \wedge \cdots \wedge B_k$; thus, by our induction assumption, each conjunct of this wff is provable, i.e., $\vdash B_1, \ldots, \vdash B_k$. By Lemma 9,

$$\vdash (B_1 \wedge \cdots \wedge B_k) \wedge B_{k+1} \rightarrow B_{k+1}$$

So, by Modus Ponens, $\vdash B_{k+1}$.

Theorem 4. $\vdash B_1 \wedge \cdots \wedge B_n$ iff $\vdash B_1, \ldots,$ and $\vdash B_n$.

Next, we shall prove that $\vee$ distributes over $\wedge$. We need some lemmas.

Lemma 10. $\vdash A \vee C \rightarrow B \vee D$ if $\vdash A \rightarrow B$ and if $\vdash C \rightarrow D$.

Dem. $\vdash C \rightarrow D$, so $\vdash A \vee C \rightarrow A \vee D$; $\vdash A \rightarrow B$, so $\vdash A \vee D \rightarrow B \vee D$. By the transitivity of $\rightarrow$, we conclude that $\vdash A \vee C \rightarrow B \vee D$.

Lemma 11. $\vdash A \wedge C \rightarrow B \wedge D$ if $\vdash A \rightarrow B$ and if $\vdash C \rightarrow D$.

Dem. $\vdash \mathsf{D}B \rightarrow \mathsf{D}A$ and $\vdash \mathsf{D}D \rightarrow \mathsf{D}C$; so by Lemma 10, $\vdash \mathsf{D}B \vee \mathsf{D}D \rightarrow \mathsf{D}A \vee \mathsf{D}C$. Thus

$$\vdash \mathsf{D}[\mathsf{D}A \vee \mathsf{D}C] \rightarrow \mathsf{D}[\mathsf{D}B \vee \mathsf{D}D]$$

and it follows that $\vdash A \wedge C \rightarrow B \wedge D$.

Lemma 12. $\vdash A \vee (B \wedge C) \rightarrow (A \vee B) \wedge (A \vee C)$ for any wffs A, B, and C.

Dem. Now $\vdash B \wedge C \rightarrow B$ and $\vdash B \wedge C \rightarrow C$. Inserting A on both sides of the arrow yields $\vdash A \vee (B \wedge C) \rightarrow A \vee B$ and $\vdash A \vee (B \wedge C) \rightarrow A \vee C$. Apply Lemma 11 to these provable wffs, and observe that $E \wedge E \equiv E$ for any wff E; we obtain

$$\vdash A \vee (B \wedge C) \rightarrow (A \vee B) \wedge (A \vee C)$$

Our aim is to show that $A \vee (B \wedge C) \equiv (A \vee B) \wedge (A \vee C)$ for any wffs A, B, and C. Lemma 12 establishes one half of this; the remaining half is more complicated.

Lemma 13. $\vdash B \wedge (A \vee C) \rightarrow A \vee (B \wedge C)$ for any wffs A, B, and C.

Dem. By Corollary 2, $\vdash B \dot{\rightarrow} C \rightarrow B \wedge C$; by Lemma 8, page 43, $\vdash C \rightarrow B \wedge C \dot{\rightarrow} A \vee C \rightarrow A \vee (B \wedge C)$. In view of the transitivity of $\rightarrow$, these provable wffs yield

$$\vdash B \dot{\rightarrow} A \vee C \rightarrow A \vee (B \wedge C)$$

By Lemma 7, this wff is equivalent to $B \wedge (A \vee C) \rightarrow A \vee (B \wedge C)$; so the latter wff is provable.

Lemma 14. $\vdash (A \vee B) \wedge (A \vee C) \rightarrow A \vee (B \wedge C)$ for any wffs A, B, and C.

Dem.

$\vdash (A \vee B) \wedge (A \vee C) \rightarrow A \vee ((A \vee B) \wedge C)$ (by Lemma 13)
$\vdash C \wedge (A \vee B) \rightarrow A \vee (C \wedge B)$ (by Lemma 13)

Insert A on both sides of the arrow in the preceding provable wff; so

$$\vdash A \vee (C \wedge (A \vee B)) \rightarrow A \vee A \vee (C \wedge B)$$

Now $A \vee A \equiv A$ and $\wedge$ is commutative; so by the Substitution Theorem for Provable Wffs,

$$\vdash A \vee ((A \vee B) \wedge C) \rightarrow A \vee (B \wedge C)$$

Taking account of the wff that begins this demonstration and the fact that $\rightarrow$ is transitive, we conclude that

$$\vdash (A \vee B) \wedge (A \vee C) \rightarrow A \vee (B \wedge C)$$

Theorem 5. $A \vee (B \wedge C) \equiv (A \vee B) \wedge (A \vee C)$ for any wffs A, B, and C.

Dem. Lemmas 12 and 14.

Applying the Principle of Duality, we easily show that $\wedge$ distributes over $\vee$.

Theorem 6. $A \wedge (B \vee C) \equiv (A \wedge B) \vee (A \wedge C)$ for any wffs A, B, and C.

Later (see Lemma 11, page 166, and Lemma 5, page 198), we shall need the following fact.

Lemma 15. If $\vdash B$ and if A is any wff, then $A \wedge B \equiv A$.

Dem. Since $A \wedge B \equiv B \wedge A$, it follows from Corollary 2 that $\vdash B \dot{\rightarrow} A \rightarrow A \wedge B$; thus $\vdash A \rightarrow A \wedge B$. By Lemma 8, $\vdash A \wedge B \rightarrow A$. Therefore, $A \wedge B \equiv A$.

Exercises

1. Prove Lemma 4.
2. Prove Lemma 5.
3. Prove Corollary 1.
4. Prove Lemma 6.
5. Prove that for any wffs A, B, and C

$$A \rightarrow (B \rightarrow C) \equiv B \rightarrow (A \rightarrow C)$$

6. Prove Theorem 6.
7. Prove that for any wffs A, B, and C

$$(A \wedge B) \vee C \equiv (A \vee C) \wedge (B \vee C)$$

8. Prove that for any wffs A, B, and C

$$(A \vee B) \wedge C \equiv (A \wedge C) \vee (B \wedge C)$$

9. Let $\vdash B$, let A be any wff, and let ν be any valuation; prove that $\nu[A \wedge B] = \nu A$.
10. Show that $A \vee (B \rightarrow C) \equiv B \rightarrow A \vee C$ for any wffs A, B, and C.
11. Show that $A \rightarrow (A \rightarrow B) \equiv A \rightarrow B$ for any wffs A and B.

12. Prove that $\vdash A \wedge B \rightarrow A \wedge C$ if $\vdash B \rightarrow C$.

13. Find the fallacy in the following argument. "If $\vdash B \rightarrow C$, then $\vdash A \wedge B \rightarrow A \wedge C$; hence, by duality, $\vdash A \vee C \rightarrow A \vee B$. Thus, $\vdash A \vee C \rightarrow A \vee B$ if $\vdash B \rightarrow C$."

14. Let B be a wff such that $\vdash \neg B$, and let A be any wff. Prove that $A \vee B \equiv A$.

15. Let A, B, and C be any wff; prove that

(a) $\vdash A \wedge (A \rightarrow B) \rightarrow B$.
(b) $\vdash (A \rightarrow B) \wedge (B \rightarrow C) \rightarrow (A \rightarrow C)$.
(c) $\vdash (A \rightarrow C) \wedge (B \rightarrow C) \dot{\rightarrow} A \vee B \rightarrow C$.
(d) $\vdash (A \rightarrow B) \wedge (C \rightarrow D) \dot{\rightarrow} A \wedge C \rightarrow B \wedge D$.

6.2. Conjunctive Normal Form

The purpose of this section is to establish a test for determining whether any wff possessing a certain specified form is provable. The same test allows us to decide whether a wff of the specified form is true. Accordingly, a wff of the specified form is provable iff it is true.

We need two concepts, the concept of a *prime* wff and the concept of a wff in *conjunctive normal form*. Recall that there are just two connectives that occur in a wff, namely $\neg$ and $\vee$; of course, the derived connectives $\wedge$, $\rightarrow$, and $\leftrightarrow$ can occur in a *name* of a wff.

Definition. We say that a wff A is *prime* if each instance of $\neg$ in A is prefixed to an atomic wff.

For example, $X \vee \neg Y \vee Z$ is prime, X is prime, and $\neg X$ is prime; but $\neg(\neg X)$ is not prime and $X \wedge Y$ is not prime.

Lemma 1. Each prime wff has the form $C_1 \vee \cdots \vee C_m$, where $m \in N$ and for each i either C_i is atomic or $C_i = \neg D_i$, where D_i is atomic.

Dem. For a prime wff, each occurrence of $\neg$ is prefixed to an atomic wff.

Each C_i is said to be a *disjunct* of the wff $C_1 \vee \cdots \vee C_m$.

The following test for provable wffs is also a test for true wffs. However, this test is not sufficiently far-reaching to achieve our goal on its own. We shall use this test to establish a more comprehensive test.

Lemma 2. Let A be any prime wff; then $\vdash A$ iff there is an atomic wff C such that both C and $\neg C$ are disjuncts of A.

Dem. 1. Assume that there is an atomic wff C such that both C and $\neg C$ are disjuncts of A. Now, $\vdash C \vee \neg C$; thus, by AS 2 and Modus Ponens, $\vdash C \vee \neg C \vee B$ for any wff B. But $A \equiv C \vee \neg C \vee B$ for some prime wff B, since $\vee$ is commutative and associative. Thus $\vdash A$.

2. Assume that $\vdash A$. Suppose that there is no atomic wff C such that both C and $\neg C$ are disjuncts of A. Let ν be the valuation such that for each atomic wff D

$$\nu[D] = \begin{cases} \mathsf{f} & \text{if } D \text{ is a disjunct of } A \\ \mathsf{t} & \text{otherwise} \end{cases}$$

It follows that ν assigns f to each disjunct of A; thus $\nu[A] = \mathsf{f}$. So, A is not true. But each provable wff is true; we conclude that A is not provable. This contradiction shows that there is an atomic wff C such that both C and $\neg C$ are disjuncts of A.

It is now an easy matter to establish the following test for true wffs; this test applies only to prime wffs.

Lemma 3. Let A be any prime wff; then A is true iff there is an atomic wff C such that both C and $\neg C$ are disjuncts of A.

We turn now to our second concept.

Definition. A wff A is said to be in *conjunctive normal form* if A is a conjunction of prime wffs.

For example, $X \wedge (X \vee \neg Y) \wedge \neg X$ is in conjunctive normal form, since the wffs X, $X \vee \neg Y$, and $\neg X$ are prime. More generally, a wff $B_1 \wedge \cdots \wedge B_m$ is in conjunctive normal form provided that each B_i is prime; each B_i is said to be a *conjunct* of $B_1 \wedge \cdots \wedge B_m$.

The importance of the concept of *conjunctive normal form* rests on the following pair of tests.

Test for Provable Wffs. A wff in conjunctive normal form, say $B_1 \wedge \cdots \wedge B_n$, is provable iff corresponding to each of its conjuncts B_i there is an atomic wff C such that both C and $\neg C$ are disjuncts of B_i.

Dem. By Theorem 4, page 78, $\vdash B_1 \wedge \cdots \wedge B_n$ iff $\vdash B_i$ for each i. But each B_i is prime; thus, by Lemma 2, $\vdash B_i$ iff there is an atomic wff C such that both C and $\neg C$ are disjuncts of B_i. This establishes our test.

This test for *provable* wffs is also a test for *true* wffs.

Test for True Wffs. A wff in conjunctive normal form, say $B_1 \wedge \cdots \wedge B_n$, is true iff corresponding to each of its conjuncts B_i there is an atomic wff C such that both C and $\neg C$ are disjuncts of B_i.

In view of these tests, it is clear that a wff in conjunctive normal form is provable iff it is true. Our plan is to prove that each wff is equivalent to a wff in conjunctive normal form. We have already shown that if $A \equiv B$, then A and B are both true or neither is true (see Corollary 1, page 46); moreover, if $A \equiv B$, then A and B are both provable or neither is provable. Accordingly, if we can show that each wff is equivalent to a wff in conjunctive normal form, then we can conclude that each wff A is true iff $\vdash A$. We shall carry out this plan in the next section.

We present one more lemma.

Lemma 4. Let A and B each be equivalent to wffs in conjunctive normal form; then $A \wedge B$ is equivalent to a wff in conjunctive normal form.

Dem. Let $A \equiv A_1$ and let $B \equiv B_1$, where A_1 and B_1 are in conjunctive normal form. By the Substitution Theorem for Wffs, $A \wedge B \equiv A_1 \wedge B_1$; note that $A_1 \wedge B_1$ is in conjunctive normal form.

Exercises

1. Which of the following wffs are prime?

   (a) $X \rightarrow \neg Y$.
   (b) $\neg X \rightarrow Y$.
   (c) $X \leftrightarrow Y$.

2. Let A and B be any prime wffs; show that $A \vee B$ is prime.

3. Use Lemma 2 to select the provable wffs from the following wffs:

   (a) $\neg Y \vee X \vee \neg Z \vee Y$.
   (b) $X \vee Y \vee Z \vee X \vee Y$.
   (c) $Z \vee \neg X \vee Z \vee Y \vee X \vee Y$.

4. Use Lemma 3 to select the true wffs from the following wffs:

   (a) $X \vee \neg Y \vee Y \vee Z$.
   (b) $Y \vee Z \vee \neg X \vee Z \vee Y$.
   (c) $\neg Y \vee X \vee Z \vee X \vee Y$.

5. Let A and B be any wffs in conjunctive normal form; show that $A \wedge B$ is in conjunctive normal form.
6. Prove Lemma 3.
7. Prove the Test for True Wffs.

6.3. Completeness

Here we shall prove that each wff is equivalent to a wff in conjunctive normal form. We shall require the following facts.

Lemma 1. $A \vee (B_1 \wedge \cdots \wedge B_n) \equiv (A \vee B_1) \wedge \cdots \wedge (A \vee B_n)$ for each $n \in N$ and any wffs $A, B_1, \ldots, B_n$.

Dem. Mathematical induction.

Lemma 2. $(A_1 \wedge \cdots \wedge A_n) \vee B \equiv (A_1 \vee B) \wedge \cdots \wedge (A_n \vee B)$ for each $n \in N$ and any wffs $A_1, \ldots, A_n, B$.

Dem. Lemma 1 and the fact that $\vee$ is commutative.

Before proceeding, we need a more general method of representing a conjunction with many conjuncts. Let I be an index set and let D_i be a wff for each $i \in I$. Then by $\bigwedge_I D_i$ we represent the conjunction of all the wffs D_i. For example, let $I = \{1, 2, 3, 4\}$ and let D_1, D_2, D_3, and D_4 be wffs; then $\bigwedge_I D_i = D_1 \wedge D_2 \wedge D_3 \wedge D_4$.

Let $C_{n,m} = \{1, \ldots, n\} \times \{1, \ldots, m\}$, a Cartesian product; so $C_{n,m}$ is the set of ordered pairs (i, j) such that $i \in \{1, \ldots, n\}$ and $j \in \{1, \ldots, m\}$. Using this notation, we present our next lemma.

Lemma 3. $(A_1 \wedge \cdots \wedge A_n) \vee (B_1 \wedge \cdots \wedge B_m) \equiv \bigwedge_{C_{n,m}} (A_i \vee B_j)$ for each $n, m \in N$ and any wffs $A_1, \ldots, A_n, B_1, \ldots, B_m$.

Dem. Apply Lemmas 1 and 2.

Similarly, let $\bigvee_I D_i$ represent the disjunction of the wff D_i, where $i \in I$. For example, let $I = \{1, 2, 3, 4\}$; then $\bigvee_I D_i = D_1 \vee D_2 \vee D_3 \vee D_4$.

We now formulate the dual of Lemma 3.

Lemma 4. $(A_1 \vee \cdots \vee A_n) \wedge (B_1 \vee \cdots \vee B_m) \equiv \bigvee_{C_{n,m}} (A_i \wedge B_j)$ for each $n, m \in N$ and any wffs $A_1, \ldots, A_n, B_1, \ldots, B_m$.

Dem. Apply the Principle of Duality to Lemma 3.

The following fact is vital to our program.

Lemma 5. If A and B are each equivalent to wffs in conjunctive normal form, then $A \vee B$ is equivalent to a wff in conjunctive normal form.

Dem. Let $A \equiv A_1 \wedge \cdots \wedge A_n$, where each A_i is prime and let $B \equiv B_1 \wedge \cdots \wedge B_m$, where each B_i is prime. By Lemma 3 and the Substitution Theorem for Wffs,

$$A \vee B \equiv \bigwedge_{C_{n,m}} (A_i \vee B_j)$$

Here, each $A_i \vee B_j$ is prime; so $\bigwedge_{C_{n,m}} (A_i \vee B_j)$ is in conjunctive normal form.

We shall prove that the N-image of each wff is equivalent to a wff in conjunctive normal form.

Theorem 1. For each wff A, $\mathsf{N}A$ is equivalent to a wff in conjunctive normal form.

Dem. We shall use a variation of the Fundamental Theorem about Wffs. Let A be the shortest wff (counting instances of $\neg$ and $\vee$) such that $\mathsf{N}A$ is *not* equivalent to a wff in conjunctive normal form. There are just three possibilities.

1. Assume that A is atomic. Then $\mathsf{N}A = A$, which is in conjunctive normal form. Thus, A is not atomic.
2. Assume that $A = \neg B$ for some wff B. Now,

$$\mathsf{N}[\neg B] = \begin{cases} \neg B & \text{if} \quad B \text{ is atomic} \\ \mathsf{N}C & \text{if} \quad B = \neg C \\ \mathsf{N}[\neg D] \wedge \mathsf{N}[\neg E] & \text{if} \quad B = D \vee E \end{cases}$$

Note that $\neg B$ is in conjunctive normal form if B is atomic; also, C is shorter than $\neg B$ (if $B = \neg C$), so $\mathsf{N}C$ is equivalent to a wff in conjunctive normal form; finally, both $\neg D$ and $\neg E$ are shorter than $\neg B$ (if $B =$

$D \vee E$), so each of N[$\neg D$] and N[$\neg E$] is equivalent to a wff in conjunctive normal form; thus by Lemma 4, page 83, N[$\neg D$] $\wedge$ N[$\neg E$] is equivalent to a wff in conjunctive normal form. Thus $\neg$ is not the main connective of A.

3. Assume that $A = C \vee D$ for some wffs C and D. Both C and D are shorter than A, so both NC and ND are equivalent to wffs in conjunctive normal form. By definition of N, N[$C \vee D$] = NC $\vee$ ND; thus, by Lemma 5, N[$C \vee D$] is equivalent to a wff in conjunctive normal form. Thus $\vee$ is not the main connective of A. This establishes Theorem 1.

Theorem 2. Each wff is equivalent to a wff in conjunctive normal form.

Dem. Let A be any wff; by Lemma 2, page 69, N$A \equiv A$. By Theorem 1, NA is equivalent to a wff in conjunctive normal form. We conclude that A is equivalent to a wff in conjunctive normal form. This establishes Theorem 2.

Theorem 3. For each wff A, A is true iff $\vdash A$.

Dem. Let A be any wff; by Theorem 2, there is a wff B such that B is in conjunctive normal form and $A \equiv B$. In view of our Test for Provable Wffs and our Test for True Wffs, B is true iff $\vdash B$. We conclude that A is true iff $\vdash A$.

This shows that the propositional calculus is *complete* in the sense that the notion of a true wff, which has been defined externally, has been characterized internally by the notion of a provable wff.

Exercises

1. Prove that $A \vee (B \wedge C \wedge D) \equiv (A \vee B) \wedge (A \vee C) \wedge (A \vee D)$ for any wffs A, B, C, and D.
2. Prove Lemma 1.
3. Prove that $(B \wedge C \wedge D) \vee A \equiv (B \vee A) \wedge (C \vee A) \wedge (D \vee A)$ for any wffs A, B, C, and D.
4. Prove Lemma 2.
5. Prove Lemma 3.

6. Prove Lemma 4.

7. Prove the generalized distributive law:

$$(B_{11} \wedge \cdots \wedge B_{1n_1}) \vee \cdots \vee (B_{m1} \wedge \cdots \wedge B_{mn_m}) \equiv \bigwedge (B_{1i_1} \vee \cdots \vee B_{mi_m})$$

$(i_1 \leq n_1, \ldots, i_m \leq n_m)$, where each B_{ij} is a wff.

8. Use Exercise 7 to prove that $\neg A$ is equivalent to a wff in conjunctive normal form, given that A is equivalent to a wff in conjunctive normal form.

9. Use the fact that $\mathsf{N}(\neg A) \equiv \neg \mathsf{N}A$ for each wff A to prove Theorem 1 of this section.

6.4. Deducibility

We begin our discussion of deducibility by generalizing the notion of a *true* wff. Recall that a wff A is true provided that μA is true for $\sum$ whenever $\sum$ is a statement system and μ is a $\sum$-interpreter. We shall generalize this concept by weakening the requirement that μA is true for $\sum$ for *each* statement system $\sum$ and *each* $\sum$-interpreter μ. Instead, we shall require that μA is true for $\sum$ for certain statement systems $\sum$ and $\sum$-interpreters μ.

Now, our purpose in this section is to characterize the notion that a wff A is *deducible* from a set of wffs K. Accordingly, we choose the family of statement systems involved in the preceding paragraph to be the models of K. This brings us to the concept of a K-*true* wff.

Definition. A wff A is said to be K-*true* if whenever $\sum$ is some model of K under some $\sum$-interpreter μ, then $\sum$ is also a model of $\{A\}$ under μ.

By this we mean that A is K-true provided that $\sum$ is a model of $\{A\}$ under μ for each statement system $\sum$ and each $\sum$-interpreter such that $\sum$ is a model of K under μ. Put more simply, A is K-true provided that μA is true for $\sum$ whenever $\sum$ is a model of K under μ.

Thus, A is K-true iff for each statement system $\sum$ and for each $\sum$-interpreter μ either: (a) there is a member of K, say B, such that μB is false for $\sum$; or (b) μA is true for $\sum$.

If A is K-true, we also say that A is a *consequence* of K.

For example, $\neg(\neg X)$ is $\{X\}$-true; each wff is $\{Y, \neg Y\}$-true; if $\vdash A$, then A is K-true for any set of wffs K (since A is true).

Of course, each true wff is K-true, where K is any set of wffs. Moreover, each statement system is a model of $\varnothing$, the empty set; so, for each wff A, A is $\varnothing$-true iff A is true.

In Section 3.3 we introduced the notion of *provable* wff in order to characterize the concept of *true* wffs within the propositional calculus. We now introduce the notion of *deducibility* in order to characterize K-true wffs within the propositional calculus.

Definition. Let K be a nonempty set of wffs and let A be any wff. We say that A is *deducible* from K (in symbols, $\mathsf{K} \vdash A$) if there is a nonempty, finite subset of K, say $\{A_1, \ldots, A_n\}$, such that

$$\vdash A_1 \wedge \cdots \wedge A_n \rightarrow A$$

Moreover, we say that *A is deducible from the empty set* (in symbols, $\varnothing \vdash A$) if $\vdash A$.

For example, $\{Z, \neg(\neg X)\} \vdash X$ since $\vdash \neg(\neg X) \rightarrow X$. Let K be the set of all atomic wffs; then $\mathsf{K} \vdash X$ and $\mathsf{K} \vdash X \wedge Y$, since $\vdash X \rightarrow X$ and $\vdash X \wedge Y \rightarrow X \wedge Y$.

In Section 6.6 we shall prove that our formal definition of *deducibility* has captured the intended idea; i.e., $\mathsf{K} \vdash A$ iff A is K-true, where K is any set of wffs and A is any wff. Bear in mind that when we assert $\mathsf{K} \vdash A$, we are thinking syntactically, i.e., in terms of the propositional calculus; on the other hand, when we assert that A is K-true, we are thinking semantically, i.e., in terms of statement systems.

Here are some facts about *deducibility*; throughout, K is any set of wffs and A and B are any wffs.

Fact 1. $\mathsf{K} \vdash A \wedge B$ if $\mathsf{K} \vdash A$ and $\mathsf{K} \vdash B$.

Fact 2. $\mathsf{K} \vdash B$ if $\mathsf{K} \vdash A$ and $A \equiv B$.

Fact 3. $\mathsf{K} \vdash B$ if $\mathsf{K} \vdash A$ and $\vdash A \rightarrow B$.

Fact 4. $\mathsf{K} \vdash A$ if $A \in \mathsf{K}$.

Fact 5. $\mathsf{K}_1 \vdash A$ if $\mathsf{K} \vdash A$ and $\mathsf{K} \subset \mathsf{K}_1$.

Fact 6. $\mathsf{K} \vdash A$ if $\vdash A$.

Fact 7. $\mathsf{K} \vdash A \vee B$ if $\mathsf{K} \vdash A$.

Fact 8. $\mathsf{K} \vdash B$ if $\mathsf{K} \vdash A$ and $\mathsf{K} \vdash A \rightarrow B$.

Our goal is to demonstrate that $\mathsf{K} \vdash A$ iff A is K-true. While it is quite easy to show that A is K-true if $\mathsf{K} \vdash A$, it is quite another matter to show that $\mathsf{K} \vdash A$ if A is K-true (we do this in Section 6.6).

Lemma 1. A is K-true if $\mathsf{K} \vdash A$.

Dem. There are two cases depending on whether K is empty or nonempty. If $\mathsf{K} = \varnothing$, we must show that A is $\varnothing$-true; i.e., we must show that A is a true wff. By definition, $\varnothing \vdash A$ iff $\vdash A$; so A is true. Next, let K be any nonempty set. By assumption, there is a finite subset of K, say $\{A_1, \ldots, A_n\}$, $n \geq 1$, such that

$$\vdash A_1 \wedge \cdots \wedge A_n \rightarrow A$$

Therefore, $A_1 \wedge \cdots \wedge A_n \rightarrow A$ is a true wff. Let $\sum$ be any model of K under μ, a $\sum$-interpreter. Then

$$\mu A_1 \wedge \cdots \wedge \mu A_n \rightarrow \mu A$$

is true for $\sum$. But each of $\mu A_1, \ldots, \mu A_n$ is true for $\sum$; so $\mu A_1 \wedge \cdots \wedge \mu A_n$ is true for $\sum$. It follows that μA is true for $\sum$. This proves that A is K-true.

We may improve our appreciation of the converse of Lemma 1 by considering the case in which K is *finitely axiomatizable*; by this we mean that there is a finite subset of K, say K_1, such that (for any $\sum$ and any $\sum$-interpreter μ) $\sum$ is a model of K_1 under μ iff it is a model of K under μ. Let $\mathsf{K}_1 = \{A_1, \ldots, A_n\}$ and let A be K-true. Then $A_1 \wedge \cdots \wedge A_n \rightarrow A$ is a true wff. But each true wff is provable; so $\vdash A_1 \wedge \cdots \wedge A_n \rightarrow A$. Thus $K \vdash A$.

The well-known *Deduction Theorem*, which normally is a rather subtle result, here is almost trivial; this is due to our formulation of deducibility in terms of provable wffs.

Deduction Theorem. $\mathsf{K} \cup \{A\} \vdash B$ iff $\mathsf{K} \vdash A \rightarrow B$.

Dem. 1. Assume that $\mathsf{K} \cup \{A\} \vdash B$. There is a finite subset of K, say $\{A_1, \ldots, A_n\}$, such that

$$\vdash A_1 \wedge \cdots \wedge A_n \wedge A \rightarrow B$$

By Lemma 7, page 77,

$$\vdash A_1 \wedge \cdots \wedge A_n \rightarrow (A \rightarrow B)$$

so $\mathsf{K} \vdash A \rightarrow B$.

2. Read "up" the first part of this proof.

Note. The above proof considers the case in which K is nonempty. If K $= \varnothing$, then the Deduction Theorem reduces to

$$\{A\} \vdash B \qquad \text{iff} \qquad \vdash A \to B$$

which is true by definition.

It is interesting to formulate *deducibility* in a manner analogous to the definition of *provability*. First, we introduce the notion of a K-proof where K is a given set of wffs.

Definition. Let K be any set of wffs. A finite sequence of wffs is called a K-*proof* if each of its terms, say E, satisfies at least one of the following conditions:

(a) $E \in$ K.
(b) E is a member of one of the three axiom schemes in Section 3.3.
(c) There is a wff D such that both D and $D \to E$ precede E in the sequence.

For example, let K $= \{X, X \to Y\}$; then

$$X, \quad X \to Y, \quad Y, \quad Y \to Y \vee X, \quad Y \vee X$$

is a K-proof.

The following criterion for deducibility is given by Rosser (1953; page 56); also see Exner and Rosskopf (1959, 1970).

Criterion for Deducibility. Let K be any set of wffs and let A be any wff; then K $\vdash A$ iff there is a K-proof whose last term is A.

Dem. First, we point out that each $\varnothing$-proof is a proof and vice versa. Moreover, $\varnothing \vdash A$ iff $\vdash A$; therefore, $\varnothing \vdash A$ iff there is a $\varnothing$-proof whose last term is A. We now assume that K is nonempty.

1. Let K $\vdash A$. Then there is a nonempty, finite subset of K, say $\{A_1, \ldots, A_n\}$, such that

$$A_1 \wedge \cdots \wedge A_n \to A \tag{1}$$

is provable. Let π_1 be a proof of (1). Certainly, π_1 is a K-proof; indeed, the sequence

$$A_1, A_2, \ldots, A_n, \pi_1$$

obtained by prefixing $A_1, \ldots, A_n$ to π_1 is a K-proof. Now, (1) is equivalent to

$$A_1 \dot{\to} A_2 \wedge \cdots \wedge A_n \to A \qquad \text{(by Lemma 7, page 77)} \tag{2}$$

so

$$(1) \rightarrow (2) \tag{3}$$

is provable. Let π_2 be a proof of (3). Then

$$A_1, \ldots, A_n, \pi_1, \pi_2, (2), A_2 \wedge \cdots \wedge A_n \rightarrow A$$

is a K-proof. Repeating this process of sloughing off the A_i from the LHS of (1), we obtain a K-proof whose last term is A.

2. Let $B_1, \ldots, B_m$ be a K-proof whose last term is A. We shall show that $\mathsf{K} \vdash A$. Let $A_1, \ldots, A_n$ be the terms of the given K-proof that are in K. We claim that

$$\vdash A_1 \wedge \cdots \wedge A_n \rightarrow A \tag{4}$$

Indeed, we shall prove, more generally, that

$$A_1 \wedge \cdots \wedge A_n \rightarrow B_i \tag{5}$$

is provable for each $i = 1, \ldots, m$. If not, there is a smallest natural number i such that (5) is not provable. Thus, $\vdash A_1 \wedge \cdots \wedge A_n \rightarrow B_j$ for each $j < i$. There are three cases to consider.

Case 1. B_i meets condition (a). Then (5) is provable by Lemma 8, page 78, and the fact that $\wedge$ is associative and commutative.

Case 2. B_i meets condition (b). Then (5) is provable by AS 2, Modus Ponens, and the fact that $\vee$ is commutative.

Case 3. B_i meets condition (c). Then there is a natural number $j, j < i$, such that $B_j \rightarrow B_i$ and B_j precede B_i in the given K-proof. Thus

$$\vdash A_1 \wedge \cdots \wedge A_n \rightarrow B_j \rightarrow B_i \qquad \text{and} \qquad \vdash A_1 \wedge \cdots \wedge A_n \rightarrow B_j$$

By Lemma 7, page 77, and Lemma 11, page 44, it follows that $\vdash A_1 \wedge \cdots \wedge A_n \rightarrow B_i$. This contradiction establishes that (5) is provable; so (4) is correct. Thus $\mathsf{K} \vdash A$. We have proven the Criterion for Deducibility.

The key to the preceding demonstration consisted in establishing (4). This fact about K-proofs is worth stating separately.

Lemma 2. Let $B_1, \ldots, B_m$ be a K-proof, and let $A_1, \ldots, A_n$ be the members of K that are terms of the given K-proof. Then

$$\vdash A_1 \wedge \cdots \wedge A_n \rightarrow B_m$$

Dem. See the demonstration of the Criterion for Deducibility.

The Criterion for Deducibility is sometimes used as the definition of deducibility in mathematics; i.e., it provides a standard for mathematical proofs. Each term of a K-proof, where K is the postulate set of a mathematical theory, may be regarded as a step in a mathematical proof. Since K-proofs tend to be lengthy, it is convenient to relax the conditions of the above criterion in an effort to reduce the number of steps in a mathematical proof.

Theorem 1. Let K be any set of wffs and let A be any wff. Then $\mathsf{K} \vdash A$ iff there is a finite sequence of wffs, with last term A, such that each of its terms, say E, satisfies at least one of the following conditions:

1. $E \in \mathsf{K}$.
2. $\vdash E$.
3. E is equivalent to a preceding term of the sequence.
4. There is a wff D such that both D and $D \rightarrow E$ precede E in the sequence.

Dem. Use the Criterion for Deducibility. Notice that each K-proof satisfies conditions 1–4. Also, each sequence that meets these conditions can be extended to a K-proof. The details are left as an exercise.

For example, let $\mathsf{K} = \{X \vee (Y \wedge Z)\}$. The sequence

$$X \vee (Y \wedge Z), \quad (X \vee Y) \wedge (X \vee Z),$$
$$(X \vee Y) \wedge (X \vee Z) \rightarrow X \vee Y, \quad X \vee Y$$

meets conditions 1–4. Thus, by Theorem 1, $\{X \vee (Y \wedge Z)\} \vdash X \vee Y$.

Another step-saving idea is to incorporate more rules of inference under condition 4 of Theorem 1.

Theorem 2. $\mathsf{K} \vdash A$ iff there is a finite sequence of wffs, with last term A, such that each of its terms, say E, satisfies at least one of the following conditions:

1. $E \in \mathsf{K}$.
2. $\vdash E$.
3. E is equivalent to a preceding term of the sequence.
4. There is a wff D such that both D and $D \rightarrow E$ precede E in the sequence.
5. There is a wff D such that both $\neg E \rightarrow \neg D$ and D precede E in the sequence.
6. There is a wff D such that $D \wedge E$ or $E \wedge D$ precedes E in the sequence.

7. $E = B \wedge C$ and B and C precede E in the sequence.
8. $E = B \rightarrow D$ and there is a wff C such that $B \rightarrow C$ and $C \rightarrow D$ precede E in the sequence.
9. $E = B \vee C \rightarrow D$ and $B \rightarrow D$ and $C \rightarrow D$ precede E in the sequence.

Dem. We can show that the rules of inference 5–9 can each be expressed in terms of a provable wff of the form $F \rightarrow E$, where F is the conjunction of the terms which are given as preceding E by the rule (e.g., see Exercise 15, page 81). The details are left as an exercise.

For example, let $\mathsf{K} = \{X \rightarrow Y, \neg X \rightarrow Z\}$. The sequence

$$X \rightarrow Y, \quad \neg Y \rightarrow \neg X, \quad \neg X \rightarrow Z, \quad \neg Y \rightarrow Z,$$
$$\neg X \vee \neg Y \rightarrow Z, \quad (X \wedge Y) \vee Z$$

meets conditions 1–9 of Theorem 2. Thus, by Theorem 2, $\{X \rightarrow Y, \neg X \rightarrow Z\} \vdash (X \wedge Y) \vee Z$.

Exercises

1. Show that $\neg(\neg X)$ is $\{X\}$-true.
2. Show that each wff is $\{Y, \neg Y\}$-true.
3. Given that $\vdash A$, show that A is K-true for any set of wffs K.
4. Prove that A is K-true for every set of wffs K, iff A is true.
5. Prove Fact 1; consider two cases, $\mathsf{K} = \varnothing$ and $\mathsf{K} \neq \varnothing$.
6. Prove Fact 2; consider two cases, $\mathsf{K} = \varnothing$ and $\mathsf{K} \neq \varnothing$. Use the Substitution Theorem for Provable Wffs.
7. Prove Fact 3.
8. Prove Fact 4.
9. Prove Fact 5.
10. Prove Fact 6.
11. Let $\sum$ be any statement system and let μ be any $\sum$-interpreter. Let

$$\mathsf{K} = \{A \in \mathsf{W} \mid \mu A \text{ is true for } \textstyle\sum\}$$

(a) Show that $A \in \mathsf{K}$ if $\vdash A$.
(b) Show that for each wff A, either $A \in \mathsf{K}$ or $\neg A \in \mathsf{K}$, but not both.

12. Let $\sum$ be any statement system and let μ be any $\sum$-interpreter. Let

$$\mathsf{K} = \{A \in \mathsf{W} \mid \mu A \text{ is false for } \sum\}$$

(a) Show that no provable wff is in K.
(b) Show that for each wff A, either $A \in \mathsf{K}$ or $\neg A \in \mathsf{K}$, but not both.

13. Let K be a set of wffs and let A be a wff such that $\mathsf{K} \cup \{A\} \vdash \neg A$. Show that $\mathsf{K} \vdash \neg A$.

14. Let K be any finite set of wffs. Prove that for each wff A, $\mathsf{K} \vdash A$ iff A is K-true.

15. Prove Fact 7.

16. Prove Fact 8. *Hint:* Use Lemma 11, page 44.

17. (a) Show that the sequence

$$X, \quad X \to X \vee Y, \quad X \vee Y$$

is an $\{X\}$-proof.
(b) Show that $\{X\} \vdash X \vee Y$.

18. Show that each $\varnothing$-proof is a proof, and each proof is a $\varnothing$-proof.

19. Prove Theorem 1.

20. (a) Let $\mathsf{K} = \{X\}$. Show that the sequence

$$X, \quad X \to X \vee Y, \quad X \vee Y, \quad Y \vee X$$

meets conditions 1–4 of Theorem 1.
(b) Show that $\{X\} \vdash Y \vee X$.

21. Prove Theorem 2.

22. (a) Let A be any wff and let $\mathsf{K} = \{X, \neg X\}$. Show that the sequence

$$X, \quad \neg X, \quad X \wedge \neg X, \quad X \wedge \neg X \to A, \quad A$$

meets the conditions 1–9 of Theorem 2.
(b) Show that $\{X, \neg X\} \vdash A$ for each wff A.

6.5. Consistent Sets and Contradictory Sets

Let K be any set of wffs and consider the set of all wffs that are deducible from K; we shall denote this set by C[K].

Definition. $\mathsf{C}[\mathsf{K}] = \{A \in \mathsf{W} \mid \mathsf{K} \vdash A\}$.

For example, $\mathsf{C}[\varnothing] = \{A \in \mathsf{W} \mid \vdash A\}$, the set of all provable wffs. Clearly, $\mathsf{C}[\mathsf{W}] = \mathsf{W}$. If K is a set of wffs such that $\mathsf{K} \vdash B$ and $\mathsf{K} \vdash \neg B$

for some wff B, then $\mathsf{K} \vdash B \wedge \neg B$; it follows that $\mathsf{K} \vdash A$ for each wff A. In this case, then, $\mathsf{C}[\mathsf{K}] = \mathsf{W}$.

We now define the terms *contradictory* and *consistent*. Throughout, K is any set of wffs.

Definition. We say that K is *contradictory* if $\mathsf{C}[\mathsf{K}] = \mathsf{W}$.

For example, $\{X, \neg X\}$ is contradictory since each wff is deducible from this set. Each superset of a contradictory set is also contradictory. Notice that the empty set is not contradictory, since X is not deducible from $\varnothing$.

Definition. We say that K is *consistent* if K is not contradictory.

Thus, a set of wffs K is consistent iff there is a wff A which is *not* deducible from K. For example, $\{X\}$ is consistent since Y is not deducible from $\{X\}$; indeed, the wff $X \to Y$ is not provable. As we have already mentioned, the empty set $\varnothing$ is consistent. Let A be any atomic wff; then both $\{A\}$ and $\{\neg A\}$ are consistent sets.

Our definition of a contradictory set is not easily applied to the task of determining whether a given set of wffs is contradictory. Fortunately, there is a simple criterion that achieves this purpose.

Criterion for Contradictory Sets. K is contradictory iff there is a wff A such that $\mathsf{K} \vdash A \wedge \neg A$.

Dem. Assume that $\mathsf{K} \vdash A \wedge \neg A$; we shall show that K is contradictory. We rely on the fact that $\vdash A \wedge \neg A \to B$ for each wff B, which is easy to verify. By assumption, there is a subset of K, say $\{A_1, \ldots, A_n\}$, such that

$$\vdash A_1 \wedge \cdots \wedge A_n \to A \wedge \neg A$$

By the transitivity of the arrow, since $\vdash A \wedge \neg A \to B$,

$$\vdash A_1 \wedge \cdots \wedge A_n \to B$$

so $\mathsf{K} \vdash B$. Thus $\mathsf{C}[\mathsf{K}] = \mathsf{W}$; this means that K is contradictory. Of course, if K is contradictory, then $X \wedge \neg X \in \mathsf{C}[\mathsf{K}]$; so $\mathsf{K} \vdash X \wedge \neg X$.

The following facts are obvious.

Fact 1. Each superset of a contradictory set is contradictory.

Fact 2. Each subset of a consistent set is consistent.

Later, we shall make use of the following observation.

Lemma 1. If A is not deducible from K, then $\mathsf{K} \cup \{\neg A\}$ is consistent.

Dem. Assume that $\mathsf{K} \cup \{\neg A\}$ is contradictory; then $\mathsf{K} \cup \{\neg A\} \vdash A$. So, there is a finite subset of K, say $\{A_1, \ldots, A_n\}$, such that

$$\vdash A_1 \wedge \cdots \wedge A_n \wedge \neg A \rightarrow A$$

Thus

$$\vdash A_1 \wedge \cdots \wedge A_n \dot{\rightarrow} \neg A \rightarrow A$$

So

$$\vdash A_1 \wedge \cdots \wedge A_n \rightarrow A$$

We conclude that $\mathsf{K} \vdash A$. This contradiction proves that $\mathsf{K} \cup \{\neg A\}$ is consistent.

Note. In the above argument, suppose that $n = 0$, i.e., $\{\neg A\}$ is the subset of $\mathsf{K} \cup \{\neg A\}$ involved. Then $\vdash \neg A \rightarrow A$, so $\vdash A$. Clearly, each provable wff is deducible from K if K is nonempty. Moreover, each provable wff is deducible from the empty set, by definition. So, in any case, $\mathsf{K} \vdash A$.

Applying the Criterion for Contradictory Sets, we verify that $\mathsf{K} \cup \{A\}$ is contradictory if $\neg A \in \mathsf{C}[\mathsf{K}]$. Therefore, $\neg A \notin \mathsf{C}[\mathsf{K}]$ if $\mathsf{K} \cup \{A\}$ is consistent. This observation allows us to strengthen Lemma 1 as follows.

Lemma 2. $\mathsf{K} \cup \{A\}$ is consistent iff $\neg A \notin \mathsf{C}[\mathsf{K}]$.

In Section 6.6 we shall prove the following important theorem, which relates our syntactical and semantical viewpoints.

Strong Completeness Theorem. K is consistent iff K has a model.

One part of the Strong Completeness Theorem is easy to establish. This we now do.

Lemma 3. If K has a model, then K is consistent.

Dem. Let $\sum$ be a model of K under μ; assume that K is contradictory. Then

$$\vdash A_1 \wedge \cdots \wedge A_n \rightarrow X \wedge \neg X$$

where $\{A_1, \ldots, A_n\}$ is a subset of K. But each provable wff is true; therefore the swff

$$\mu[A_1 \wedge \cdots \wedge A_n \rightarrow X \wedge \neg X]$$

is true for $\sum$. But $\mu A_1 \wedge \cdots \wedge \mu A_n$ is true for $\sum$, since $\sum$ is a model of K under μ. Thus $\mu X \wedge \neg \mu X$ is true for $\sum$; certainly, this swff is false for $\sum$. This contradiction proves that K is consistent.

It is much more difficult to prove that each consistent set has a model. We postpone a proof of this fact to the next section.

Here is a useful fact.

Lemma 4. K is consistent iff each finite subset of K is consistent.

Dem. 1. Assume that K is consistent; then each subset of K is consistent.

2. Assume that each finite subset of K is consistent. If K is contradictory, then there is a wff B such that $K \vdash B \wedge \neg B$. Therefore, there is a finite subset of K, say $\{A_1, \ldots, A_n\}$, such that

$$\vdash A_1 \wedge \cdots \wedge A_n \rightarrow B \wedge \neg B$$

So the finite subset $\{A_1, \ldots, A_n\}$ is contradictory (see the Criterion for Contradictory Sets earlier in this section). This contradiction proves that K is consistent.

Exercises

1. Show that W, the set of all wffs, is contradictory.
2. Prove Fact 1.
3. Prove Fact 2.
4. Show that $\{A\}$ is consistent if $\neg A$ is not a true wff; here A is a wff.
5. Let K be the set of all provable wffs.
   (a) Show that K is consistent.
   (b) Show that for each wff B, $K \vdash B$ iff $B \in K$.
   (c) Prove that $K \cup \{X\}$ is consistent.
   (d) Prove that $K \cup \{\neg X\}$ is consistent.

6. Let K be the set of all atomic wffs.

   (a) Show that K is consistent.
   (b) Show that $K \cup \{\neg X\}$ is contradictory.
   (c) Let $\vdash A$; is $K \cup \{A\}$ necessarily consistent?

7. Let $\sum$ be any statement system and let μ be any $\sum$-interpreter. Let

$$K = \{A \in W \mid \mu A \text{ is true for } \textstyle\sum\}$$

   (a) Show that K is consistent.
   (b) Show that $K \cup \{B\}$ is contradictory if B is a wff not in K.
   (c) Show that each proper superset of K is contradictory.
   (d) Show that for each wff A, either $A \in K$ or $\neg A \in K$ (but not both).

8. Let $\sum$ be any statement system and let μ be any $\sum$-interpreter. Let

$$K = \{A \in W \mid \mu A \text{ is false for } \textstyle\sum\}$$

   (a) Show that $K \neq W$.
   (b) Show that K is contradictory.
   (c) Show that for each wff A, either $A \in K$ or $\neg A \in K$ (but not both).

9. Prove that $K \cup \{B\}$ is consistent if $K \cup \{A\}$ is consistent and $\vdash A \rightarrow B$.

6.6. Maximal-Consistent Sets

We now present a concept which we need to prove the Strong Completeness Theorem.

Definition. A set of wffs K is said to be *maximal-consistent* if K is consistent and if each proper superset of K is contradictory.

To illustrate this concept we present a method of constructing a statement system $\sum$ from a given propositional calculus. Let $\sum$ be the statement system such that:

1. dom $\sum$ is the set of all propositions.
2. The truth-value of each statement in dom $\sum$ is "true."

Let ι be the identity interpreter and form the set of wffs

$$K = \{A \in W \mid \iota A \text{ is true for } \textstyle\sum\}$$

We claim that K is maximal-consistent.

Dem. 1. By construction, $\sum$ is a model of K under ι. By Lemma 3, page 96, K is consistent.

2. Let K_1 be any proper superset of K, and let $A \in K_1 - K$. Then the swff ιA is false for $\sum$ (since $A \notin K$). Thus $\iota[\neg A]$ is true for $\sum$; so

$\neg A \in \mathsf{K}$, thus $\mathsf{K} \vdash \neg A$. Now consider $\mathsf{K} \cup \{A\}$. Clearly, $\mathsf{K} \cup \{A\} \vdash A$, so $\mathsf{K} \cup \{A\} \vdash A \wedge \neg A$. Thus $\mathsf{K}_1 \vdash A \wedge \neg A$. By our Criterion for Contradictory Sets, K_1 is contradictory. We have proved that each proper superset of K is contradictory. So K is maximal-consistent, as claimed.

As we have indicated in the preceding argument, we can decide whether a set of wffs is maximal-consistent by applying the following criterion.

Criterion for Maximal-Consistent Sets. A consistent set of wffs, say K, is maximal-consistent iff $\mathsf{K} \cup \{B\}$ is contradictory for each wff B not in K.

Dem. Show that L is contradictory whenever L is a proper superset of K; this is left as an exercise.

Here are some useful properties of maximal-consistent sets.

Lemma 1. If K is maximal-consistent and if $\mathsf{K} \vdash A$, then $A \in \mathsf{K}$.

Dem. Assume that $A \notin \mathsf{K}$. Then $\mathsf{K} \cup \{A\}$ is contradictory. Thus $\mathsf{K} \cup \{A\} \vdash \neg A$ and it follows that $\mathsf{K} \vdash \neg A$. But $\mathsf{K} \vdash A$, so $\mathsf{K} \vdash A \wedge \neg A$; thus K is contradictory. This contradiction establishes our lemma.

Lemma 2. If K is maximal-consistent and if $B \notin \mathsf{K}$, then $\neg B \in \mathsf{K}$.

Dem. Since K is maximal-consistent, $\mathsf{K} \cup \{B\}$ is contradictory. Thus

$$\mathsf{K} \cup \{B\} \vdash \neg B$$

and it follows that $\mathsf{K} \vdash \neg B$. By Lemma 1, $\neg B \in \mathsf{K}$.

Corollary 1. If K is maximal-consistent and if B is any wff, then either $B \in \mathsf{K}$ or else $\neg B \in \mathsf{K}$.

Dem. Apply Lemma 2.

Lemma 3. If K is maximal-consistent, if $A \in \mathsf{K}$, and if $\vdash A \rightarrow B$, then $B \in \mathsf{K}$.

Dem. Clearly $\mathsf{K} \vdash B$; so, by Lemma 1, $B \in \mathsf{K}$.

The notion of maximal-consistent sets is especially important because each maximal-consistent set K has a model; indeed, a model of K can be constructed from K in a simple and direct fashion. Let $\sum$ be the statement system such that $P \in \text{dom} \sum$ iff P is a proposition; moreover, define the truth-value of each atomic swff (P) of $\sum$ to be "true" iff $(P) \in$ K. There is a close connection between the maximal-consistent set K and the truth-values of the swffs of $\sum$; consider the following lemma, where ι is the identity interpreter.

Lemma 4. ιA is true for $\sum$ iff $A \in$ K.

Dem. Let $S = \{A \in \mathsf{W} \mid \iota A \text{ is true for } \sum \text{ iff } A \in \mathsf{K}\}$; we shall apply the Fundamental Theorem about Wffs to show that $S = \mathsf{W}$, the set of all wffs.

1. A atomic. By construction of $\sum$, $A \in S$.
2. Assume that $B \in S$; we shall show that $\neg B \in S$.

(a) Assume that $\iota[\neg B]$ is true for $\sum$. Then ιB is false for $\sum$; but $B \in S$, so $B \notin$ K, thus by Lemma 2, $\neg B \in$ K.

(b) Assume that $\neg B \in$ K. Then $B \notin$ K; so ιB is false for $\sum$, thus $\iota[\neg B]$ is true for $\sum$. We conclude that $\neg B \in S$.

3. Assume that $C, D \in S$; we shall show that $C \vee D \in S$.

(a) Assume that $\iota[C \vee D]$ is true for $\sum$. Then ιC is true for $\sum$ or ιD is true for $\sum$. In the former case, $C \in$ K, so $C \vee D \in$ K; in the latter case, $D \in$ K, so $C \vee D \in$ K.

(b) Assume that $C \vee D \in$ K. If $C \in$ K or if $D \in$ K, then $\iota C \vee \iota D$ is true for $\sum$ (since $C, D \in S$), and it follows that $C \vee D \in S$. Accordingly, we shall assume that $C \notin$ K and that $D \notin$ K. Then $\neg C \in$ K and $\neg D \in$ K; so $\neg C \wedge \neg D \in$ K, and it follows that $C \vee D \notin$ K. This contradiction proves that $C \in$ K or $D \in$ K, and it follows that $\iota[C \vee D]$ is true for $\sum$; so $C \vee D \in S$. We conclude that $S = \mathsf{W}$. This establishes Lemma 4.

In view of Lemma 4, it is clear that $\sum$ is a model of K under the identity interpreter ι. This establishes the following important fact.

Theorem 1. Each maximal-consistent set has a model.

We shall make use of Theorem 1 in Section 6.7.

Exercises

1. Prove the Criterion for Maximal-Consistent Sets.
2. Let K be a maximal-consistent set, let ν be the valuation such that for each atomic wff (P), $\nu(P) = \mathbf{t}$ iff $(P) \in$ K. Prove that for each wff A, $A \in$ K iff $\nu A = \mathbf{t}$.

3. Let K be a consistent set of wffs such that for each atomic wff (P), either $(P) \in \mathsf{K}$ or $\mathsf{K} \cup \{(P)\}$ is contradictory. Let $\sum$ be the statement system such that:

   (1) $P \in \operatorname{dom} \sum$ iff P is a proposition.
   (2) For each P in dom $\sum$, P is true for $\sum$ iff $(P) \in \mathsf{K}$.

   Show that $\sum$ is a model of K under the identity interpreter ι. *Hint:* Prove that for each wff A, $\mathsf{K} \cup \{A\}$ is consistent iff ιA is true for $\sum$. To this purpose, let B be the shortest wff such that $\mathsf{K} \cup \{B\}$ is contradictory iff ιB is true for $\sum$; show that B does not exist.

4. Let $\sum$ be the statement system such that:

   (1) $P \in \operatorname{dom} \sum$ iff P is a proposition.
   (2) For each P in dom $\sum$, P is false for $\sum$.

   Let ι be the identity interpreter and form the set of wffs

   $$\mathsf{K} = \{A \in \mathsf{W} \mid \iota A \text{ is true for } \textstyle\sum\}$$

   Prove that K is maximal-consistent.

5. Let ν be any valuation and let $\sum$ be the statement system such that:

   (1) $\operatorname{dom} \sum = \{P \mid P \text{ is a proposition of the propositional calculus}\}$.
   (2) For each P in dom $\sum$,

   $$\textstyle\sum P = \begin{cases} \text{true} & \text{if } \nu(P) = \mathsf{t} \\ \text{false} & \text{if } \nu(P) = \mathsf{f} \end{cases}$$

   Let ι be the identity interpreter and form the set of wffs

   $$\mathsf{K} = \{A \in \mathsf{W} \mid \iota A \text{ is true for } \textstyle\sum\}$$

   Prove that K is maximal-consistent.

6.7. Strong Completeness Theorem

Our objective here is to prove the Strong Completeness Theorem. Of course, no contradictory set of wffs has a model; thus, it is enough to prove that each consistent set has a model. In view of Theorem 1, page 100, we have only to show that each consistent set can be extended to a maximal-consistent set.

Theorem 1. Each consistent set possesses a maximal-consistent superset.

Dem. Let K be any consistent set of wffs. In view of the axiom of choice, which we assume, W can be well-ordered. Choose an ordering of W and call it the standard ordering. Let B be the first wff in the standard ordering such that $\mathsf{K} \cup \{B\}$ is consistent. Let C be the next wff in the standard ordering such that $\mathsf{K} \cup \{B, C\}$ is consistent. Keep going. We claim that the resulting set $\mathsf{K} \cup \{B, C, \ldots\} = \mathsf{K}_1$ is maximal-consistent. First, we shall show that K_1 is consistent. If not, K_1 possesses a finite subset that is contradictory. But K is consistent, so each of its finite subsets is consistent; moreover, the union of K and any finite subset of $\{B, C, \ldots\}$ is consistent by construction. We conclude that $\mathsf{K} \cup \{B, C, \ldots\}$ is consistent. Next, let A be a wff such that $\mathsf{K}_1 \cup \{A\}$ is consistent; we shall prove that $A \in \mathsf{K}_1$. Notice that each subset of $\mathsf{K}_1 \cup \{A\}$ is consistent. Now, A occurs in the standard ordering of W; so A is considered at some point in the construction of K_1; the subset of K_1 constructed at that point, say K', is certainly consistent. By assumption, $\mathsf{K}' \cup \{A\}$ is consistent; so $A \in \mathsf{K}_1$ by the definition of K_1. We conclude that if B is a wff such that $B \notin \mathsf{K}_1$, then $\mathsf{K}_1 \cup \{B\}$ is contradictory. Thus K_1 is maximal-consistent. This establishes Theorem 1.

Note. For a more rigorous proof of this theorem, see Henkin (1949).

Considering Theorem 1, page 100, we conclude that each consistent set has a model. As we have observed above, this verifies our next statement.

Strong Completeness Theorem. K is consistent iff K has a model.

We now present the important *Compactness Theorem.*

Compactness Theorem. K has a model iff each finite subset of K has a model.

Dem. If $\sum$ is a model of K under μ, a $\sum$-interpreter, then $\sum$ is a model of each subset of K under μ. We must show that K has a model if each finite subset of K has a model. Now, each consistent set has a model; so if K has no model, then K is contradictory. This means that there is a finite subset of K, say $\{A_1, \ldots, A_n\}$, such that

$$\vdash A_1 \wedge \cdots \wedge A_n \rightarrow X \wedge \neg X$$

Thus $\{A_1, \ldots, A_n\}$ is contradictory, so does not have a model. We conclude from this contradiction that K has a model. This establishes the Compactness Theorem.

Here is another corollary of the Strong Completeness Theorem.

Theorem 2. A is K-true iff $\mathsf{K} \vdash A$.

Dem. 1. Assume that $\mathsf{K} \vdash A$. Then A is K-true by Lemma 1, page 89.

2. Assume that A is K-true. We want to show that $\mathsf{K} \vdash A$; assume that this is false. Then $\mathsf{K} \cup \{\neg A\}$ is consistent by Lemma 1, page 96. By the Strong Completeness Theorem, $\mathsf{K} \cup \{\neg A\}$ has a model, say Σ, under an interpreter μ; thus $\mu[\neg A]$ is true for Σ, a model of K. We conclude that μA is false for Σ; so A is not K-true. This contradiction proves that $\mathsf{K} \vdash A$.

PART II
Semantical Systems and Predicate Calculus

7

Semantical Systems

7.1. Relational Systems

The predicate calculus, in its classical form, serves as the blueprint for the language of any relational system. In this book we shall develop a more expressive predicate calculus, with the aim of providing a language better suited to the needs of mathematics than the language of relational systems.

However, the language of relational systems is easy to grasp, and so provides a good introduction to the language that primarily concerns us—the language of semantical systems. Accordingly, we now briefly introduce relational systems and the language of a relational system.

By a *relational system* we mean a system of the form $\langle S, R_1, R_2, R_3, \ldots \rangle$, where S is a nonempty set and each R_i is a relation on S, i.e., a set of tuples whose terms are in S. There must be at least one relation in the system. Of course, mathematical systems usually involve operations as well as relations; e.g., the system of natural numbers $\langle N, +, \cdot, < \rangle$, where $N = \{1, 2, 3, \ldots\}$ is the set of all natural numbers, $+$ and $\cdot$ represent the operations of addition and multiplication, and $<$ denotes the *less than* relation on N. It is well known, however, that any operation on a set can be regarded as a relation on that set. For example, a binary operation on a set A associates a member of A with each ordered pair whose terms are in A; however, mathematicians usually identify a map of $A \times A$ into A with the corresponding set of triples, which is a subset of $A \times A \times A$, and so is a relation on A. In more practical terms, we note that the operation of addition can be interpreted as the relation $+$ such that $(a, b, c) \in +$ iff c is the sum of a and b.

As we have indicated, we regard the relations of a relational system $\mathscr{R} = \langle S, R_1, R_2, R_3, \ldots \rangle$ as being sets, indeed sets of tuples. Here S is the *supporting* set of the relational system $\mathscr{R}$ in the sense that the remaining components of $\mathscr{R}$, namely its relations, are built up from S by way of forming tuples whose terms are in S. From the viewpoint of naive set theory, a set is known once we know of each object x whether or not x is in that set. Accordingly, once we have specified the supporting set of $\mathscr{R}$, which we regard as the universe of discourse for $\mathscr{R}$, the most basic question we can put regarding $\mathscr{R}$ is whether a tuple α is in a relation of $\mathscr{R}$, say R_i. The statement $\alpha \in R_i$, which may be true or false, is called an *atomic* statement, or atomic swff, of $\mathscr{R}$. The language of $\mathscr{R}$ is built up from its atomic statements by utilizing the connectives of symbolic logic, namely $\neg$ (*not*), $\vee$ (*or*), and $\forall$ (*for each*). The remaining connectives of symbolic logic can be expressed in terms of $\neg$, $\vee$, and $\forall$. Moreover, the truth-value of each compound statement built up from the atomic statements of $\mathscr{R}$ is determined by the truth-values of the atomic statements involved in the compound statement.

The relations of a relational system are usually regarded as mathematical objects of a higher type than the members of its supporting set S. In particular, the usual convention is that no relation of $\mathscr{R}$ is a member of S; indeed, the relations of $\mathscr{R}$ do not possess names within the relational system itself; rather, their names lie outside the system. On the other hand, the classical predicate calculus provides names for relations (namely, *predicates*), and names for members of S (namely, *individuals*); but the dichotomy between these objects is maintained in the sense that no predicate is also an individual.

Under the classical view, any two members of a relation have the same number of terms (i.e., the same *length*). We have dropped this restriction as the first step toward developing a language which is sufficiently flexible and expressive to exploit the potential of nonstandard analysis.

For example, let T be the set of all finite tuples whose terms are real numbers; then T is a relation on the set of all real numbers. Moreover, the following subset of T:

$$\{\alpha \in T \mid \text{the last term of } \alpha \text{ is the sum of its other terms}\}$$

is a relation on the set of all real numbers. This relation, which we denote by S, represents the operation of summing the terms of a finite tuple. Let us agree that the sum of no numbers is 0; then $(0) \in S$ and $(1) \notin S$. Clearly, for each natural number n there is a tuple in S whose length is n; we do not regard 0 as a natural number.

Exercises

1. Show that the system of integers, which involves addition, multiplication, and the *less than* relation, can be represented by a relational system.

2. Show that the additive identity of the system of integers can be represented by a unary relation.

3. Characterize the language of a relational system

$$\mathscr{R} = \langle S, R_1, R_2, R_3, \ldots \rangle$$

and define inductively the truth-value of each statement of this language.

7.2. Semantical Systems

We have seen that each atomic statement of a relational system $\mathscr{R}$ has the form "$\alpha \in R$," where α is a tuple whose terms are in the basic set of $\mathscr{R}$ and R is a relation of $\mathscr{R}$. To avoid various philosophical problems associated with the notion of a *relation*, we now present a somewhat more fundamental method of looking at relational systems; our approach yields systems more comprehensive than relational systems, called *semantical systems*.

First, let us agree to represent a statement of the form "$\alpha \in R$" by writing "$R\alpha$." Now, α is a tuple, say $(a_1, \ldots, a_n)$; as an abbreviating device, we shall denote this tuple by the string "$a_1 \cdots a_n$" (i.e., we delete commas and parentheses). So the string "$Ra_1 \cdots a_n$," which has $n + 1$ terms, represents the informal statement

$$(a_1, \ldots, a_n) \in R \tag{1}$$

i.e., "the tuple $(a_1, \ldots, a_n)$ is in the set R." Of course, the truth of a statement of this sort depends upon the membership of R. Since the relations of a relational system are regarded as known, we can decide the truth-value of (1).

Our plan is to carry matters a step further. We shall regard each term of the string "$Ra_1 \cdots a_n$" as a linguistic entity or object, and we shall base a language on these linguistic entities. In particular, we shall regard the first term of each of our strings merely as a symbol, not a relation or set. Moreover, a language must be capable of assigning truth-values to its statements. Accordingly, each initial string

$$Ra_1 \cdots a_n$$

must have "true" or "false" associated with it.

In summary, our plan is to develop a language based on given strings, each having at least two terms; moreover, "true" or "false" is associated with each given string. To handle this situation we introduce the map $\sum$ whose domain is the set of given strings, and whose range is {true, false}, where $\sum$ associates "true" with a given string so as to reflect the given situation. We say that any map of this sort is a *semantical system.*

Definition. By a *semantical system* we mean any map whose domain is a nonempty set of strings, each with at least two terms, and whose range is included in {true, false}.

For each string $\beta \in \operatorname{dom} \sum$ we say that β *is true for* $\sum$ if $\sum$ associates "true" with β; we say that β *is false for* $\sum$ if $\sum$ associates "false" with β.

Each term of a string in dom $\sum$ is called an *object*; the first term of each string in dom $\sum$ is called a *relation symbol*; all other terms of these strings are called *constants*, i.e., a constant is an ith term, $i > 1$, of some string in dom $\sum$. An object can be both a relation symbol and a constant. We say that a string α is an *n-string* if α has exactly n terms.

We want to associate a relation (i.e., a set of certain tuples) with each relation symbol of $\sum$. First, we decide the *length* of a string that is a candidate for membership in the relation associated with any relation symbol, say R. Accordingly, we now associate a set of natural numbers with R, called its *type*.

Definition. For each relation symbol of $\sum$, say R,

$$\text{type } R = \{n \in N \mid \text{there is an } n\text{-string } \alpha \text{ such that } R\alpha \in \operatorname{dom} \textstyle\sum\}$$

Notice that for each relation symbol of $\sum$, say R, type R is a nonempty subset of N; this subset can have just one member, several members, or it may even be N itself.

An n-string whose terms are constants of $\sum$ is called an *R-string* provided that R is a relation symbol of $\sum$ and $n \in$ type R. The relation which we wish to associate with a relation symbol R is the following set of R-strings:

$$\{\alpha \mid R\alpha \text{ is true for } \textstyle\sum\}$$

Breaking down the condition for membership in this set, we require that: (i) $R\alpha \in \operatorname{dom} \sum$; (ii) $R\alpha$ is true for $\sum$.

We shall denote the relation associated with a relation symbol R by "$\underline{R}$" (or by R itself, if there is no ambiguity).

Definition. For each relation symbol of $\sum$, say R,

$$\underline{R} = \{\alpha \mid R\alpha \text{ is true for } \textstyle\sum\}$$

We say that $\underline{R}$ is a *relation of* $\sum$.

In this way, each relation symbol of $\sum$ provides a name for the corresponding relation of $\sum$. Notice that each term of each string in a relation of $\sum$ is a constant of $\sum$; however, a constant may also be a relation symbol. This is one sense in which we have generalized the usual notion of a relational system. Moreover, the strings in a relation $\underline{R}$ are not necessarily of the same length; indeed, $\underline{R}$ can be empty, can contain many strings of different lengths, or even all strings whose terms are constants of $\sum$.

Another useful idea is the notion of the *diagram* of $\sum$ (denoted by "diag $\sum$"). This is the subset of dom $\sum$ that consists of the strings that are true for $\sum$.

Definition. Let $\sum$ be any semantical system; then

$$\text{diag} \sum = \{\beta \in \text{dom} \sum \mid \beta \text{ is true for } \sum\}$$

Each relation of $\sum$, say $\underline{R}$, can be obtained from diag $\sum$ by forming the following subset of diag $\sum$:

$$\{\beta \in \text{diag} \sum \mid R \text{ is the first term of } \beta\} \tag{2}$$

and then deleting the first term of each string in (2).

It is useful to display the components of a semantical system $\sum$. In particular, we want to exhibit the constants, relation symbols, and relations of $\sum$. Accordingly, we shall sometimes represent $\sum$ by writing

$$\langle C, \text{RS}, \underline{R}_1, \underline{R}_2, \underline{R}_3, \ldots \rangle$$

where $C = \{t \mid t \text{ is a constant of } \sum\}$ and $\text{RS} = \{R \mid R \text{ is a relation symbol of } \sum\}$. Here $C \cup \text{RS} = \text{Ob}$, the set of all objects of $\sum$. The set of constants C is sometimes called the *supporting set* (or *basic set*) of $\sum$ since, as we shall see, all quantifiers refer to C; moreover, the relations of $\sum$ involve only constants (i.e., each term of each string in each relation of $\sum$ is a constant of $\sum$). Notice that names for these relations are provided by RS, the set of all relation symbols. The empty set may be a relation of $\sum$, depending on the nature of $\sum$. In fact, several relations of $\sum$ may be the empty set, where these relations are yielded by distinct relation symbols of $\sum$. In general, several relation symbols of $\sum$ may yield the same relation.

If dom $\sum$ = diag $\sum$, we can spell out the semantical system $\sum$ by specifying its constants, relation symbols, and relations.

We sometimes focus on the constants and relations of a semantical system

$$\sum = \langle C, \text{RS}, \underline{R}_1, \underline{R}_2, \underline{R}_3, \ldots \rangle$$

ignoring its relation symbols. This yields the relational system

$$\langle C, \underline{R}_1, \underline{R}_2, \underline{R}_3, \ldots \rangle$$

Example 1. Let $\sum$ be the semantical system such that

$$\text{dom} \sum = \{<12,\ <21,\ <13,\ <31,\ <23,\ <32\}$$

and

$$\text{diag} \sum = \{<12,\ <13,\ <23\}$$

Here type $< \; = \{2\}$. Then

$$\sum = \langle \{1, 2, 3\}, \{<\}, \{12, 13, 23\} \rangle$$

and the corresponding relational system is $\langle \{1, 2, 3\}, \{12, 13, 23\} \rangle$.

Example 2. Let $\sum$ be the semantical system such that

$$\text{dom} \sum = \{+000,\ +101,\ +011,\ +110\}$$

and diag $\sum$ = dom $\sum$. Examining dom $\sum$, we see that 0 and 1 are the constants of $\sum$ and that $+$ is its only relation symbol. Clearly, $\{000, 011, 101, 110\}$ is the corresponding relation of $\sum$, i.e.,

$$+ \; = \{000, 011, 101, 110\}$$

(Note that here, as elsewhere, we dispense with the underlining.) Here type $+ \; = \{3\}$. Thus

$$\sum = \langle \{0, 1\}, \{+\}, \{000, 101, 011, 110\} \rangle$$

and the corresponding relational system is $\langle \{0, 1\}, \{000, 101, 011, 110\} \rangle$.

Example 3. Let $\sum$ be the semantical system such that

$$\text{dom} \sum = \{+000,\ +101,\ +011,\ +110,\ -00,\ -11\}$$

and diag $\sum$ = dom $\sum$. Here $C = \{0, 1\}$, RS $= \{+, -\}$, $+ = \{000, 101, 011, 110\}$, and $- = \{00, 11\}$. Clearly type $+ = \{3\}$ and type $- = \{2\}$. Thus

$$\sum = \langle \{0, 1\}, \{+, -\}, \{000, 101, 011, 110\}, \{00, 11\} \rangle$$

and the corresponding relational system is $\langle \{0, 1\}, \{000, 101, 011, 110\}, \{00, 11\} \rangle$.

Example 4. Let $\sum$ be the semantical system such that dom $\sum$ = $\{S\alpha \mid \alpha$ is a string of real numbers of length at least one$\}$ and diag $\sum$ =

$\{S\alpha \mid$ the last term of α is the sum of its other terms$\}$. So $S\alpha \in \text{diag} \sum$ if $\alpha = \beta x$, where β is a string of real numbers and x is the sum of the terms of β. The constants of $\sum$ are real numbers, and S is its only relation symbol. Let us interpret the sum of no numbers as zero; then $S0 \in \text{diag} \sum$. Thus type $S = N$. If we restrict diag $\sum$ so that $S\alpha \in \text{diag} \sum$ iff α has at least two terms and its last term is the sum of its other terms, then type $S = N - \{1\}$.

The constants of a semantical system need not be of the same sort. This is convenient, for example, to bring out the properties of the generalized summation of Example 4; this involves manipulating tuples of real numbers. The point is that the objects under discussion are tuples of real numbers, not just real numbers.

Example 5. Let $\sum$ be the semantical system whose constants are real numbers and tuples of real numbers of finite length. Let $\sum$ have three relation symbols $+$, S, and C. Finally, let dom $\sum$ = diag $\sum$, where each member of diag $\sum$ has one of the forms $+xyz$, $S\beta x$, or $C\alpha\beta\gamma$ and

$+xyz \in \text{diag} \sum$	iff	$x, y, z \in R$ and $x + y = z$
$S\beta x \in \text{diag} \sum$	iff	β is a tuple of real numbers, $x \in R$, and x is the sum of the terms of β
$C\alpha\beta\gamma \in \text{diag} \sum$	iff	α and β are tuples of real numbers and γ is the tuple formed by adjoining the terms of β to the terms of α

For example, $S(4, 1, 2)7 \in \text{diag} \sum$, $+314 \in \text{diag} \sum$, and

$$C(1, 2)(5, 6, 7)(1, 2, 5, 6, 7) \in \text{diag} \textstyle\sum.$$

With relations, these statements simplify to: $(4, 1, 2)7 \in S$, $314 \in +$, and $(1, 2)(5, 6, 7)(1, 2, 5, 6, 7) \in C$. In this system we can express the following connection between $+$, S, and C:

$$S\alpha x \wedge S\beta y \wedge S\gamma z \wedge C\alpha\beta\gamma \rightarrow +xyz \tag{3}$$

for each $x, y, z \in R$ and for any tuples α, β, and γ.

Example 6. Let $\sum$ be the semantical system whose constants are natural numbers and finite tuples of real numbers; let L be the only relation of $\sum$ where $\alpha n \in L$ iff α is a tuple of real numbers with n terms. Here we regard αn as a string with two terms, α and n, where α itself is a tuple and $n \in N$; e.g., $L(5, 2, 7)3 \in \text{diag} \sum$, i.e., $(5, 2, 7)3 \in L$. Let T be the set of all tuples of finite length whose terms are real numbers; then $\sum = \langle N \cup T, \{L\}, L\rangle$ and the corresponding relational system is $\langle N \cup T, L\rangle$.

Example 7. Let $\sum$ be the semantical system whose constants are real numbers and tuples of real numbers of finite length. Let R, N, T, L, and t be the relations of $\sum$, where

$\alpha \in T$ iff α is a tuple of real numbers of finite length
$\alpha n \in L$ iff $\alpha \in T$ and α has n terms
$a \in R$ iff a is a real number
$n \in N$ iff n is a natural number
$n\alpha a \in t$ iff $n \in N$, $\alpha \in T$, $a \in R$, and a is the nth term of α

Here the relational system yielded by $\sum$ is

$$\langle R \cup T, R, N, T, L, t\rangle$$

For example, let α be a tuple such that $\alpha \in T$, $\alpha 3 \in L$, $1\alpha 5 \in t$, $2\alpha 2 \in t$, $3\alpha 7 \in t$; then $\alpha = (5, 2, 7)$. Finally, we mention that type $L = \{2\}$, type $t = \{3\}$, and type $T =$ type $R =$ type $N = \{1\}$.

We return to fundamentals to describe the semantical system of our next example. First, we partially describe dom $\sum$ by announcing the constants and relation symbols of $\sum$, and then we provide the values of the map $\sum$ by characterizing its diagram. We complete our description of dom $\sum$ by announcing that dom $\sum =$ diag $\sum$.

Example 8. Let $\sum$ be the semantical system whose constants are real numbers and sets of real numbers (and each real number and each set of real numbers is a constant of $\sum$), and let R, S, M, C, $\bigcup$, $\bigcap$, $\leq$, and $+$ be the relation symbols of $\sum$, where the type of R and S is $\{1\}$, the type of M, C, and $\leq$ is $\{2\}$, the type of $+$ is $\{3\}$, and the type of $\bigcup$ and $\bigcap$ is $\{n \in N \mid n \geq 3\}$. Let dom $\sum =$ diag $\sum$, where diag $\sum$ is characterized by the following statements:

1. $Ra \in$ diag $\sum$ iff a is a real number.
2. $Sa \in$ diag $\sum$ iff a is a set of real numbers.
3. $Mab \in$ diag $\sum$ iff $a \in b$, a is real, and b is a set of real numbers.
4. $Can \in$ diag $\sum$ iff a is a set of n real numbers.
5. For each $n \in N$, $n \geq 3$, $\bigcup a_1 \cdots a_n \in$ diag $\sum$ iff $a_1 \cup \cdots \cup a_{n-1} = a_n$ and each a_i is a set of real numbers.
6. For each $n \in N$, $n \geq 3$, $\bigcap a_1 \cdots a_n \in$ diag $\sum$ iff $a_1 \cap \cdots \cap a_{n-1} = a_n$ and each a_i is a set of real numbers.

7. $\leq xy \in \text{diag} \sum$ iff $x \leq y$ and x and y are real numbers.
8. $+xyz \in \text{diag} \sum$ iff $x + y = z$ and x, y, and z are real numbers.

The associated relational system is

$$\langle R \cup S, R, S, M, C, \bigcup, \bigcap, \leq, +\rangle$$

where $R \cup S$ is the set of constants of $\sum$.

To illustrate, let the following statements be true about $\sum$: Sa, $M4a$, $M5a$, $Ca2$. Then $a = \{4, 5\}$. Also, the following statements are true about this system: For any sets a, b, and c and for any real numbers x, y, z, and u

$$\bigcup abc \leftrightarrow \forall t(Rt \rightarrow (Mtc \leftrightarrow Mta \vee Mtb)) \tag{4}$$

$$\bigcup abc \wedge Cax \wedge Cby \wedge Ccz \wedge +xyu \rightarrow \leq zu \tag{5}$$

There are two special conditions that many semantical systems will meet. Let R be any relation symbol of a semantical system $\sum$ and let $n \in \text{type } R$. The conditions are as follows.

1. There is an R-string of length n, say α, such that $R\alpha \in \text{diag} \sum$.
2. For each constant t of $\sum$, there is a string β such that t is the ith term of β, $i > 1$, and $\beta \in \text{diag} \sum$.

If these conditions are met, then diag $\sum$ tells the whole story of $\sum$. Indeed, from diag $\sum$ we can read off the relation symbols of $\sum$ and the type of each relation symbol; diag $\sum$ also discloses the constants of $\sum$ and the relations of $\sum$. Any semantical system *not* meeting these two conditions will be called *pathological.*

We now present a semantical system for which some relation symbols are also constants.

Example 9. Let $\sum$ be the semantical system whose constants are real numbers and functions, and let R, F, $+$, and each function be the relation symbols of $\sum$. The type of R and F is $\{1\}$, the type of $+$ is $\{3\}$, and the type of each function is $\{2\}$. Let dom $\sum = \text{diag} \sum$, where diag $\sum$ is characterized by the following statements:

1. $Ra \in \text{diag} \sum$ iff a is a real number.
2. $Ff \in \text{diag} \sum$ iff f is a function (i.e., a map of the set of all real numbers into the set of real numbers).
3. $+xyz \in \text{diag} \sum$ iff $x + y = z$ and either x, y, and z are functions or x, y, and z are real numbers.
4. For each function f,

   $fab \in \text{diag} \sum$ iff $f(a) = b$ and a and b are real numbers.

Although it is customary to regard addition of real numbers and addition of functions as distinct mathematical operations, we have represented addition by a single relation symbol, which in turn yields a single relation of the corresponding relational system. Here is a fact about this generalized notion.

For any functions f, g, and h

$$+fgh \leftrightarrow \forall yzuv[fyz \wedge gyu \wedge +zuv \rightarrow hyv] \tag{6}$$

Notice that f, g, and h appear as constants on the LHS of (6) and as relation symbols on the RHS of (6).

A constant that is also a relation symbol of a semantical system $\sum$ is said to be a *mixed* constant of $\sum$; a constant of $\sum$ that is not a relation symbol of $\sum$ is called a *pure* constant of $\sum$. The pure constants of $\sum$ have a special role in developing the language of $\sum$, as we shall soon see.

Exercises

1. Let $\sum$ be the semantical system such that

$$\text{dom } \sum = \{121, 101, 1862, 146, 16452, 0298, 04513, 017, 08, 029, 211\}$$

and $\text{diag } \sum = \{121, 1862, 16452, 08, 017, 029\}$.

(a) Exhibit the relation symbols of $\sum$.
(b) Exhibit the constants of $\sum$.
(c) State the type of each relation symbol of $\sum$.
(d) Exhibit each relation of $\sum$.
(e) Is $\sum$ pathological?

2. Let $\sum$ be the semantical system such that

$$\begin{aligned}\text{dom } \sum = {} & \{>\alpha \mid \alpha \text{ is a finite tuple of real numbers with at least one term}\} \\ & \cup \{<\alpha \mid \alpha \text{ is a finite tuple of real numbers with at least one term}\}\end{aligned}$$

and

$$\begin{aligned}\text{diag } \sum = {} & \{>\alpha \mid \text{the last term of } \alpha \text{ is greater than its other terms}\} \\ & \cup \{<\alpha \mid \text{the last term of } \alpha \text{ is less than its other terms}\}\end{aligned}$$

(a) Exhibit the relation symbols of $\sum$.
(b) Exhibit the constants of $\sum$.
(c) State the type of each relation symbol of $\sum$.
(d) Exhibit each relation of $\sum$.
(e) Is $\sum$ pathological?

7.3. Language of a Semantical System

Each semantical system, say $\sum$, has its own language, which is built up from the relation symbols and constants of $\sum$ by means of the logical connectives *not* ($\neg$), the *inclusive or* ($\vee$), and *for each* ($\forall$); in conjunction with the universal quantifier $\forall$ we shall require placeholders. Each statement of this language is called a *semantical well-formed formula*, or *swff* for short.

First, we define the atomic swffs of $\sum$, the basic statements from which each swff of $\sum$ is obtained. Let R be any relation symbol of $\sum$, and let α be any R-string of $\sum$ (i.e., α is an n-string, where $n \in$ type R, and each term of α is a constant of $\sum$); we say that $(R\alpha)$ is an *atomic* swff of $\sum$. Accordingly, we regard the expression "$(R\alpha)$" as a statement; we may interpret this statement as meaning that the tuple α is an element of the relation $\underline{R}$, i.e., $\alpha \in \underline{R}$. Alternatively, even preferably, we regard each atomic swff $(R\alpha)$ as a basic linguistic component of the language of $\sum$, and concern ourselves only with its truth-value in $\sum$. Of course, $(R\alpha)$ is true for $\sum$ iff $\sum$ associates "true" with $R\alpha$. More precisely, we define the truth-value of each atomic swff of $\sum$ as follows.

Definition. Let $(R\alpha)$ be any atomic swff of $\sum$; then we say that $(R\alpha)$ is *true for* $\sum$ if $\sum$ associates "true" with $R\alpha$, and we say that $(R\alpha)$ is *false for* $\sum$ if $\sum$ does not associate "true" with $R\alpha$.

In this way, we obtain the truth-value of each atomic swff of $\sum$ directly from the map $\sum$ itself.

The compound swffs of $\sum$ are built up from atomic swffs by connecting a finite number of them with $\neg$, $\vee$, or $\forall$. The universal quantifier also involves placeholders $x, y, z, \ldots$. For example, let c and d be pure constants of $\sum$, let x and y be placeholders, and let R be a relation symbol of $\sum$ such that $3 \in$ type R. Then $(Rcdc)$ is an atomic swff of $\sum$. The result of quantifying the c's in this atomic swff

$$(\forall x(Rxdx)) \tag{1}$$

will also be a swff. But we point out that $Rxdx$ is *not* a swff. It is, however, a quasi-swff, used as a component in the formation of (1). Similarly we can choose another placeholder, e.g., y, and form the swff

$$(\forall y(\forall x(Rxyx))) \tag{2}$$

from the quasi-swff $Rxyx$.

The alphabet of the language of $\sum$ is derived mainly from dom $\sum$, which yields the relation symbols and constants of the language. We also require connectives, parentheses, and placeholders.

Alphabet of the Language of $\sum$:
Relation Symbols: Each first term of each string in dom $\sum$.
Constants: Each ith term, $i > 1$, of each string in dom $\sum$.
Connectives: $\neg$, $\vee$, $\forall$.
Parentheses: (,).
Placeholders: $x, y, z, \ldots$. (This list is infinite and does not include any of the above symbols.)

Before officially defining the swffs of $\sum$, we will define a broader class, the *quasi-swffs* of $\sum$. We begin by extending the notion of *R-string* to cover strings (of length n where $n \in$ type R) consisting of constants and placeholders.

An *expression* is any finite string of symbols from the alphabet of the language of $\sum$. A placeholder t is said to be *free* in an expression φ if t occurs in φ and "$\forall t$" does not occur in φ. We say that t is *bound* in φ if "$\forall t$" occurs in φ.

Definition of Quasi-Swff of $\sum$:

(a) $(R\alpha)$ is a quasi-swff of $\sum$ if R is a relation symbol and α is an R-string.

(b) $(\neg A)$ is a quasi-swff of $\sum$ if A is a quasi-swff of $\sum$.

(c) $(A \vee B)$ is a quasi-swff of $\sum$ if A and B are quasi-swffs of $\sum$ and furthermore no placeholder is both free in A and bound in B, or vice versa.

(d) $(\forall tC)$ is a quasi-swff of $\sum$ if C is a quasi-swff of $\sum$ and t is a placeholder free in C.

(e) Each quasi-swff of $\sum$ contains only a finite number of instances of connectives.

Definition. A *swff* of $\sum$ is a quasi-swff of $\sum$ in which no placeholder is free.

We are interested only in the swffs of $\sum$; the quasi-swffs are useful only because they lead to swffs.

For example, let R and S be relation symbols of a semantical system $\sum$, and let a, b, and c be pure constants of $\sum$. Let $1 \in$ type R and let $2 \in$ type S; then both (Rb) and (Sca) are atomic swffs of $\sum$, by (a). By (b), $(\neg(Rb))$ is a swff of $\sum$. By (c), $((\neg(Rb)) \vee (Sca))$ is a swff of $\sum$. Also $((\neg(Rb)) \vee (Sxa))$ is a quasi-swff of $\sum$, so, by (d), $(\forall x((\neg(Rb)) \vee (Sxa)))$ is a swff of $\sum$.

We obtain the truth-value of each atomic swff of a semantical system, say $\sum$, directly from the map $\sum$ itself. The truth-value of each compound swff of $\sum$ is obtained by considering the significance of the connectives $\neg$, $\vee$, and $\forall$; in short, by the truth-table method. Bearing this in mind, we formulate the following definition.

Definition of Truth-Values:

1. An atomic swff $(R\alpha)$ is *true for* $\sum$ if $\sum$ associates "true" with $R\alpha$; $(R\alpha)$ is *false for* $\sum$ if $\sum$ does not associate "true" with $R\alpha$.
2. $(\neg A)$ is *true for* $\sum$ if A is false for $\sum$; $(\neg A)$ is *false for* $\sum$ if A is true for $\sum$. Here A is any swff of $\sum$.
3. $(A \vee B)$ is *true for* $\sum$ if A is true for $\sum$, B is true for $\sum$, or both A and B are true for $\sum$; $(A \vee B)$ is *false for* $\sum$ if both A and B are false for $\sum$. Here A and B are any swffs of $\sum$.
4. $(\forall tE)$ is *true for* $\sum$ if $\mathrm{S}_t^a[E]$ is true for $\sum$ for each constant (either pure or mixed) a of $\sum$ ($\mathrm{S}_t^a[E]$ is the expression obtained by substituting a for each t in E). $(\forall tE)$ is *false for* $\sum$ if there is a constant of $\sum$, say b, such that $\mathrm{S}_t^b[E]$ is false for $\sum$.

We can verify that each swff of $\sum$ has a unique truth-value under the above definition, by using the analog of the Construction Theorem for Wffs from Part I (cf. p. 150).

In Section 1.3 we introduced *names* for swffs of a statement system, including various parentheses-omitting conventions. We shall freely apply these conventions to assist us in naming swffs of a semantical system. Thus, we shall omit parentheses whenever this does not produce an ambiguous expression. For example, the outermost pair of parentheses of any swff and the pair of parentheses involved in each atomic swff appearing in a compound swff can usually be suppressed without harm.

Short names for certain swffs are obtained in terms of the logical connectives $\wedge$, $\rightarrow$, $\leftrightarrow$, and $\exists$ (*there is at least one*). Our agreement is that for any swffs A, B, and C, and for any placeholder t,

$(A \wedge B)$	is a name for	$(\neg((\neg A) \vee (\neg B)))$
$(A \rightarrow B)$	is a name for	$((\neg A) \vee B)$
$(A \leftrightarrow B)$	is a name for	$((A \rightarrow B) \wedge (B \rightarrow A))$
$(\exists tC)$	is a name for	$(\neg(\forall t(\neg C)))$

We emphasize that $\neg$, $\vee$, and $\forall$ are the basic connectives of our language, whereas the connectives $\wedge$, $\rightarrow$, $\leftrightarrow$, and $\exists$ are defined in terms of the basic connectives.

Moreover, we shall freely drop the outermost pair of parentheses of a *name* for a swff. For example, "$\exists xSxa$" is a name for the swff named by "$\neg\forall x\neg Sxa$," namely the swff $(\neg(\forall x(\neg(Sxa))))$.

We shall attribute a built-in bracketing power, or reach, to the connectives, other than the quantifiers $\forall$ and $\exists$, in the following order:

$$\neg, \vee, \wedge, \rightarrow, \leftrightarrow$$

where weakest connectives appear first. As for statement systems, the reach of a connective is strengthened by placing a dot over the connective.

These conventions for naming swffs apply also to quasi-swffs. But we have no permanent interest in quasi-swffs; they are useful—indeed necessary—in the construction of swffs.

We have a special convention for the quantifiers $\forall$ and $\exists$. Let $t_1, \ldots, t_n$ be n distinct placeholders free in a quasi-swff C. Let Q_i be a quantifier, either $\forall$ or $\exists$, for $i = 1, \ldots, n$. Then "$Q_1t_1 \cdots Q_nt_nC$" is a name for the swff $(Q_1t_1 \cdots (Q_nt_nC) \cdots)$. Of course, if C is atomic, we may drop the pair of parentheses of C. The name "$Q_1t_1 \cdots Q_nt_nC$" is said to be in *prenex normal form* provided that no quantifier occurs in C. For example, "$\forall x \exists y Gxy$" is a name for the swff $(\forall x(\exists y Gxy))$.

We now illustrate some of the ideas of this section.

Example 1. Let $\sum = \langle\{0, 1\}, \{+\}, \{000, 011, 101, 110\}\rangle$ be the semantical system of Example 2, page 112. Each of the following expressions is a true swff of $\sum$: $\forall x(+x0x)$, $\forall x\exists y(+xy0)$, $+101 \wedge \neg +111$. The swff $\forall x(+x1x)$ is false for $\sum$. Since $\{3\}$ = type $+$, the expressions $(+11)$ and $(+1111)$ are not swffs of $\sum$.

Example 2. Let $\sum$ be the semantical system of Example 5, page 113, augmented by two additional relation symbols R and T, and let diag $\sum$ be augmented so that $Ra \in \operatorname{diag} \sum$ iff a is a real number, and $T\alpha \in \operatorname{diag} \sum$ iff α is a finite tuple of real numbers. We also include α, β, and γ among our placeholders. Then statement (3), page 113, is expressed by the following swff:

$$\forall xyz\alpha\beta\gamma(Rx \wedge Ry \wedge Rz \wedge T\alpha \wedge T\beta \wedge T\gamma \\ \wedge S\alpha x \wedge S\beta y \wedge S\gamma z \wedge C\alpha\beta\gamma \rightarrow +xyz)$$

Example 3. Let $\sum$ be the semantical system of Example 8, page 114. Since there is a set with no members (the empty set), the following expression is a true swff of $\sum$: $\exists y(Sy \wedge \forall x(\neg Mxy))$. Moreover, statements (4) and (5), page 115, are expressed by the following true swffs of $\sum$, where a, b, and c are placeholders of $\sum$:

$$\forall abc(Sa \wedge Sb \wedge Sc \dot{\rightarrow} \bigcup abc \leftrightarrow \forall x(Rx \rightarrow (Mxc \leftrightarrow Mxa \vee Mxb)))$$

$$\forall abcxyzu(Sa \wedge Sb \wedge Sc \wedge Rx \wedge Ry \wedge Rz \wedge Ru \\ \dot{\rightarrow} \bigcup abc \wedge Cax \wedge Cby \wedge Ccz \wedge +xyu \rightarrow \leq zu)$$

Example 4. Let $\sum$ be the semantical system of Example 9, page 115, and let f, g, and h be placeholders. Then statement (6), page 116, is expressed by the following:

$$\forall fgh(Ff \wedge Fg \wedge Fh \dot{\rightarrow} +fgh \leftrightarrow \forall xyzu(fxy \wedge gxz \wedge +yzu \rightarrow hxu)) \quad (3)$$

We mention that (3) is *not* a swff of $\sum$, since it is obtained by quantifying mixed constants of $\sum$.

Example 4 illustrates the fact that we have restricted the language of semantical systems by prohibiting quantification over mixed constants. The point is that we can establish powerful theorems about this basic language (see Section 12.3).

Here is another statement which expresses a fact about the semantical system of Example 9, page 115, but is not a swff:

$$\forall fxy(Ff \wedge fxy \rightarrow Rx \wedge Ry) \quad (4)$$

The standard method of expressing this fact by means of swffs is to formulate a family of swffs, each of the form

$$\forall xy(Fg \wedge gxy \rightarrow Rx \wedge Ry) \quad (5)$$

where g is a specific mixed constant of $\sum$ (namely, a function). Of course, since g is a function there is no need to include the conjunct Fg (which asserts that g is a function) in (5). So the family of statements (5) simplifies to the family

$$\forall xy(gxy \rightarrow Rx \wedge Ry) \quad (6)$$

where there is one statement in this family for each mixed constant of $\sum$.

Here is a fact about the swffs of a semantical system $\sum$ which we shall need in Section 7.4.

Lemma 1. Let A be any swff of $\sum$; there is a swff of $\sum$ which has the form "$\exists tE$" such that A is true for $\sum$ iff $\exists tE$ is true for $\sum$.

Dem. Let t be a placeholder that does not occur in A. Let B be a quantifier-free quasi-swff of $\sum$ such that $\exists tB$ is a swff true for $\sum$. Then

$$A \wedge \exists tB \quad (7)$$

is true for $\sum$ iff A is true for $\sum$. So

$$\exists t(A \wedge B) \quad (8)$$

is true for $\sum$ iff A is true for $\sum$. This proves our lemma.

Exercises

1. Let *a* and *b* be included among the relation symbols of a semantical system $\sum$, and let *a*, *b*, *c*, *d*, *e* be included among its constants. Given that *x* and *y* are placeholders, simplify the following:

   (a) $\mathrm{S}^{y}_{d}[(adea)]$.
   (b) $\mathrm{S}^{y}_{a}[(adea)]$.
   (c) $\mathrm{S}^{y}_{b}[(\forall x(abxb) \vee (bb))]$.

2. Under the assumptions of Exercise 1, simplify the following:

   (a) $\mathrm{S}^{a}_{x}[(\forall y(aya \vee bxa))]$.
   (b) $\mathrm{S}^{b}_{y}[(bxy \vee \neg(ayxb))]$.

3. Let $(\forall xA)$ be a swff of a semantical system.

   (a) Prove that *x* occurs in *A* and prove that no occurrence of *x* in *A* is immediately preceded by a LH parenthesis.
   (b) Prove that "$\forall x$" does not occur in *A*.

4. Let *R* be a relation symbol, with type *N*, of a semantical system $\sum$, and let *a*, *b*, and *R* be the constants of $\sum$. Prove:

   (a) Each of the following expressions is a swff of $\sum$: (Rab), (RR), $(RRRaa)$, $(\forall x(Rx))$, $(\forall y(Ray))$, $(\forall x(\forall y((Rx) \vee (Ry))))$, $(\forall x(\forall y(Rxy)))$.
   (b) $Ra \wedge Rb$ is true for $\sum$ iff (Ra) is true for $\sum$ and (Rb) is true for $\sum$.
   (c) $(\forall x(Rx))$ is true for $\sum$ iff $(RR) \wedge (Ra) \wedge (Rb)$ is true for $\sum$.
   (d) $(\exists y(Ry))$ is true for $\sum$ iff $(RR) \vee (Ra) \vee (Rb)$ is true for $\sum$.
   (e) $(\exists y(Ry))$ is true for $\sum$ iff at least one of the swffs (RR), (Ra), (Rb) is true for $\sum$.

5. Let $\sum$ be the semantical system of Example 1 of this section. Verify that:

   (a) $+111$ is false for $\sum$.
   (b) $+101 \wedge \neg(+111)$ is true for $\sum$.
   (c) $\forall x(+x0x)$ is true for $\sum$.
   (d) $\forall x(+x1x)$ is false for $\sum$.
   (e) $\exists x(+x1x)$ is false for $\sum$.
   (f) $\forall x(\neg(+x1x))$ is true for $\sum$.

7.4. Extensions; Elementary Extensions

The following ideas are vital to the applications of logic that we shall consider in Part III. Two semantical systems $\sum$ and $\sum'$ are said to be *similar* if they have the same relation symbols and each relation symbol has the same type in both systems. For example, the natural number system $\mathscr{N} = \langle N, +, \cdot, < \rangle$ and the system of integers $\mathscr{I} = \langle I, +, \cdot, < \rangle$

are similar; each of the relation symbols $+$, $\cdot$, and $<$ has the same type in both of these semantical systems.

Here is the important notion of an *extension*, or *subsystem*, of a semantical system. Throughout, $\sum$ and $\sum'$ are semantical systems.

Definition. We say that $\sum'$ is an *extension* of $\sum$, or that $\sum$ is a *subsystem* of $\sum'$, provided that:

1. $\sum$ and $\sum'$ are similar.
2. For each atomic swff A of $\sum$, A is true for $\sum'$ iff A is true for $\sum$.

By condition 2 of this definition we mean that each atomic swff of $\sum$ is an atomic swff of $\sum'$ and the following hold:

(a) Each atomic swff of $\sum$ that is true for $\sum$ is true for $\sum'$.
(b) Each atomic swff of $\sum$ that is false for $\sum$ is false for $\sum'$.

Since each atomic swff of $\sum$ is also an atomic swff of $\sum'$, each constant of $\sum$ is a constant of $\sum'$. Also, from (a), each relation of $\sum$ is also a subset of the corresponding relation of $\sum'$.

For example, the system of integers is an extension of the natural number system; the rational number system is an extension of the system of integers; the real number system is an extension of the rational number system; the complex number system is an extension of the real number system (with $<$ deleted).

We now present some facts about extensions.

Lemma 1. Let $\sum'$ be an extension of $\sum$. Then for each quantifier-free swff of $\sum$, say A,

$$A \text{ is true for } \textstyle\sum \text{ iff } A \text{ is true for } \textstyle\sum' \tag{1}$$

Dem. If the lemma is false, there is a shortest quantifier-free swff A, counting connectives, for which (1) is false. By assumption, (1) is true for each atomic swff. Clearly, if (1) is true for a swff B, then (1) is true for the swff $\neg B$. Moreover, it is easy to verify that if (1) is true for swffs C and D, then (1) is true for the swff $C \vee D$. We conclude that (1) is true for each quantifier-free swff of $\sum$.

Our next two lemmas show that truth-values are preserved for swffs having a special form.

Lemma 2. Let $\sum'$ be an extension of $\sum$. If E is quantifier-free and if $\exists sE$ is true for $\sum$, then $\exists sE$ is true for $\sum'$.

Dem. By assumption, there is a constant of $\sum$, say a, such that $S_s^a[E]$ is true for $\sum$. By Lemma 1, $S_s^a[E]$ is true for $\sum'$; therefore, $\exists sE$ is true for $\sum'$.

Lemma 3. Let $\sum'$ be an extension of $\sum$. If E is quantifier-free and if $\forall sE$ is true for $\sum'$, then $\forall sE$ is true for $\sum$.

Dem. If $\forall sE$ is false for $\sum$, there is a constant of $\sum$, say a, such that $S_s^a[E]$ is false for $\sum$. By Lemma 1, $S_s^a[E]$ is false for $\sum'$; therefore, $\forall sE$ is false for $\sum'$. This contradiction proves that $\forall sE$ is true for $\sum$, given that $\forall sE$ is true for $\sum'$.

It is tempting to conjecture that if $\sum'$ is an extension of $\sum$, then each swff of $\sum$ that has the form $\forall sE$ (where E is not necessarily quantifier-free) is true for $\sum$ provided that it is true for $\sum'$. We state emphatically that this conjecture is false! For example, the swff $\forall x \exists y(y^2 = x)$ is true for the complex number system, but is false for the real number system.

We now present the notion of an elementary extension.

Definition. $\sum'$ is said to be an *elementary extension* of $\sum$ provided that:

1. $\sum'$ is an extension of $\sum$.
2. For each swff A of $\sum$, A is true for $\sum$ iff A is true for $\sum'$.

Trivially, each semantical system is an elementary extension of itself. In Part III we shall introduce a method for obtaining elementary extensions, due to Abraham Robinson. Some insight into the nature of elementary extensions may be provided by considering extensions that are *not* elementary extensions. The system of integers is *not* an elementary extension of the natural number system, since the swff $\forall x(1 \leq x)$ is true for the natural number system, but is false for the system of integers. The rational number system is *not* an elementary extension of the system of integers, since the swff $\forall x(2x \neq 1)$ is true for the integers, but is false for the rationals. The complex number system is *not* an elementary extension of the real number system (with $<$ deleted), since the swff $\forall x(x^2 \neq -1)$ is true for the real numbers, but is false for the complex numbers.

We are interested in the connection between the notions of *extension* and *elementary extension*. Of course, if $\sum'$ is an elementary extension of $\sum$, then $\sum'$ is an extension of $\sum$. What we need is a condition that ensures that an extension is elementary. As a first step toward developing a suitable condition, we present the following fact.

Lemma 4. Let $\sum'$ be an extension of $\sum$. Then $\sum'$ is *not* an elementary extension of $\sum$ iff there is a swff of $\sum$, say A, such that A is true for $\sum'$ and is false for $\sum$.

Dem. If A is a swff of $\sum$ which is true for $\sum'$ and is false for $\sum$, then $\sum'$ is not an elementary extension of $\sum$. On the other hand, if $\sum'$ is not an elementary extension of $\sum$, then there is a swff of $\sum$, say B, such that

$$B \text{ is true for } \textstyle\sum \text{ iff } B \text{ is false for } \textstyle\sum'$$

i.e., either B is true for $\sum$ and is false for $\sum'$, or B is false for $\sum$ and is true for $\sum'$. If the latter, there is nothing to prove. If the former, then $\neg B$ is true for $\sum'$ and is false for $\sum$. This completes our proof of Lemma 4.

Bearing Lemma 4 in mind, we shall prove that $\sum'$ is an elementary extension of $\sum$ if $\sum'$ is an extension of $\sum$ and if each swff of $\sum$ that has the form $\exists tE$ and is true for $\sum'$ is also true for $\sum$.

Test for Elementary Extensions. $\sum'$ is an elementary extension of $\sum$ iff:

1. $\sum'$ is an extension of $\sum$.
2. Each swff of $\sum$ that has the form $\exists tE$ and is true for $\sum'$ is also true for $\sum$.

Dem. If $\sum'$ is an elementary extension of $\sum$, then conditions 1 and 2 are necessary. We must show that these conditions are sufficient for an elementary extension. Assume conditions 1 and 2. We shall prove that A is true for $\sum$ if A is a swff of $\sum$ that is true for $\sum'$. By Lemma 1, page 121, there is a swff of $\sum$, say $\exists tE$, that has the same truth-value in $\sum$ as A. Taking $\exists tE$ to be $\exists t(A \wedge B)$, where B is quantifier-free, it follows from condition 1 that $\exists tE$ and A have the same truth-value in $\sum'$. By assumption, A is true for $\sum'$; so $\exists tE$ is true for $\sum'$. By condition 2, $\exists tE$ is true for $\sum$; thus A is true for $\sum$. We conclude from Lemma 4 that $\sum'$ is an elementary extension of $\sum$. This establishes our test.

A similar test, which follows, refers to truth only in $\sum'$.

Test for Elementary Extensions. $\sum'$ is an elementary extension of $\sum$ iff:

1. $\sum'$ is an extension of $\sum$.
2. For each swff of $\sum$ that has the form $\exists tE$ and is true for $\sum'$, there is a constant of $\sum$, say a, such that $S_t^a[E]$ is true for $\sum'$.

Dem. It is easy to verify that the conditions are necessary for an elementary extension. We shall prove that the conditions are sufficient. Assume conditions 1 and 2 and suppose that $\sum'$ is not an elementary extension of $\sum$. Let A be the *shortest* (i.e., contains the smallest number of connectives $\neg$, $\vee$, and $\forall$) swff of $\sum$ such that

$$A \text{ is true for } \textstyle\sum' \text{ iff } A \text{ is false for } \textstyle\sum \tag{2}$$

Since $\sum'$ is an extension of $\sum$, A is not atomic. Therefore, $A = \neg B$, $A = C \vee D$, or $A = \forall tE$, where these are swffs of $\sum$.

Case 1: $A = \neg B$. Since B is shorter than A, B has the same truth-value in $\sum$ and $\sum'$. It follows that $\neg B$ has the same truth-value in $\sum$ and $\sum'$. So A does not have the form "$\neg B$."

Case 2: $A = C \vee D$. Both C and D are shorter than A; so C has the same truth-value in $\sum$ and $\sum'$, and D has the same truth-value in $\sum$ and $\sum'$. It follows that $C \vee D$ has the same truth-value in $\sum$ and $\sum'$. So A does not have the form "$C \vee D$."

Case 3: $A = \forall tE$. Let a be any constant of $\sum$; then $S_t^a[E]$ is shorter than A. Thus $S_t^a[E]$ has the same truth-value in $\sum$ and $\sum'$.

(a) Assume that $\forall tE$ is true for $\sum'$. Then $S_t^a[E]$ is true for $\sum'$ for each constant a of $\sum$. Thus $\forall tE$ is true for $\sum$.

(b) Assume that $\forall tE$ is false for $\sum'$. Then $\exists t(\neg E)$ is true for $\sum'$. By condition 2, there is a constant of $\sum$, say b, such that $S_t^b[\neg E]$ is true for $\sum'$. So, $S_t^b[E]$ is false for $\sum'$. Since $S_t^b[E]$ is shorter than $\forall tE$, it follows that $S_t^b[E]$ is false for $\sum$. Thus, $\forall tE$ is false for $\sum$. We conclude that $\forall tE$ has the same truth-value in $\sum$ and $\sum'$. This proves that (2) above is impossible; so $\sum'$ is an elementary extension of $\sum$.

This test can be modified still further.

Lemma 5. $\sum'$ is an elementary extension of $\sum$ iff:

1. $\sum'$ is an extension of $\sum$.
2. For each true swff of $\sum'$, say A, that involves just one constant b that is not in $\sum$, there is a constant of $\sum$, say a, such that $S_b^a[A]$ is true for $\sum'$.

Dem. It is easy to verify that the conditions are necessary for an elementary extension. We shall prove that the conditions are sufficient.

Assume conditions 1 and 2. Let $\exists tE$ be a swff of $\sum$ that is true for $\sum'$. There is a constant of $\sum'$, say b, such that $\mathsf{S}_t^b[E]$ is true for $\sum'$. If b is not a constant of $\sum$, there is a constant of $\sum$, say a, such that $\mathsf{S}_t^a[E]$ is true for $\sum'$. We conclude, by the preceding test, that $\sum'$ is an elementary extension of $\sum$.

Note. If b is a constant of $\sum'$ that is not in $\sum$, then b is pure.

The second requirement of the definition of an elementary extension focuses attention on the swffs of $\sum$; we require that each swff of $\sum$ is true for $\sum'$ iff it is true for $\sum$. From the first condition of the definition, $\sum'$ is similar to $\sum$; thus, a swff of $\sum'$ that is not a swff of $\sum$ necessarily involves a finite number of constants of $\sum'$ that are not constants of $\sum$. Replacement of these constants by constants of $\sum$ yields a swff of $\sum$. In our next lemma we point out that if A is true for $\sum'$, an elementary extension of $\sum$, then so is the swff obtained from A by replacing each of its constants not in $\sum$ by appropriately chosen constants of $\sum$.

Lemma 6. Let $\sum'$ be an elementary extension of $\sum$, let A be a true swff of $\sum'$ and let $b_1, \ldots, b_n$ be the constants in A that are not constants of $\sum$. Then there are constants of $\sum$, say $a_1, \ldots, a_n$, such that $\mathsf{S}_{b_1}^{a_1} \cdots \mathsf{S}_{b_n}^{a_n}[A]$ is a swff of $\sum$ that is true for $\sum'$.

Dem. By assumption, $\exists x_1 \cdots \exists x_n \mathsf{S}_{b_1}^{x_1} \cdots \mathsf{S}_{b_n}^{x_n}[A]$ is a swff of $\sum$ that is true for $\sum'$; therefore, this swff is true for $\sum$. It follows that there are constants of $\sum$, say $a_1, \ldots, a_n$, such that $\mathsf{S}_{b_1}^{a_1} \cdots \mathsf{S}_{b_n}^{a_n}[A]$ is true for $\sum$. Thus, $\mathsf{S}_{b_1}^{a_1} \cdots \mathsf{S}_{a_n}^{a_n}[A]$ is true for $\sum'$.

Note. We point out that $b_1, \ldots, b_n$ are *pure* constants of $\sum'$.

Exercises

1. Prove the Test for Extensions.
2. Complete the proof of Lemma 1.
3. Let $\sum'$ be an extension of $\sum$, and let $\exists sE$ be a swff of $\sum$ that is true for $\sum'$, where E is quantifier-free. Does it follow that $\exists sE$ is true for $\sum$?
4. Let $\sum'$ be an extension of $\sum$, and let $\forall sE$ be a swff of $\sum$ that is true for $\sum$, where E is quantifier-free. Does it follow that $\forall sE$ is true for $\sum'$?

5. Let $\sum'$ be an extension of $\sum$. Prove that $\sum'$ is *not* an elementary extension of $\sum$ iff there is a swff of $\sum$, say A, such that A is true for $\sum$ and is false for $\sum'$.

6. Complete the proof of the second Test for Elementary Extensions (page 125).

7. Complete the proof of Lemma 5.

8. Prove that $\sum'$ is an elementary extension of $\sum$ if:

   (1) $\sum'$ is an extension of $\sum$.
   (2) Each swff of $\sum$ that has the form $\forall tE$ and is true for $\sum$ is true for $\sum'$.

8

Predicate Calculus

8.1. Well-Formed Formulas

A *predicate calculus* is a language based on a set $\prod$ of strings, each of length greater than one. This language involves *variables*, *predicates*, and *individuals*, which are defined as follows.

Variables. Each term of each string in $\prod$.
Predicates. Each first term of each string in $\prod$.
Individuals. Each ith term, $i > 1$, of each string in $\prod$.

We point out that a predicate may also be an individual.

We associate a set of natural numbers with each predicate, called its *type*; this is defined as follows. Let P be any predicate of $\prod$; then

$$\text{type } P = \{n \in N \mid \text{there is an } n\text{-string } \alpha \text{ such that } P\alpha \in \prod\}$$

We agree to call any finite list of variables a *string*; so we are not confined to the strings in $\prod$. A string β is said to be a P-string provided that:

1. length $\beta \in$ type P.
2. Each term of β is an individual of $\prod$.

The alphabet of the language of $\prod$ consists of the variables of $\prod$ and certain other symbols, namely connectives and parentheses. We now exhibit our alphabet.

Alphabet of $\prod$:
Variables. The predicates and individuals of $\prod$.
Connectives. $\neg$, $\vee$, $\forall$.
Parentheses. (,).

Before characterizing the well-formed formulas of $\prod$ we need the notion of an *expression* and what it means for an individual to be *free* or *bound* in an expression. An expression is any finite string of symbols from the alphabet of $\prod$. For example, "$yF(\neg xxyv$" and "$v\forall))yGGF\neg$" are expressions provided that x, y, F, and G are variables of $\prod$. Hereafter, let x, y, z, and w be individuals of $\prod$ that are not predicates, and let F and G be predicates of $\prod$ that may also be individuals of $\prod$.

An individual, say t, is *free* in an expression φ if:

1. t occurs in φ.
2. t is not a predicate.
3. "$\forall t$" does not occur in φ.

An individual, say t, is *bound* in an expression φ if "$\forall t$" occurs in φ.

For example, x is free in the expression "$\vee \neg yy(Fyx)\neg$" x is not free in either of the expressions "$\vee \neg yy(Fy)\neg$" or "$\vee \neg\forall xF))xy$." Moreover, x is bound in "$\vee \neg\forall x\neg yyx$"; x is not bound in "$\vee \neg\forall yx\neg yyy$."

Having defined expressions, we now select certain expressions which we shall call well-formed formulas, or *wffs* for short. First, we shall define the *atomic* wffs of $\prod$.

Definition of Atomic Wff. Let P be any predicate of $\prod$ and let α be any P-string; then the expression "$(P\alpha)$" is said to be an atomic wff of $\prod$.

For example, (Fyx) is an atomic wff provided that $2 \in \text{type } F$.

The well-formed formulas of $\prod$ are generated in the usual way from its atomic wffs by applying the connectives $\neg$, $\vee$, and $\forall$. This is carried out as follows.

Definition of Wff:

1. Each atomic wff is a wff.
2. $(\neg A)$ is a wff if A is a wff.
3. $(A \vee B)$ is a wff provided that A and B are wffs, and no individual is both free in A and bound in B, or vice versa.
4. $(\forall tD)$ is a wff provided that D is a wff and t is free in D.
5. Each wff contains only a finite number of instances of connectives.

Notice that each wff is an expression. For example, let $1 \in \text{type } F$ and let $2 \in \text{type } G$. Then

$$((\neg(Fx)) \vee (Gyx)) \qquad \text{and} \qquad ((Fz) \vee (\forall y(Gxy)))$$

are wffs, whereas none of

$$Fx, \quad (Gx, \quad (\forall y(Gxz)), \quad ((Fx) \vee (\forall x(Gxz)))$$

is a wff. Although we are using capital letters for predicates and lower case letters for individuals, it is important to bear in mind that a variable can be both a predicate and an individual.

A wff is atomic if no connective occurs in the wff. We now say that a wff is *composite* if it is not atomic; thus a composite wff contains at least one of the connectives $\neg$, $\vee$, or $\forall$. By the *length* of a wff we mean the number of instances of these connectives in the wff. The length of each atomic wff is zero; the length of each composite wff is a positive integer. For example, the length of $((\neg(\neg(Fx))) \vee (\forall y(Gyy)))$ is 4.

We now present a general method of proving that each wff of $\prod$ possesses a stated property. For concreteness we get at the property involved by considering the set of all wffs that have the property. Essentially this means that the notion of *property* is reduced to membership in a specified set of wffs. Here is our method.

Fundamental Theorem about Wffs. A set of wffs, say S, is the set of all wffs, iff:

1. $A \in S$ for each atomic wff A.
2. $(\neg A) \in S$ provided that $A \in S$.
3. $(A \vee B) \in S$ provided that $A \in S$, $B \in S$, and $(A \vee B)$ is a wff.
4. $(\forall tD) \in S$ provided that $D \in S$ and that $(\forall tD)$ is a wff.

Dem. See the proof of the Fundamental Theorem about Wffs for the propositional calculus.

Let P be a property which is meaningful for wffs of $\prod$; i.e., it is true or false that a specific wff has the property. By forming the set of all wffs that possess property P and applying the Fundamental Theorem about Wffs to that set, we easily derive the following algorithm.

Algorithm to Prove That Each Wff Has Property P:

1. Prove that each atomic wff has property P.

2. Prove that $(\neg A)$ has property P, assuming that A is a wff with property P.

3. Prove that $(A \vee B)$ has property P, assuming that A and B are wffs with property P and that $(A \vee B)$ is a wff.

4. Prove that $(\forall tD)$ has property P, assuming that D has property P and that $(\forall tD)$ is a wff.

Notice that if $(\forall tD)$ is a wff, then t is free in D, so t is not a predicate. Thus, if an individual t is bound in a wff, then t is not a predicate. Any individual which is not a predicate is called a *pure* individual. Thus, only pure individuals can be free or bound in a wff.

We shall denote the set of all wffs of a predicate calculus $\prod$ by "W"; i.e., W $= \{A \mid A \text{ is a wff of } \prod\}$.

Note that the alphabet of $\prod$ does not include placeholders, even though placeholders are part of the alphabet of the language of a semantical system. We point out that a pure individual, say t, performs the function of a placeholder when t is bound in a wff.

Definition. A variable t is said to be a *placeholder of a wff A* if t is bound in A.

Thus t is a placeholder of a wff A iff t is a pure individual and "$\forall t$" occurs in A. For example, y and z are the placeholders of

$$((\forall y(Gxy)) \vee (\forall z(Fz)))$$

x is not a placeholder of this wff.

We will generally want $\prod$ to have infinitely many pure individuals, so that in building wffs we will never run out of symbols that can be used as placeholders.

Moreover, we say that a variable t is *interpretable in a wff A* if t occurs in A and t is not a placeholder of A. Thus, a variable t is interpretable in A if t is a predicate that occurs in A, or t is a pure individual that is free in A. The point is that the placeholders of A are not interpretable in A.

For example, the variables F, G, and x are interpretable in the wff $((\forall y(Gxy)) \vee (\forall z(Fz)))$; the variables y and z are not interpretable in this wff.

Thus, we regard "predicate or free" and "interpretable" as synonyms, and "bound" and "placeholder" as synonyms.

Whereas a predicate P either occurs in a wff A or does not occur in A, a pure individual t either does not occur in A, is free in A, or is bound in A.

Exercises

1. Let $\prod$ be a predicate calculus for which F is a predicate with type $\{1\}$ and G is a predicate with type $\{1, 2\}$. Let x, y, z, and F be the individuals of $\prod$. Which of the following statements are correct?

(a) xy is a G-string.
(b) Fy is a G-string.
(c) F is an F-string.
(d) x is free in both Fx and $\forall yGxy$.
(e) x is bound in $\neg\forall x(Fx)$.
(f) y is bound in $(\forall x(Gxy))$.
(g) y is free in $((Gxx) \vee (Fz))$.
(h) F is free in $Gxy \vee Fx$.
(i) F is free in (GxF).
(j) F is bound in $\forall F(Gxy)$.

2. Which of the following expressions are wffs of the predicate calculus of Exercise 1?

(a) Fy.
(b) $((GxF) \vee (\neg(FF)))$.
(c) $(\forall x(GxF)) \vee (Fy)$.
(d) $((\forall x(GxF)) \vee (Fy))$.
(e) $((Gxy) \vee (\forall x(Fx)))$.
(f) $(\forall y(\forall x(Gxy)))$.
(g) $(\forall F(GxF))$.
(h) (GF).
(i) $(\forall G(FG))$.
(j) $(\forall F(FF))$.

8.2. Parentheses

The purpose of introducing a pair of parentheses at each construction step in our definition of wff is to impose a certain structure on wffs. In order to study this structure effectively, we now present the notion of the *mate* of a parenthesis in a wff. We here repeat the discussion in Section 2.2.

Definition of Mate. The *mate* of a particular left-hand (LH) parenthesis occurring in a wff is the first right-hand (RH) parenthesis to its right such that an equal number of left-hand parentheses and right-hand parentheses occur in between. The *mate* of a particular right-hand parenthesis occurring in a wff is the first left-hand parenthesis to its left such that an equal number of left-hand parentheses and right-hand parentheses occur in between.

For example, the mate of each parenthesis of the following wff is indicated by placing the same letter below a parenthesis and its mate:

$$\underset{a}{(}\neg\underset{b}{(}\underset{c}{(}\forall x\underset{d}{(}Fx\underset{d}{)}\underset{c}{)} \vee \underset{e}{(}\neg\underset{f}{(}Gyy\underset{f}{)}\underset{e}{)}\underset{b}{)}\underset{a}{)}$$

The following facts about parentheses and mates are easy to verify by applying the Fundamental Theorem about Wffs, page 131.

Lemma 1. Each wff has an equal number of LH and RH parentheses.

Dem. Let $S = \{A \in \mathsf{W} \mid$ the number of LH parentheses in A is the same as the number of RH parentheses in $A\}$; apply the Fundamental Theorem about Wffs to show that $S = \mathsf{W}$.

Lemma 2. Counting from left to right, the number of instances of LH parentheses in a wff is greater than the number of instances of RH parentheses until the final RH parenthesis is reached.

Dem. Let S be the set of all wffs that satisfy this lemma; apply the Fundamental Theorem about Wffs to prove that $S = \mathsf{W}$.

Corollary 1. The rightmost RH parenthesis of a wff is the mate of its leftmost LH parenthesis.

Dem. Consider Lemmas 1 and 2.

Corollary 2. The leftmost LH parenthesis of a wff is the mate of its rightmost RH parenthesis.

Dem. The result of interchanging each instance of "left" and "right" throughout a statement concerning parenthesis is called the *dual* of the statement. We point out that the dual of Lemma 2 is true. Using Lemma 1 and the dual of Lemma 2, we easily establish Corollary 2.

Lemma 3. Each LH parenthesis of a wff has a mate.

Dem. Let $S = \{A \in \mathsf{W} \mid$ each LH parenthesis of A has a mate$\}$; apply the Fundamental Theorem about Wffs to prove that $S = \mathsf{W}$.

Lemma 4. Each parenthesis of a wff has a mate.

Dem. First, we must establish the dual of Lemma 3. Let

$$S = \{A \in \mathsf{W} \mid \text{each RH parenthesis of } A \text{ has a mate}\}$$

Applying the Fundamental Theorem about Wffs, we easily prove that $S = \mathsf{W}$; so each RH parenthesis of a wff has a mate. By Lemma 3, each

LH parenthesis of a wff has a mate; thus each parenthesis of a wff has a mate.

Lemma 5. No two LH parentheses of a wff have the same mate.

Dem. Let $S = \{A \in \mathsf{W} \mid \text{no two LH parentheses of } A \text{ have the same mate}\}$; applying the Fundamental Theorem about Wffs, we can prove that $S = \mathsf{W}$.

Lemma 6. No two parentheses of a wff have the same mate.

Dem. It is easy to prove the dual of Lemma 5; i.e., no two RH parentheses of a wff have the same mate.

Since a pair of parentheses is introduced into a wff A at each construction step involved in building up A from its atomic wffs, we are tempted to conjecture that each RH parenthesis of A is the mate of a LH parenthesis of A iff the LH parenthesis is the mate of the RH parenthesis. The following definition brings out the basic idea more clearly.

Definition. A parenthesis of a wff is said to be ***mated*** if it is the mate of its mate.

For example, each parenthesis of an atomic wff is mated; each parenthesis of the wff displayed on page 133 is mated.

We want to prove that each parenthesis of each wff is mated. First, we need the following fact.

Lemma 7. The leftmost LH parenthesis of each wff is mated; the rightmost RH parenthesis of each wff is mated.

Dem. Apply Corollaries 1 and 2.

Here is our main result in this connection.

Theorem 1. Each parenthesis of each wff is mated.

Dem. See the demonstration of Theorem 1, page 19.

Exercises

1. Find the mate of the fourth symbol of

$$(\neg(((\neg(\forall y(Gxy))) \vee (Fx)) \vee (Fx)))$$

2. Prove Lemma 1.
3. Prove Lemma 2.
4. State and prove the dual of Lemma 2.
5. Prove Corollary 1.
6. Prove Corollary 2.
7. Prove Lemma 3.
8. State and prove the dual of Lemma 3.
9. Prove Lemma 5.
10. State and prove the dual of Lemma 5.
11. Prove Lemma 7.
12. Prove Theorem 1.
13. Show directly that the first three LH parentheses of the wff of Exercise 1 are mated.

8.3. Main Connective of Wffs

As we shall soon see, our discussion of the parentheses of a wff has an impact on the notion of a *main* connective of a composite wff, which we now present.

Definition of Main Connective. Let A be any composite wff. An occurrence of a connective in A is said to be a *main* connective of A if:

1. $A = (\neg B)$, where B is a wff. In this case, the displayed instance of the connective $\neg$ is said to be a *main* connective of A.
2. $A = (C \vee D)$, where C and D are wffs. In this case, the displayed instance of the connective $\vee$ is said to be a *main* connective of A.
3. $A = (\forall tE)$, where E is a wff. In this case, the displayed instance of the connective $\forall$ is said to be a *main* connective of A.

For example, the tenth symbol (counting from left to right) of $((\forall z(Fz)) \vee (Gxx))$ is a main connective of this wff.

Of course, it is conceivable that a specific wff A has several main connectives. This will be the case if there are wffs, B, C, and D such that $A = (\neg B)$ and $A = (C \vee D)$; or if there are wffs B, C, D, and E such that $A = (B \vee C)$ and $A = (D \vee E)$, where $B \neq D$. In the former case, both $\neg$ and $\vee$ are main connectives of A; in the latter case, two instances of $\vee$ in A are main connectives of A. We shall prove that this cannot happen. Indeed, we shall prove that each composite wff has exactly one main connective. Of course, each composite wff has at least one main connective. We must show that no composite wff has more than one main connective.

Theorem 1. No composite wff has two main connectives.

Dem. Assume that $\neg$ is a main connective of a composite wff A; then $A = (\neg B)$, where B is a wff. Thus "$\neg$" is the second symbol of A. But if $\forall$ is a main connective of A, then "$\forall$" is the second symbol of A. If $\vee$ is a main connective of A, then $A = (C \vee D)$, where C and D are wffs; notice that the first symbol of C is a LH parenthesis, so the second symbol of A is a LH parenthesis. So, neither $\forall$ nor $\vee$ is a main connective of A. Of course, no other instance of $\neg$ in A, except the second symbol of A, is a main connective of A. A similar argument shows that if $\forall$ is a main connective of a wff A, then that instance of $\forall$ in A is its only main connective. The situation is a little different in case an instance of $\vee$ is a main connective of a wff A; at first sight, it is conceivable that another occurrence of $\vee$ in this wff is also a main connective of A. This means that there are wffs B and C, and wffs D and E, such that

$$A = (B \vee C) = (D \vee E)$$

where $B \neq D$ and $C \neq E$. Now, the mate of the second symbol of A is the last symbol of B, since B is a wff. Since D is a wff, we see that the mate of the second symbol of A is the last symbol of D. But each parenthesis of A has a unique mate; therefore, the last symbol of D is also the last symbol of B, and it follows that $B = D$. So, the two displayed occurrences of $\vee$ in A are actually the same occurrence of $\vee$. We conclude that each composite wff has at most one main connective.

Corollary 1. Each composite wff has a unique main connective.

The problem, now, is to locate the main connective of a given wff. Each wff is a finite string of symbols, some of which may be connectives. We need an algorithm that will mechanically exhibit the main connective of a composite wff. Here is a suitable algorithm.

Algorithm for Main Connective. Counting from left to right, the first connective reached for which the number of LH parentheses is one greater than the number of RH parentheses is the main connective of the wff.

Dem. Let $S = \{A \in \mathsf{W} \mid A$ is atomic or the algorithm is true for $A\}$. Apply the Fundamental Theorem about Wffs to show that $S = \mathsf{W}$.

To illustrate this algorithm, consider the wff

$$((\forall y(Fy)) \vee (Gxx))$$

The tenth symbol of this wff, counting from left to right, is the first connective that is preceded by one more LH parenthesis than RH parenthesis. We conclude that the only instance of $\vee$ in this wff is its main connective; of course, this means that the given wff has the form "$(A \vee B)$."

In operational terms this algorithm can be expressed as follows.

Operational Form of Algorithm for Main Connective. Add 1 for a LH parenthesis and -1 for a RH parenthesis, and take subtotals starting at the leftmost symbol of a given wff A. Then the connective reached when the subtotal is 1 is the main connective of A.

Applying this algorithm to the wff $((\forall y(Fy)) \vee (Gxx))$, notice that we reach its tenth symbol on the count of 1. So this instance of $\vee$ is its main connective.

Our algorithm provides us with a simple method of determining whether a given expression is a wff. First, we must decide whether the given expression is an *atomic* wff. If not, then we apply the *algorithm for main connective*. Either (i) the algorithm fails, or (ii) the algorithm yields a unique connective of the expression. In the first case, we conclude that the given expression is *not* a wff. In the second case, our problem is reduced to considering one or two expressions each shorter than the given expression. Repeating this program sufficiently often, we either prove that the given expression is not a wff, or else we obtain a list of atomic wffs from which the expression was constructed. In the latter case, the expression is a wff.

To illustrate, consider the expression

$$(\forall y(\neg(\forall x(\neg((\neg((Fy) \vee (Gxx))) \vee (Fz)))))) \tag{1}$$

Our algorithm points out the first "$\forall$" of (1); so we delete its first three symbols and its last symbol, obtaining the expression

$$(\neg(\forall x(\neg((\neg((Fy) \vee (Gxx))) \vee (Fz))))) \tag{2}$$

Our algorithm points out the first "$\neg$" of (2); deleting this "$\neg$" and the associated pair of parentheses, we obtain

$$(\forall x(\neg((\neg((Fy) \vee (Gxx))) \vee (Fz)))) \tag{3}$$

In two more steps we obtain

$$((\neg((Fy) \vee (Gxx))) \vee (Fz)) \tag{4}$$

Our algorithm points out the second "$\vee$" of (4); we delete this "$\vee$" and the associated pair of parentheses to obtain the following two expressions:

$$(\neg((Fy) \vee (Gxx))) \tag{5}$$

$$(Fz) \tag{6}$$

Certainly, (6) is atomic. Examining (5), we obtain

$$((Fy) \vee (Gxx)) \tag{7}$$

which splits into (Fy) and (Gxx), which are both atomic wffs. We conclude that (1) is a wff; moreover, this wff is built up from the atomic wffs (Fy), (Gxx), and (Fz).

Exercises

Let $\prod$ be a predicate calculus such that $1 \in$ type F, $2, 3 \in$ type G, and x, y, and z are pure individuals.

1. Use the Definition of Main Connective to show that the 25th symbol of the wff

$$((\neg((\forall x(Fx)) \vee (\forall x(Gxx)))) \vee (\forall x((Fx) \vee (Gxx))))$$

is a main connective of this wff. Does this wff have another main connective?

2. Use the Definition of Main Connective to show that the 12th symbol of the wff

$$((\forall x(Gxyy)) \vee ((\neg(\forall z(Gzz))) \vee (\forall x(Fx))))$$

is a main connective of this wff. Does this wff have another main connective?

3. Prove the Algorithm for Main Connective.

4. Use the Algorithm for Main Connective to find the main connective of the wff of Exercise 1.

5. Use the Algorithm for Main Connective to find the main connective of the wff of Exercise 2.

Determine which of the following expressions are wffs of $\prod$.

6. $((\forall x(Gxyy)) \vee ((\neg(\forall z(Gzz)))) \vee (\forall x(Fx)))$.

7. $((\neg(\forall x(\neg(Gxxx)))) \vee (\neg(Gyyy)))$.

8. $((\neg(\forall x(\neg(Gxxx))) \vee (\neg(Gyyy))))$.

9. $(\forall x((\forall y((Fy) \vee (Gxy))) \vee (\neg(Gxxx))))$.

8.4. Names for Wffs; Principal Connective of Name

As we have shown, the role of parentheses in imposing a certain structure on wffs is of fundamental importance. On the other hand, it is difficult to read a specific wff when it involves many parentheses. Just as for the swffs of a semantical system, we can obtain the best of both worlds by introducing parentheses-omitting conventions. This is achieved by utilizing *names* for wffs. Of course, we must avoid ambiguity; i.e., each of our names must name a unique wff.

The conventions introduced in Section 7.3, which yield short names for swffs, can be used to obtain short names for wffs. Let A be any wff of $\prod$; there is no ambiguity in suppressing the outermost pair of parentheses of A; moreover, we can safely suppress the pair of parentheses involved in each atomic wff $(R\alpha)$ which occurs in A.

Just as for swffs, short names for certain wffs are obtained by using the connectives $\wedge$, $\rightarrow$, $\leftrightarrow$, and $\exists$. Let A, B, and C be any wffs and let t be any pure individual free in C; then

$(A \wedge B)$	is a name for	$(\neg((\neg A) \vee (\neg B)))$
$(A \rightarrow B)$	is a name for	$((\neg A) \vee B)$
$(A \leftrightarrow B)$	is a name for	$((A \rightarrow B) \wedge (B \rightarrow A))$
$(\exists tC)$	is a name for	$(\neg(\forall t(\neg C)))$

Moreover, we shall usually drop the outermost pair of parentheses in a *name* for a wff. For example, "$\exists xGxy$" is a name for the wff denoted by "$\neg\forall x(\neg Gxy)$," namely the wff $(\neg(\forall x(\neg(Gxy))))$.

We attribute a built-in bracketing power, or reach, to the connectives $\neg$, $\vee$, $\wedge$, $\rightarrow$, and $\leftrightarrow$ in this order (weakest connectives appear first). As usual, the reach of any of these connectives is strengthened by placing a dot over the connective.

Also, we shall carry over the notion of *prenex normal form* (see page 120) from swffs to wffs. Let $t_1, \ldots, t_n$ be n distinct individuals that are free in a quantifier-free wff A. For each $i = 1, \ldots, n$ let Q_i be a quantifier, either $\forall$ or $\exists$. Then we say that

$$Q_1t_1 \cdots Q_nt_nA \tag{1}$$

is a name for the wff

$$(Q_1t_1 \cdots (Q_nt_nA) \cdots) \tag{2}$$

The name (1) is said to be the *prenex normal form* of the wff (2). Moreover, if A is atomic, we usually omit the parentheses of A in (1). For example, "$\forall x\exists yGxy$" is an abbreviation for "$(\forall x(\exists y(Gxy)))$," which in turn denotes the wff

$$(\forall x(\neg(\forall y(\neg(Gxy))))))$$

Similarly, "$\exists y \forall z \forall w \exists x(Fy \vee Gwx \rightarrow Fz)$" is an abbreviation for

$$(\exists y(\forall z(\forall w(\exists x(Fy \vee Gwx \rightarrow Fz)))))$$

which, in turn, denotes the wff

$$(\neg(\forall y(\neg(\forall z(\forall w(\neg(\forall x(\neg((\neg((Fy) \vee (Gwx))) \vee (Fz))))))))))$$

The sequence of quantifiers and individuals "$Q_1t_1 \cdots Q_nt_n$" is said to be the ***prefix*** of the wff $Q_1t_1 \cdots Q_nt_nA$; A is said to be the ***matrix*** of $Q_1t_1 \cdots Q_nt_nA$. If several consecutive Q's in a prefix are the same (i.e., all $\forall$'s or all $\exists$'s), then we write the first Q, deleting the others. Thus

$$\forall xyzA = \forall x \forall y \forall zA$$

and

$$\forall x \exists y_1 y_2 \forall z_1 z_2 z_3 A = \forall x \exists y_1 \exists y_2 \forall z_1 \forall z_2 \forall z_3 A$$

The quantifiers $\forall$ and $\exists$ have the shortest reach of any connective. Thus

$$(\forall yFy) \vee (Gxx) = \forall yFy \vee Gxx$$

whereas

$$(\forall y((Fy) \vee (Gxx))) = \forall y(Fy \vee Gxx)$$

The basic connectives of a predicate calculus $\prod$ are $\neg$, $\vee$, and $\forall$. The defined connectives are $\wedge$, $\rightarrow$, $\leftrightarrow$, and $\exists$. Of course, our conventions for abbreviating names of wffs are concerned with ***names***, so deal with defined connectives. From this viewpoint, it is convenient to utilize the notion of the ***principal*** connective of a ***name*** of a wff. Remember that ***main*** connectives are concerned with wffs, not their names; principal connectives refer to names, rather than wffs.

Definition of Principal Connective. Let N be a name for a wff, let c be one of $\vee$, $\wedge$, $\rightarrow$, or $\leftrightarrow$, and let Q be a $\forall$ or $\exists$. Then

1. c is the principal connective of N if $N = (N_1)\ c\ (N_2)$, where N_1 and N_2 are names for wffs.
2. $\neg$ is the principal connective of N if $N = \neg(N_1)$, where N_1 is a name for a wff.

3. Q is the principal connective of N if $N = Qt(N_1)$, where t is an individual and N_1 is a name for a wff in which t is free.

For example, $\exists$ is the principal connective of $\exists yFy$; here we are looking at a name of a wff. Notice that $\neg$ is the main connective of this wff.

We mention that each name for a composite wff has exactly one principal connective.

Exercises

Write down a short name for each of the following wffs.

1. $((\neg(Fx)) \vee (\forall y(Gxy)))$.
2. $((\neg(\forall y(\neg(Fy)))) \vee (Gxx))$.
3. $(\forall z((\neg(Fz)) \vee (Fy)))$.
4. $(\forall x(\forall y((\neg(Gyx)) \vee (\forall z(Fz)))))$.
5. $(\forall y(\forall x(\forall z(\forall w((Gxy) \vee (Gzw))))))$.

Exhibit the wffs denoted by each of the following (use only the connectives $\neg$, $\vee$, or $\forall$).

6. $\exists xFx \rightarrow Fy$.
7. $\exists x(Fx \rightarrow Fy)$.
8. $\forall x\exists yGxy$.
9. $\forall x\exists y\forall z(Gxy \vee Fz)$.
10. $\forall yGyy \wedge \exists xFx$.
11. $\forall xFx \leftrightarrow \forall xGxx$.

Indicate the principal connective of each of the following, by inserting one or two pairs of parentheses.

12. $Fx \vee Fy \rightarrow Fz$.
13. $\forall y(Fy \vee Gxy) \rightarrow \exists zFz$.
14. $\forall x(Fy \vee Gxx) \leftrightarrow Fy \vee \forall xGxx$.
15. $\forall z(Fz \rightarrow Gxx) \vee Fy$.
16. $\forall z\exists yGyz$.
17. $\exists yx(Fx \leftrightarrow Fy)$.

18. $\forall xyGxy \leftrightarrow Fz$.

19. $\forall yFy \dashrightarrow Gxx \leftrightarrow Fx$.

20. $Gxy \dashrightarrow Fx \rightarrow \forall zFz$.

8.5. Syntactical Transforms

As for the propositional calculus, the introduction of syntactical transforms facilitates our development of the predicate calculus; by a *syntactical transform* we mean any map of W into W. As usual, "T[A]" denotes the wff that a syntactical transform T associates with a wff A; we shall suppress the brackets, writing "TA," but only if no ambiguity results.

We point out that a map is a syntactical transform if and only if its domain is W and its range is a subset of W. For example, let T be the map such that for each wff A,

$$\mathsf{T}[A] = \begin{cases} \forall xA & \text{if } x \text{ is free in } A \\ A & \text{otherwise} \end{cases}$$

Then T is a syntactical transform.

As illustrated, we can characterize a specific syntactical transform by presenting a rule for determining the image of A, where A is any wff. Just as for the propositional calculus, we can specify a syntactical transform without providing an explicit rule that directly yields the image of each wff under the transform. To this purpose, it is enough to provide the following information about the transform, say T:

(a) Define TA in terms of A for each atomic wff A.
(b) Define T[$\neg B$] in terms of B and the T-images of a finite number of wffs, each shorter than $\neg B$ (here B is any wff).
(c) Define T[$C \vee D$] in terms of C, D, and the T-images of a finite number of wffs each shorter than $C \vee D$ (here C and D are any wffs such that $C \vee D$ is a wff).
(d) Define T[$\forall tE$] in terms of E and the T-images of a finite number of wffs each shorter than $\forall tE$ (here E is any wff and t is any individual free in E).

We shall prove this in a moment. First, we establish the following basic fact about maps.

Lemma 1. A map T of S into W is a syntactical transform if S is a set of wffs such that:

1. $A \in S$ for each atomic wff A.
2. $\neg B \in S$ provided that $B \in S$.
3. $C \vee D \in S$ provided that $C, D \in S$ and $C \vee D$ is a wff.
4. $\forall tE \in S$ provided that $E \in S$ and $\forall tE$ is a wff.

Dem. By the Fundamental Theorem about Wffs, page 131, dom $\mathsf{T} = \mathsf{W}$. Thus T is a syntactical transform.

We come now to our main result. Note that we define $\mathsf{T}[\neg B]$, $\mathsf{T}[C \vee D]$, and $\mathsf{T}[\forall tE]$ in terms of certain T-images. In a sense, then, we define T in terms of T. Of course, this is generally unacceptable; here, our procedure is sound because of the requirement that $\mathsf{T}[\neg B]$ is defined in terms of the T-images of wffs *shorter* than $\neg B$, $\mathsf{T}[C \vee D]$ is defined in terms of the T-images of wffs *shorter* than $C \vee D$, and $\mathsf{T}[\forall tE]$ is defined in terms of the T-images of wffs shorter than $\forall tE$.

Theorem 1. Let T be any syntactical transform; then T is characterized by providing the following information.

1. Announce $\mathsf{T}A$ in terms of A; here A is any atomic wff. The definition may involve A but may not involve a T-image.
2. Announce $\mathsf{T}[\neg B]$ in terms of B and the T-images of a finite number of wffs, each shorter than $\neg B$; here B is any wff.
3. Announce $\mathsf{T}[C \vee D]$ in terms of C, D, and the T-images of a finite number of wffs, each shorter than $C \vee D$; here C and D are any wffs such that $C \vee D$ is a wff.
4. Announce $\mathsf{T}[\forall tE]$ in terms of E and the T-images of a finite number of wffs, each shorter than $\forall tE$; here E is any wff and t is any individual free in E.

Dem. See the proof of Theorem 1, page 60.

In Section 8.6 we shall use Theorem 1 to exhibit two families of syntactical transforms, the *interchange* transforms I_t^s and the *substitution* transforms S_t^s. In Section 11.2 we shall follow this procedure to characterize the *normal* transforms N, M, R, and D.

Exercises

1. Is $\{(A, \forall yA) \mid A \in \mathsf{W}\}$ a syntactical transform?
2. Is $\{(A, A \vee Fx) \mid A \in \mathsf{W}\}$ a syntactical transform?

3. Prove Theorem 1.

4. Let N be a map whose domain is a subset of W and whose range is a subset of W, such that:

(1) $\mathsf{N}A = A$ for each atomic wff A.

$$(2)\quad \mathsf{N}[\neg A] = \begin{cases} \neg A & \text{if } A \text{ is atomic} \\ \mathsf{N}B & \text{if } A = \neg B \\ \mathsf{N}[\neg C] \wedge \mathsf{N}[\neg D] & \text{if } A = C \vee D \\ \exists t(\mathsf{N}[\neg E]) & \text{if } A = \forall tE. \end{cases}$$

(3) $\mathsf{N}[C \vee D] = \mathsf{N}C \vee \mathsf{N}D$ if $C \vee D$ is a wff.

(4) $\mathsf{N}[\forall tE] = \forall t(\mathsf{N}E)$ if $\forall tE$ is a wff.

(a) Prove directly, i.e., without using Lemma 1, that dom $\mathsf{N} = \mathsf{W}$.

(b) Prove directly, i.e., without using Lemma 1, that N is a syntactical transform.

(c) Prove directly, i.e., without using Theorem 1, that there is just one syntactical transform that satisfies conditions 1–4.

5. Let S be a map whose domain is a subset of W and whose range is a subset of W, such that:

(1) $\mathsf{S}A = Fx$ for each atomic wff A.
(2) $\mathsf{S}[\neg A] = \neg \mathsf{S}A$ for each wff A.
(3) $\mathsf{S}[A \vee B] = \mathsf{S}A \vee \mathsf{S}B$ for each wff $A \vee B$.
(4) $\mathsf{S}[\forall tE] = \forall t(\mathsf{S}E)$ for each wff $\forall tE$.

(a) Compute $\mathsf{S}[\exists y(Fy \wedge Gyy) \rightarrow \forall xGxx]$.
(b) Show that $\mathsf{S}[\mathsf{S}A] = \mathsf{S}A$ for each wff A.
(c) Describe the map S in intuitive terms.

8.6. Interchange and Substitution Transforms

We now present two families of syntactical transforms, $\mathbf{I}_t^s$ (read "interchange s and t") and $\mathbf{S}_t^s$ (read "substitute s for t"), where s and t are individuals of a predicate calculus $\prod$. For the $\mathbf{I}_t^s$ transforms we require that neither s nor t is a predicate; for the $\mathbf{S}_t^s$ transforms we require that t is not a predicate. We shall use the $\mathbf{S}_t^s$ transforms to assist us in defining the notion of a provable wff; the main role of the $\mathbf{I}_t^s$ transforms is to simplify the task of defining $\mathbf{S}_t^s$.

By "$\mathbf{I}_t^s$" we denote a transform that associates with each wff A the wff obtained from A by interchanging the individuals s and t throughout A; i.e., each occurrence of s is replaced by t, and each occurrence of t is replaced by s. This defines the image of A under $\mathbf{I}_t^s$; nonetheless, it is helpful to define the syntactical transforms $\mathbf{I}_t^s$ in the step-by-step manner that apes the permitted construction steps. We do this to prepare the way for our definition of the syntactical transforms $\mathbf{S}_t^s$, which cannot be defined so simply. The role of $\mathbf{S}_t^s$ is to substitute s for each free instance of t in the wff involved. This looks simple enough; however, we must announce what $\mathbf{S}_t^s$ does to a wff in case either s or t is bound in the wff. Because of this complication it is necessary to define $\mathbf{S}_t^s$ in the usual step-by-step manner based on Theorem 1, Section 8.5. For pedagogical reasons, therefore, we shall characterize $\mathbf{I}_t^s$ this way as well.

Definition of $\mathbf{I}_t^s$. Here s and t are pure individuals. Let P be any predicate, let $x_1 \cdots x_n$ be any P-string, and let A and B be any wffs. Then

1. $\mathbf{I}_t^s[Px_1 \cdots x_n] = Pz_1 \cdots z_n$, where $z_i = \begin{cases} s & \text{if } x_i = t \\ t & \text{if } x_i = s \\ x_i & \text{otherwise} \end{cases}$
2. $\mathbf{I}_t^s[\neg A] = \neg \mathbf{I}_t^s[A]$.
3. $\mathbf{I}_t^s[A \vee B] = \mathbf{I}_t^s[A] \vee \mathbf{I}_t^s[B]$.
4. $\mathbf{I}_t^s[\forall u A] = \begin{cases} \forall s \mathbf{I}_t^s[A] & \text{if } u = t \\ \forall t \mathbf{I}_t^s[A] & \text{if } u = s \\ \forall u \mathbf{I}_t^s[A] & \text{otherwise} \end{cases}$

In fact, we have defined a family of syntactical transforms, since s and t may be any individuals that are not predicates. Applying the Fundamental Theorem about Wffs, we can prove that the four properties of $\mathbf{I}_t^s$ displayed above characterize a unique map of $\mathbf{W}$ into $\mathbf{W}$ for each s and t.

Let us illustrate our definition:

$$\begin{aligned} \mathbf{I}_x^y[\forall xGxz \vee \forall yFy] &= \mathbf{I}_x^y[\forall xGxz] \vee \mathbf{I}_x^y[\forall yFy] && \text{(by 3)} \\ &= \forall yGyz \vee \forall xFx && \text{(by 4)} \end{aligned}$$

Clearly, $\mathbf{I}_t^s = \mathbf{I}_s^t$ provided that s and t are pure individuals; moreover, $\mathbf{I}_s^s$ is the identity map for each pure individual s.

We turn now to the family of substitution transforms $\mathbf{S}_t^s$, our main concern here. The role of $\mathbf{S}_t^s$ is to substitute s for each free occurrence of t in a wff. We now come to the complications. If s is bound in a wff A, then $\mathbf{S}_t^s[A]$ is formed by replacing the bound s's in A by some other individual; we must avoid obtaining an expression that is not a wff (remember that each free t in A is replaced by s); moreover, if a free t of A occurs in the scope of a bound s, substituting s for t will change the meaning of A, when interpreted in a semantical system, in an unintended manner. Accordingly, we shall substitute t for each bound s in A (note that there are no free t's in $\mathbf{S}_t^s[A]$—they have all been replaced by s's). Finally, in case t is bound in A (so there are no free t's in A) we agree to do essentially nothing (i.e., at most we shall replace bound s's by bound t's, and we shall replace bound t's by bound s's).

Certainly, this substitution transform is too complicated to characterize in a phrase; accordingly, we present a formal definition which follows the pattern of our definition of the interchange transforms.

Definition of S_t^s. Here s and t are individuals and t is pure. Let P be any predicate, let $x_1 \cdots x_n$ be any P-string, and let A and B be any wff. Then

1. $\mathrm{S}_t^s[Px_1 \cdots x_n] = Pz_1 \cdots z_n$, where $z_i = \begin{cases} s & \text{if } x_i = t \\ x_i & \text{otherwise} \end{cases}$
2. $\mathrm{S}_t^s[\neg A] = \neg \mathrm{S}_t^s[A]$.
3. $\mathrm{S}_t^s[A \vee B] = \mathrm{S}_t^s[A] \vee \mathrm{S}_t^s[B]$.
4. $\mathrm{S}_t^s[\forall uA] = \begin{cases} \forall uA & \text{if } u = t \\ \mathrm{I}_t^s[\forall uA] & \text{if } u = s \\ \forall u\mathrm{S}_t^s[A] & \text{otherwise} \end{cases}$

Note the impact of S_t^s on a wff A; each free instance of t in A is replaced by s; each bound s not in the scope of $\forall t$ is replaced by t; each bound t in the scope of $\forall s$ is replaced by s.

To illustrate, we shall use the wff of the preceding example:

$$\begin{aligned} \mathrm{S}_x^y[\forall xGxz \vee \forall yFy] &= \mathrm{S}_x^y[\forall xGxz] \vee \mathrm{S}_x^y[\forall yFy] && \text{(by 3)} \\ &= \forall xGxz \vee \forall xFx && \text{(by 4)} \end{aligned}$$

Clearly $\mathrm{S}_t^s \neq \mathrm{S}_s^t$ unless $s = t$; of course, S_s^s is the identity transform provided that s is pure. The following lemmas are easy to verify; throughout, s and t are any individuals and t is pure.

Lemma 1. $\mathrm{S}_t^s[A \wedge B] = \mathrm{S}_t^s[A] \wedge \mathrm{S}_t^s[B]$ provided $A \wedge B$ is a wff.

Lemma 2. $\mathrm{S}_t^s[A \to B] = \mathrm{S}_t^s[A] \to \mathrm{S}_t^s[B]$ provided $A \to B$ is a wff.

Lemma 3. $\mathrm{S}_t^s[A \leftrightarrow B] = \mathrm{S}_t^s[A] \leftrightarrow \mathrm{S}_t^s[B]$ provided $A \leftrightarrow B$ is a wff.

Lemma 4.

$$\mathrm{S}_t^s[\exists uA] = \begin{cases} \exists tA & \text{if } u = t \\ \exists t\mathrm{I}_t^s[A] & \text{if } u = s, \\ \exists u\mathrm{S}_t^s[A] & \text{otherwise} \end{cases} \quad \text{provided } \exists uA \text{ is a wff}$$

Finally, we present two facts about the transforms of this section.

Theorem 1. $\mathrm{I}_t^s[A]$ is a wff provided that A is a wff, where s and t are pure individuals.

Dem. Apply the Fundamental Theorem about Wffs.

Theorem 2. $\mathrm{S}_t^s[A]$ is a wff provided that A is a wff, where s and t are individuals and t is pure.

Dem. Apply the Fundamental Theorem about Wffs.

Now that the substitution transforms S_t^s are available to us, we can present a form of the Fundamental Theorem about Wffs (see page 131) which is sometimes useful, even indispensable. Note that the fourth induction assumption (involving $\forall$) is considerably strengthened.

Strong Fundamental Theorem about Wffs. Let S be a set of wffs; then $S = \mathsf{W}$ iff:

1. $A \in S$ for each atomic wff A.
2. $\neg A \in S$ provided that $A \in S$.
3. $A \vee B \in S$ provided that $A \in S$, $B \in S$, and $A \vee B$ is a wff.
4. $\forall tE \in S$ provided that $\forall tE$ is a wff and $\mathrm{S}_t^s[E] \in S$ for each individual s.

Dem. Assume the theorem is false. Then there is a proper subset of W, say S, that meets the four conditions of the theorem. Let A be the shortest wff that is not in S (count instances of $\neg$, $\vee$, and $\forall$). By condition 1, A is composite; therefore, A has a main connective. In view of conditions 2–4 this is impossible. This contradiction establishes our theorem.

Exercises

Simplify each of the following:

1. $\mathrm{I}_y^x[\forall xGxz \vee \forall yFy]$.
2. $\mathrm{I}_z^y[\forall xGxz \vee \forall yFy]$.
3. $\mathrm{S}_y^x[\forall xGxz \vee \forall yFy]$.
4. $\mathrm{S}_z^y[\forall xGxz \vee \forall yFy]$.
5. $\mathrm{S}_z^y[\forall xy\exists z(Gxz \vee Fy)]$.
6. $\mathrm{S}_z^y[\forall yzGyz]$.

7. $\mathsf{S}_z^y[\forall zyGyz]$.
8. Prove Theorem 1.
9. Prove Theorem 2.
10. Prove that $\mathsf{S}_t^s \neq \mathsf{S}_s^t$ if $s \neq t$.
11. Show that $\mathsf{S}_t^s[\mathsf{S}_s^t[A]] = A$ for any wff A in which s and t are free. What if $s = t$?
12. Complete the proof of the Strong Fundamental Theorem about Wffs.

8.7. Valuations

Our procedure for constructing the wffs of a predicate calculus $\prod$ is based on the definition of the swff of a semantical system, but is somewhat simpler since *placeholders* are not explicitly included in the alphabet of $\prod$.

Later we shall want to construct a semantical system from a given predicate calculus $\prod$. It is important, therefore, that we develop a technique for injecting truth-values into $\prod$. As for a propositional calculus, we introduce symbols "t" and "f," which stand for *true* and *false*, respectively. Let σ be a map whose domain is the set of all atomic wffs of $\prod$ and whose range is $\{\mathsf{t}, \mathsf{f}\}$; then σ is said to be an *assignment*. The idea is that σ assigns a truth-value to each atomic wff of $\prod$. The purpose of an assignment is to reflect the basic feature of each semantical system, namely that each of its atomic swffs has a unique truth-value.

In the same way that compound swffs receive unique truth-values in terms of the truth-values of their components, we use an assignment σ to provide each composite wff of $\prod$ with a unique truth-value. In fact, we extend σ to ν, a map of W into $\{\mathsf{t}, \mathsf{f}\}$ defined as follows.

Definition of ν:

1. $\nu[A] = \sigma[A]$ for each atomic wff A.

2. $$\nu[\neg B] = \begin{cases} \mathsf{t} & \text{if } \nu[B] = \mathsf{f} \\ \mathsf{f} & \text{if } \nu[B] = \mathsf{t} \end{cases}$$

3. $$\nu[C \vee D] = \begin{cases} \mathsf{f} & \text{if } \nu[C] = \nu[D] = \mathsf{f} \\ \mathsf{t} & \text{otherwise} \end{cases}$$

4. $$\nu[\forall tE] = \begin{cases} \mathsf{t} & \text{if } \nu[\mathsf{S}_t^s[E]] = \mathsf{t} \text{ for each individual } s \\ \mathsf{f} & \text{otherwise} \end{cases}$$

The map ν is called a *valuation*; for each wff A, $\nu[A]$ is said to be the truth-value of A.

We must prove that our definition of a valuation does indeed associate a unique truth-value with each wff. For that purpose we use the next theorem.

Construction Theorem for Wffs. In order to define an operation on all wffs, it suffices to give the following.

(a) The definition of the operation on the atomic wffs.
(b) The definition of the value of the operation at $(\neg B)$, possibly using its value at B.
(c) The definition of the value of the operation at $(A \vee B)$, possibly using its values at A and B.
(d) The definition of the value of the operation at $\forall tE$, possibly using its values at wffs $\mathsf{S}_t^s[E]$ for individuals s.

Dem. This parallels the Construction Theorem for Wffs in Part I. The proof given there can be extended to cover the present case. The details are left to Exercise 3.

Lemma 1. Each valuation is a map of W into $\{\mathsf{t}, \mathsf{f}\}$.

Dem. Apply the Construction Theorem for Wffs.

The following facts about valuations can be established just as for the propositional calculus (see page 28). Throughout, ν is any valuation and A and B are any wffs of $\prod$.

Fact 1.

$$\nu[A \wedge B] = \begin{cases} \mathsf{t} & \text{if } \nu A = \nu B = \mathsf{t} \\ \mathsf{f} & \text{otherwise} \end{cases}$$

Fact 2.

$$\nu[A \to B] = \begin{cases} \mathsf{t} & \text{if } \nu A = \mathsf{t} \text{ or } \nu B = \mathsf{t} \\ \mathsf{f} & \text{if } \nu A = \mathsf{t} \text{ and } \nu B = \mathsf{f} \end{cases}$$

Fact 3.

$$\nu[A \leftrightarrow B] = \begin{cases} \mathsf{t} & \text{if } \nu A = \nu B \\ \mathsf{f} & \text{if } \nu A \neq \nu B \end{cases}$$

Here is a new fact concerning valuation.

Fact 4.

$$\nu[\exists tA] = \begin{cases} \mathsf{t} & \text{if } \nu[\mathrm{S}_t^s[A]] = \mathsf{t} \text{ for some individual } s \\ \mathsf{f} & \text{if } \nu[\mathrm{S}_t^s[A]] = \mathsf{f} \text{ for each individual } s \end{cases}$$

Dem. We point out that $\exists tA = \neg(\forall t(\neg A))$.

We shall be specially interested in wffs A such that $\nu[A] = \mathsf{t}$ for each valuation ν. Here are some examples: $\forall xFx \rightarrow Fy$, $\forall x(Fy \wedge Gxx) \rightarrow Fy \wedge \forall x(Gxx)$, $\exists x(Fx \wedge Gxy) \rightarrow \exists xFx \wedge \exists xGxy$.

Exercises

1. Show that for each valuation ν:

   (a) $\nu[\forall xFx \rightarrow Fy] = \mathsf{t}$.
   (b) $\nu[\forall x(Fy \wedge Gxx) \rightarrow Fy \wedge \forall xGxx] = \mathsf{t}$.
   (c) $\nu[\exists x(Fx \wedge Gxy) \rightarrow \exists xFx \wedge \exists xGxy] = \mathsf{t}$.
   (d) $\nu[\forall xFx \wedge \neg Fy] = \mathsf{f}$.

2. Exhibit a valuation ν such that

$$\nu[\exists xFx \wedge \exists xGxy \rightarrow \exists x(Fx \wedge Gxy)] = \mathsf{f}$$

3. Prove the Construction Theorem for Wffs.

4. Prove Fact 4.

5. Let ν be the valuation such that $\nu A = \mathsf{t}$ for each atomic wff A. Compute each of the following:

   (a) $\nu[\forall xFx \rightarrow \exists yGyy]$.
   (b) $\nu[\forall xFx \rightarrow \exists y(\neg Gyy)]$.
   (c) $\nu[\forall x(Fx \wedge \neg Gxx)]$.

9

Provable Wffs

9.1. $\sum$-Interpreters

Let $\prod$ be a predicate calculus and let $\sum$ be some semantical system. A map μ which associates a unique object of $\sum$ with each variable of $\prod$ is said to be a $\sum$-*interpreter* provided that:

1. μ associates a relation symbol of $\sum$ with each predicate of $\prod$, and μP has the same type as P for each predicate P.
2. μ associates a constant of $\sum$ with each individual of $\prod$.

Note that if P is both a predicate and an individual of $\prod$, then μP is both a relation symbol and a constant of $\sum$.

If $\sum$ is sufficiently rich in objects, then there will be some choice for the image of each variable of $\prod$; in this case, there are many $\sum$-interpreters. However, $\prod$ does not possess a $\sum$-interpreter if there is a predicate such that no relation symbol of $\sum$ has its type.

In order that a semantical system $\sum$ can be linked to $\prod$ via a $\sum$-interpreter, it is necessary that $\sum$ match $\prod$ in two ways:

(a) For each predicate of $\prod$ there is a relation symbol of $\sum$ with the same type.
(b) For each predicate of $\prod$ that is also an individual of $\prod$, there must be a relation symbol of $\sum$, with the same type as the predicate, that is also a constant of $\sum$.

We say that $\sum$ is *matched* to $\prod$ if both conditions are met.

The idea is that a $\sum$-interpreter, say μ, translates each wff A of $\prod$ into a swff of $\sum$, denoted by "μA"; this swff is obtained from A by replacing each interpretable variable of A by its image under μ. Recall that a variable t is interpretable in a wff A if t occurs in A and is not bound in A (so t is either a predicate or a pure individual that is free in A). We point out that the image, under μ, of a pure individual can be a *mixed* constant of $\sum$.

We emphasize that individuals that are bound in A are not affected by μ. Moreover, all connectives and parentheses occurring in A are ignored by the $\sum$-interpreter μ in making the translation. Only predicates and free individuals are interpreted by μ. Since we usually use $x, y, z, \ldots$ for the individuals of $\prod$, the μ-image of a wff will automatically have $x, y, z, \ldots$ as its placeholders.

Example 1. Let $\prod$ be a predicate calculus with predicates F and G, where type $F = \{1\}$ and type $G = \{2\}$, and individuals $x, y, z, \ldots$. Let $\sum$ be a semantical system with relation symbols T and R, where type $T = \{1\}$ and type $R = \{2\}$, and constants T and R. Let μ be the $\sum$-interpreter such that $\mu F = T$, $\mu G = R$, and $\mu t = R$ for each individual t of $\prod$. Let $A = \forall xFx \rightarrow Fz \vee \forall x\exists yGxy$; then

$$\mu A = \forall xTx \rightarrow TR \vee \forall x\exists yRxy$$

a swff of $\sum$. In this example we are not involved with diag $\sum$.

We now introduce a procedure for constructing a semantical system $\sum$ from a given predicate calculus $\prod$ and a valuation ν. First, we define the domain of the map $\sum$; next, we announce which member of {true, false} $\sum$ associates with a given member of dom $\sum$.

Definition of $\sum$:

1. dom $\sum = \{P\alpha \mid (P\alpha)$ is an atomic wff of $\prod\}$, i.e., dom $\sum$ is the set of all expressions obtained by deleting the outermost pair of parentheses from each atomic wff of $\prod$.
2. $\sum$ associates "true" with a member of its domain, say $P\alpha$, if $\nu[(P\alpha)] = \mathsf{t}$; $\sum$ associates "false" with $P\alpha$ if $\nu[(P\alpha)] = \mathsf{f}$.

By construction, the swffs of $\sum$ are precisely the wffs of $\prod$. To avoid confusion, we shall use the identity interpreter ι to indicate a swff; i.e., if A is a wff of $\prod$, then ιA is the corresponding swff of $\sum$.

Utilizing these conventions, we now formulate the following important relation between ν and truth-values in $\sum$.

Lemma 1. For each wff A of $\prod$, ιA is true for $\sum$ iff $\nu[A] = \mathsf{t}$.

Dem. Let $S = \{A \in \mathsf{W} \mid \iota A$ is true for $\sum$ iff $\nu[A] = \mathbf{t}\}$. We can show that S meets the four conditions of the Strong Fundamental Theorem about Wffs; thus $S = \mathsf{W}$. The details are left as an exercise.

Exercises

1. Let $\prod$, $\sum$, and μ be as in Example 1.
   (a) Exhibit $\mu[\forall y Gxy \vee \neg Fx]$.
   (b) Given that diag $\sum = \{TR, RTR, RRT\}$, determine the truth-value of the swff of part (a).
   (c) Given that diag $\sum = \{TR, RRR\}$, determine the truth-value of the swff of part (a).
2. Complete the proof of Lemma 1.
3. Let $\prod$ be any predicate calculus and let ν be the valuation such that $\nu[A] = \mathbf{t}$ for each atomic wff A. Exhibit the semantical system $\sum$ constructed from ν and $\prod$ by applying the definition which precedes Lemma 1. Verify, in this case, that for each wff A,

$$\iota A \text{ is true for } \textstyle\sum \text{ iff } \nu[A] = \mathbf{t}$$

   where ι is the identity interpreter.

9.2. True Wffs

Continuing our study of wffs, we now consider the important family of wffs referred to earlier, i.e., the *true* wffs of $\prod$. In this section we shall characterize true wffs in two ways; first, via semantical systems, and second, via valuations. Our purpose is to motivate the work of Section 9.3, where we shall use the axiomatic method to characterize this concept within our formal language.

We now define the notion of a *true* wff of $\prod$.

Definition of True Wff. A wff A is *true* provided that for each semantical system $\sum$ matched to $\prod$, and for each $\sum$-interpreter μ, μA is true for $\sum$.

For example, $\forall x Fx \to Fy$ is a true wff of $\prod$. To see this, let $\sum$ be any semantical system matched to $\prod$ and let μ be any $\sum$-interpreter. Let $\mu F = R$ and let $\mu y = a$; then

$$\mu[\forall x Fx \to Fy] = \forall x Rx \to Ra$$

Clearly, the swff $\forall xRx \to Ra$ is true for $\sum$. We conclude that $\forall xFx \to Fy$ is a true wff.

It is worth noting that our notions of wff and swff are matched to the point that the image of any wff is a swff.

Lemma 1. Let $\sum$ be any semantical system matched to $\prod$ and let μ be any $\sum$-interpreter. For each wff A, μA is a swff of $\sum$.

Here are two important facts about true wffs. First, we present the familiar rule of inference Modus Ponens. Next, we establish a powerful rule of inference which involves introducing a quantifier.

Modus Ponens. If A and $A \to B$ are true wffs, so is B.

Dem. Let $\sum$ be any semantical system matched to $\prod$ and let μ be any $\sum$-interpreter; we must show that μB is true for $\sum$. By assumption, μA and $\mu[A \to B]$ are both true for $\sum$; i.e.,

$$\mu A \qquad \text{and} \qquad \mu A \to \mu B$$

are both true for $\sum$. Thus, μB is true for $\sum$. We conclude that B is true.

Here is an extremely fertile rule of inference which we shall call "I$\forall$R," *introduce $\forall$ to the right.*

Lemma 2. $A \to \forall tB$ is true if:
1. $A \to B$ is true.
2. t is free in B.
3. t does not occur in A.

Dem. Let $\sum$ be any semantical system matched to $\prod$ and let μ be any $\sum$-interpreter; we must show that $\mu[A \to \forall tB]$ is true for $\sum$. There are two cases.

(i) Assume that μA is false for $\sum$. Then $\mu A \to \mu[\forall tB]$ is true for $\sum$.

(ii) Assume that μA is true for $\sum$. We must show that $\mu[\forall tB]$ is true for $\sum$. Let $\mu[\forall tB] = \forall tC$ and let a be any constant of $\sum$; we shall prove that $S_t^a[C]$ is true for $\sum$. To this purpose, let λ be the $\sum$-interpreter such that for each variable v,

$$\lambda v = \begin{cases} a & \text{if} \quad v = t \\ \mu v & \text{if} \quad v \neq t \end{cases}$$

i.e., λ is constructed from μ by replacing the image of t by a (notice that there is no matching problem since t is not a predicate). Now, $A \to B$ is true; so $\lambda A \to \lambda B$ is true for $\sum$. Also, $\lambda A = \mu A$ since t does not occur in A; thus λA is true for $\sum$; therefore λB is true for $\sum$. By construction, $\lambda B = \mathrm{S}_t^a[C]$; so $\mathrm{S}_t^a[C]$ is true for $\sum$. Since a is any constant of $\sum$, it follows that $\forall tC$ is true for $\sum$; i.e., $\mu[\forall tB]$ is true for $\sum$. Thus $\mu[A \rightarrow \forall tB]$ is true for $\sum$. We conclude that $A \rightarrow \forall tB$ is a true wff. This establishes Lemma 2.

We point out that if $A \rightarrow B$ and $A \rightarrow \forall tB$ are both wffs, then t is free in B and t does not occur in A. Accordingly, the conditions of Lemma 2 can be modified as follows.

Lemma 3. $A \rightarrow \forall tB$ is true if

1. $A \rightarrow B$ is true.
2. $A \rightarrow \forall tB$ is a wff.

A predicate calculus is sometimes defined by other authors so that $A \vee B$ is a wff even if there is an individual that is free in A and is bound in B, or vice versa. For this reason, we prefer to express the preceding rule of inference as given by Lemma 2, rather than Lemma 3.

To see the necessity for condition 3 of Lemma 2, consider the following example, which applies to a predicate calculus that allows the same individual to be free in one disjunct and bound in another disjunct of a wff.

Example 1. Now, $Fx \rightarrow Fx$ is a true wff, since $\mu Fx \rightarrow \mu Fx$ is true for $\sum$ if μ is a $\sum$-interpreter. Let us show that $Fx \rightarrow \forall xFx$, which we pretend for the moment is a wff, is not true. Let $\sum$ be a semantical system that matches the given predicate calculus. Let R be a relation symbol of $\sum$ such that $\underline{R} = \{a\}$, and let b be another constant of $\sum$. Take $\mu F = R$, $\mu x = a$, and $\mu t = b$ for each individual t different from x. Then

$$\mu[Fx \rightarrow \forall xFx] = Ra \rightarrow \forall xRx$$

But Ra is true for $\sum$, and $\forall xRx$ is false for $\sum$. We conclude that the wff $Fx \rightarrow \forall xFx$ is *not* true.

Here is the vital notion of a *model* of a set of wffs. Let K be any set of wffs of $\prod$; a semantical system $\sum$ matched to $\prod$ is said to be a model of K under μ provided that:

1. μ is a $\sum$-interpreter.
2. For each wff $A \in \mathsf{K}$, μA is true for $\sum$.

Here is an example.

Example 2. Let $\prod$ be the predicate calculus with predicates F and G, where type $F = \{1\}$ and type $G = \{1, 2\}$, and with individuals

x, y, and z. Let $\sum$ be the semantical system with relation symbols T and R, where type $T = \{1\}$ and type $R = \{1, 2\}$, and constants T, a, and b. Moreover, let diag $\sum = \{Ta, RT, Rab\}$. Let μ be the $\sum$-interpreter for which

$$\mu F = T, \qquad \mu G = R, \qquad \mu x = T, \qquad \mu y = a, \qquad \mu z = b$$

Let $\mathsf{K} = \{Fy, Gx, \neg Gxy, \exists xGxz\}$; then $\sum$ is a model of K under μ. Indeed, $\mu Fy = Ta$, $\mu Gx = RT$, $\mu[\neg Gxy] = \neg RTa$, and $\mu[\exists xGxz] = \exists xRxb$. Each swff in this list is true for $\sum$; the swff $\exists xRxb$ is true for $\sum$ since Rab is true for $\sum$.

It is important that we characterize the set of all true wffs of a predicate calculus $\prod$ internally; i.e., without referring to semantical systems or interpreters. This we shall do in Section 9.3. In Chapter 12 we shall show that the Strong Completeness Theorem (see page 102) also applies to $\prod$. This fact is the bridge that connects mathematical logic and mathematics, and so provides many applications to mathematics.

We now present a fact about true wffs that involves valuations.

Theorem 1. If B is a true wff then $\nu[B] = \mathrm{t}$ for each valuation ν.

Dem. Assume that B is a true wff. Let ν be any valuation and let $\sum$ be the semantical system constructed from $\prod$ and ν by following the prescription in the definition preceding Lemma 1, page 154. Since B is a true wff, ιB is true for $\sum$. By Lemma 1, page 154, $\nu[B] = \mathrm{t}$.

Exercises

1. Let $\prod$, $\sum$, and μ be as in Example 2.
   (a) Exhibit μA, where $A = \forall y(Fx \lor Gxy) \to Fx \lor \forall yGxy$.
   (b) Determine the truth-value of μA in $\sum$, where A is the wff of part (a).
2. Show that $\forall y(Fx \lor Gxy) \to Fx \lor \forall yGxy$ is a true wff of $\prod$, provided it is a wff. Use the definition of true wff.
3. Show that $Fx \lor \forall yGxy \to \forall y(Fx \lor Gxy)$ is a true wff of $\prod$, provided it is a wff. Use the definition of true wff.
4. Show that $\exists xFx \to \forall xFx$ is not a true wff of $\prod$.
5. Let $\prod$, $\sum$, and μ be as in Example 2. Show that $\sum$ is a model of $\{\neg Fz, Fz \to Gxx, \exists z(Fz), \forall x(Fx \to Gxz)\}$ under μ.

6. Prove that each of the following wffs of $\prod$ is true:

   (a) $\forall xFx \rightarrow Fy$.
   (b) $\forall xFx \rightarrow \exists xFx$.
   (c) $\forall x(Fx \wedge Gxx) \leftrightarrow \forall xFx \wedge \forall xGxx$.

9.3. Proofs and Provable Wffs

Our purpose in this section is to characterize the set of all true wffs without referring to semantical systems or interpreters; instead, we shall apply the axiomatic method.

First, we note that each member of the following four sets of wffs is true:

AS 1. $\{A \vee A \rightarrow A \mid A \in \mathsf{W}\}$.
AS 2. $\{A \rightarrow A \vee B \mid A \vee B \in \mathsf{W}\}$.
AS 3. $\{A \rightarrow B \dot{\rightarrow} C \vee A \rightarrow B \vee C \mid A \vee (B \vee C) \in \mathsf{W}\}$.
AS 4. $\{\forall tA \rightarrow \mathrm{S}_t^s[A] \mid \forall tA \rightarrow \mathrm{S}_t^s[A] \in \mathsf{W}\}$.

Referring to AS 4, let $\forall tA \rightarrow \mathrm{S}_t^s[A]$ be a wff; then certainly t is free in A. Moreover, if s is bound in A, then $\mathrm{S}_t^s[A] = \mathrm{I}_t^s[A]$; so s is free in $\mathrm{S}_t^s[A]$ and is bound in $\forall tA$. In this case, $\forall tA \rightarrow \mathrm{S}_t^s[A]$ is not a wff. Again, if $s = t$, then s is bound in $\forall tA$ and is free in $\mathrm{S}_t^s[A]$; so $\forall tA \rightarrow \mathrm{S}_t^s[A]$ is not a wff. We conclude that $\forall tA \rightarrow \mathrm{S}_t^s[A]$ is a wff iff $A \in \mathsf{W}$, t is free in A, and s is not bound in $\forall tA$.

We recall two rules of inference for true wffs discussed in Section 9.2, Modus Ponens and I∀R.

Modus Ponens. B is true if both A and $A \rightarrow B$ are true wffs.

I∀R. $A \rightarrow \forall tB$ is true if $A \rightarrow B$ is true and $A \rightarrow \forall tB$ is a wff.

Our notion of *proof*, which we now present, is based on the preceding observations concerning true wffs.

Definition of Proof. A finite sequence of wffs is called a *proof* if each of its terms, say E, satisfies at least one of the following conditions:

(a) E is a member of one of the four sets of wffs displayed above.
(b) There is a wff D such that both D and $D \rightarrow E$ precede E in the sequence.
(c) E has the form $A \rightarrow \forall tB$ and is preceded in the sequence by $A \rightarrow B$.

The four sets of wffs involved in our definition of *proof* are called *axiom schemes*; notice that our definition involves two rules of inference, Modus Ponens and I∀R.

Here is an example.

Example 1. The following sequence is a proof: $\forall xFx \to Fy$, $\forall xFx \to \forall yFy$. The first term of this sequence is a member of AS 4; the second term is justified by the rule of inference (c).

Notice that the above definition of *proof* includes the three axiom schemes involved in the corresponding concept of the propositional calculus (see page 37); one of the two rules of inference appearing in our definition is Modus Ponens, the only rule of inference involved in the definition of proof for the propositional calculus. Therefore, each proof in the propositional calculus yields a proof in our predicate calculus $\prod$ by merely replacing its atomic wffs by atomic wffs of $\prod$. Accordingly, the results about proofs in Section 3.3 carry over to $\prod$. For convenience, we state some of these results here.

Lemma 1. Let π_1 and π_2 be proofs; then the sequence π_1, π_2 obtained by adjoining the terms of π_2 to the terms of π_1 is a proof.

Lemma 2. Let π_1 be a proof whose last term is A, and let π_2 be a proof whose last term is $A \to B$; then the sequence π_1, π_2, B is a proof.

Using condition (c) of the above definition, we can establish the following fact, which applies only to a predicate calculus.

Lemma 3. Let π be a proof whose last term is $A \to B$, where t is free in B and t does not occur in A; then the sequence π, $A \to \forall tB$ is a proof.

Next, we introduce the notion of a *provable* wff.

Definition. A wff, say C, is said to be *provable* if there is a proof whose last term is C.

We shall abbreviate the statement "C is provable" by writing "$\vdash C$." The last term of the proof of Example 1 is $\forall xFx \to \forall yFy$; so $\vdash \forall xFx \to \forall yFy$, i.e., the wff $\forall xFx \to \forall yFy$ is provable.

Let C be a provable wff; then there exists a proof, say π, whose last term is C. In this case we say that π is a *proof of C*. So, the proof of Example 1 is a proof of $\forall xFx \to \forall yFy$.

Analogous to the Fundamental Theorem about Wffs is the useful Fundamental Theorem about Provable Wffs, which we now present.

Fundamental Theorem about Provable Wffs. A set of provable wffs, say S, is the set of all provable wff provided that:

1. $A \in S$ if A is a member of an axiom scheme.
2. $B \in S$ if there is a wff such that both $A \rightarrow B$ and A are in S.
3. $A \rightarrow \forall tB \in S$ if $A \rightarrow B \in S$ and if $A \rightarrow \forall tB$ is a wff.

Dem. See page 39.

Our purpose in introducing the notion of a provable wff is to characterize the concept of a *true* wff within our formal language, i.e., without referring to valuations, semantical systems, or interpreters. Accordingly, we must show that the set of provable wffs *is* the set of true wffs. We are now in a position to prove that each provable wff is true; i.e., the set of all provable wffs is a subset of the set of all true wffs.

Lemma 4. Each provable wff is true.

Dem. Let $S = \{A \in W \mid A \text{ is true and } \vdash A\}$. Applying the Fundamental Theorem about Provable Wffs, we conclude that each provable wff is in S; i.e., each provable wff is true. For more details see page 40.

Later (see Section 12.3) we shall prove that each true wff is provable.

Corollary 1. No atomic wff is provable.

Dem. No atomic wff is true.

We can say a little more in this direction.

Corollary 2. The length of each provable wff is at least two.

Dem. No wff of length one (i.e., with exactly one basic connective) is true; so, by Lemma 4, no wff of length one is provable.

Following the pattern of the proof of Example 1, page 38, we can easily verify that $\vdash Fx \vee \neg Fx$. This provable wff has length two; so Corollary 2 is the best we can do in this direction.

Continuing our reference to the proof of Example 1, page 38, we point out that we can construct a proof of the wff $A \vee \neg A$, where A is any wff of our predicate calculus, by substituting A for the atomic wff (X) throughout the proof of Example 1, page 38. This establishes the following fact.

Lemma 5. $\vdash A \vee \neg A$ for any wff A.

Exercises

1. Prove Lemma 1.
2. Prove Lemma 2.
3. Given that $\vdash A$ and that $A \vee B$ is a wff, show that $\vdash A \vee B$.
4. Given that $\vdash A \to B$ and that $A \to \forall tB$ is a wff, prove that $\vdash A \to \forall tB$.
5. Prove Lemma 4.
6. Prove that $\vdash \mathsf{S}_t^s[A] \to \exists tA$ provided that $A \in \mathsf{W}$, t is free in A, and s is an individual which is not bound in $\forall tA$.

9.4. Equivalent Wffs

Just as for the propositional calculus, we need the notion of *equivalent* wffs.

Definition. We shall say that wffs A and B are *equivalent* (in symbols $A \equiv B$) if $\vdash A \to B$ and $\vdash B \to A$.

We have observed that $\vdash \forall xFx \to \forall yFy$; similarly, we can show that $\vdash \forall yFy \to \forall xFx$. Thus $\forall xFx \equiv \forall yFy$.

Although we read "$\equiv$" as "is equivalent to," we should realize that this binary relation on W is *not* an equivalence relation. Certainly, $\equiv$ is symmetric; moreover, by Lemma 2, page 165, $\equiv$ is reflexive. However, $\equiv$ is not transitive. For example,

$$Fx \vee \neg Fx \equiv \forall yFy \to Fz \qquad \text{and} \qquad \forall yFy \to Fz \equiv \forall xFx \to Fz$$

yet

$$Fx \vee \neg Fx \equiv \forall xFx \to Fz$$

is false since

$$Fx \vee \neg Fx \dot{\to} \forall xFx \to Fz$$

is not a wff.

We can, however, establish the following weakened form of the transitive law for $\equiv$; here, we rely on Lemma 3, page 165.

Lemma 1. $A \equiv C$ if $A \equiv B$, $B \equiv C$, and $A \rightarrow C$ is a wff.

The point is that the transitive law for $\equiv$ fails only if the expression involved, here $A \rightarrow C$, is not a wff. This happens only if some individual is bound in A and free in C or vice versa.

We mention that if $A_1, \ldots, A_n$ are wffs such that $A_1 \equiv A_2$, $A_2 \equiv A_3, \ldots, A_{n-1} \equiv A_n$ and if $A_i \rightarrow A_j$ is a wff for each i and j between 1 and n, then $A_1 \equiv A_n$. So, even though $\equiv$ is not an equivalence relation, it can be used to carry out a chain of simplifications provided that no two wffs in the chain involve an individual that is free in one wff but bound in the other.

Hereafter, we assume that our predicate calculus $\prod$ has infinitely many pure individuals $x, y, z, \ldots$. In view of the fact that each wff is a finite string of symbols, we conclude that corresponding to each wff A there is a pure individual that does not occur in A.

Later, we shall prove that $\forall tA \equiv \forall s\mathsf{S}_t^s[A]$ provided that $\forall tA \rightarrow \forall s\mathsf{S}_t^s[A]$ is a wff (see Theorem 1, page 177). The first step toward proving this fact is to establish the following lemma.

Lemma 2. $\vdash \forall tA \rightarrow \forall s\mathsf{S}_t^s[A]$ if $\forall tA$ is a wff and s is a pure individual that does not occur in A.

Dem. By AS 4, $\vdash \forall tA \rightarrow \mathsf{S}_t^s[A]$; thus, by Lemma 3, page 160, $\vdash \forall tA \rightarrow \forall s\mathsf{S}_t^s[A]$.

Corollary 1. $\forall tA \equiv \forall s\mathsf{S}_t^s[A]$ if $\forall tA$ is a wff and s is a pure individual that does not occur in A.

Dem. By assumption, t is free in A and s does not occur in A; so t is a pure individual that does not occur in $\mathsf{S}_t^s[A]$. Therefore, by Lemma 2,

$$\vdash \forall s\mathsf{S}_t^s[A] \rightarrow \forall t\mathsf{S}_s^t[\mathsf{S}_t^s[A]]$$

But $\mathsf{S}_s^t[\mathsf{S}_t^s[A]] = A$; so $\vdash \forall s\mathsf{S}_t^s[A] \rightarrow \forall tA$. We conclude, from Lemma 2, that $\forall tA \equiv \forall s\mathsf{S}_t^s[A]$.

We now present some connections between equivalent wffs, valuations, truth-values in semantical systems, and true wffs.

Lemma 3. If $A \equiv B$, then $\nu A = \nu B$ for each valuation ν.

Dem. See Lemma 4, page 46.

The following lemma is closely related to the preceding one, but involves the concept of a true wff.

Lemma 4. If $A \equiv B$, then both A and B are true wffs or else neither is true.

Dem. See Corollary 1 on page 46.

We will later prove the Completeness Theorem (see the discussion on page 213; the proof is completed on page 223). The Completeness Theorem asserts that for each wff A, A is provable if and only if A is true. But at this point, we have established only the weaker property that all of the provable wffs are true (Lemma 4, page 161). This latter property, sometimes called the *soundness* property, is clearly essential; we would not want a notion of proof that allowed us to derive any wff that was not true.

We turn now to the connection with semantical systems.

Lemma 5. Let $A \equiv B$, let $\sum$ be any semantical system matched to $\prod$, and let μ be any $\sum$-interpreter; then μA and μB have the same truth-value in $\sum$.

Dem. See Lemma 6, page 46.

Lemma 6. If $A \equiv B$, then $A \leftrightarrow B$ is a true wff.

Dem. See Lemma 7, page 46.

We mention that Lemma 5 follows from Lemma 6.

Finally, we observe that Lemma 7, page 77, carries over to our predicate calculus.

Lemma 7. Let $A \vee (B \vee C) \in \mathsf{W}$; then:
(a) $A \dot{\rightarrow} B \rightarrow C \equiv A \wedge B \rightarrow C$.
(b) $A \dot{\rightarrow} B \rightarrow C \equiv B \wedge A \rightarrow C$.
(c) $A \dot{\rightarrow} B \rightarrow C \equiv B \dot{\rightarrow} A \rightarrow C$.

Exercises

1. Prove Lemma 1.
2. Let $A_1 \equiv A_2, A_2 \equiv A_3, \ldots, A_{n-1} \equiv A_n$ and let $A_i \rightarrow A_j$ be a wff for each i and j between 1 and n. Show that $A_1 \equiv A_n$.

3. Prove Lemma 3.

4. Prove Lemma 4.

5. Prove Lemma 5.

6. Prove Lemma 6.

7. Let $A \leftrightarrow B$ be a true wff. Is it necessarily the case that $A \equiv B$? Justify your answer.

8. Prove Lemma 7.

9.5. Rules of Inference

The various provable wffs and rules of inference established in Section 3.4 (see pages 41 to 43) carry over to our predicate calculus $\prod$; to see this, simply interpret A, B, and C as wffs of $\prod$ and repeat each argument in the context of $\prod$. In the same way, the results about equivalent wffs established in Section 3.5 (see page 45) and in Section 6.1 (see pages 75–80) carry over to $\prod$. For convenience, we now list some of these results about $\prod$.

Lemma 1. $\vdash A \rightarrow B$ iff $\vdash \neg B \rightarrow \neg A$.

Lemma 2. $\vdash A \rightarrow A$ for each wff A.

Lemma 3. $\vdash A \rightarrow C$ if $\vdash A \rightarrow B$, $\vdash B \rightarrow C$ and if $A \rightarrow C$ is a wff.

Lemma 4. $\vdash A_1 \wedge \cdots \wedge A_n$ iff $\vdash A_1, \ldots, \vdash A_n$ and $A_1 \wedge \cdots \wedge A_n$ is a wff.

Lemma 5. $A \vee B \equiv B \vee A$ provided that $A \vee B$ is a wff.

Lemma 6. $A \equiv C$ if $A \equiv B$, $B \equiv C$, and if $A \rightarrow C$ is a wff.

Lemma 7. $A \vee (B \vee C) \equiv (A \vee B) \vee C$ provided $A \vee (B \vee C)$ is a wff.

Lemma 8. $A \wedge (B \wedge C) \equiv (A \wedge B) \wedge C$ provided $A \wedge (B \wedge C)$ is a wff.

Lemma 9. $A \vee (B \wedge C) \equiv (A \vee B) \wedge (A \vee C)$ provided $A \vee (B \wedge C)$ is a wff.

Lemma 10. $A \wedge (B \vee C) \equiv (A \wedge B) \vee (A \wedge C)$ provided $A \wedge (B \vee C)$ is a wff.

Lemma 11. $A \wedge B \equiv A$ if $\vdash B$ and if $A \wedge B$ is a wff.

Lemma 12. $\neg(\neg A) \equiv A$ for each wff A.

Lemma 13. $A \equiv B$ iff $\neg A \equiv \neg B$.

Lemma 14. Let $A \equiv B$ and let C be a wff such that both $A \vee C$ and $B \vee C$ are wffs; then $A \vee C \equiv B \vee C$ and $C \vee A \equiv C \vee B$.

As for the propositional calculus, the rule of inference Modus Ponens is yielded by Lemma 2, page 160.

Modus Ponens. $\vdash B$ if $\vdash A$ and $\vdash A \rightarrow B$.

Our next rule of inference, I∀R, is based on Lemma 3, page 160.

Lemma 15 (I∀R). $\vdash A \rightarrow \forall tB$ provided that $\vdash A \rightarrow B$, t is free in B, and t does not occur in A.

The result of prefixing "$\forall t$" to a provable wff in which t is free is provable. This is known as the *Principle of Generalization.*

Lemma 16. $\vdash \forall tA$ provided that $\vdash A$ and t is free in A.

Dem. Here we shall use our assumption that $\prod$ has infinitely many pure individuals. By this assumption, there is a pure individual, say s, that does not occur in A. Since $\vdash A$, it follows that $\vdash Fs \vee \neg Fs \rightarrow A$; by Lemma 15, $\vdash Fs \vee \neg Fs \rightarrow \forall tA$. But $\vdash Fs \vee \neg Fs$; so by Modus Ponens, $\vdash \forall tA$.

Here is an interesting rule of inference.

Lemma 17. $\vdash \forall tA \rightarrow \forall tB$ provided that $\vdash A \rightarrow B$ and t is free in both A and B.

Dem. Let s be a pure individual that occurs in neither A nor B. Then $\vdash \forall s\mathrm{S}^s_t[A] \rightarrow A$, since this wff is in AS 4. But $\vdash A \rightarrow B$, so $\vdash \forall s\mathrm{S}^s_t[A] \rightarrow B$ by Lemma 3. By Lemma 15, $\vdash \forall s\mathrm{S}^s_t[A] \rightarrow \forall tB$. From Lemma 2, page 163, $\vdash \forall tA \rightarrow \forall s\mathrm{S}^s_t[A]$; thus by Lemma 3, $\vdash \forall tA \rightarrow \forall tB$.

Corollary 1. $\forall tA \equiv \forall tB$ if $A \equiv B$ and if t is free in both A and B.

Lemma 18. $\vdash \mathrm{S}_t^s[A]$ provided that $\vdash \forall tA$ and s is an individual that is not bound in $\forall tA$.

Dem. By AS 4, $\vdash \forall tA \to \mathrm{S}_t^s[A]$; but $\vdash \forall tA$, so by Modus Ponens $\vdash \mathrm{S}_t^s[A]$.

We consider, now, the converse of Lemma 16.

Lemma 19. $\vdash A$ if $\vdash \forall tA$.

Dem. Let s be a pure individual that does not occur in A. By Lemma 2, page 163, $\vdash \forall tA \to \forall s\mathrm{S}_t^s[A]$. But $\vdash \forall tA$; thus by Modus Ponens, $\vdash \forall s\mathrm{S}_t^s[A]$. Again by Lemma 2, page 163, $\vdash \forall s\mathrm{S}_t^s[A] \to \mathrm{S}_s^t[\mathrm{S}_t^s[A]]$; i.e., $\vdash \forall s\mathrm{S}_t^s[A] \to A$. So, by Modus Ponens, $\vdash A$.

We recall three more rules of inference for the propositional calculus which carry over to $\prod$, namely Lemma 9, page 43, its corollary, and Lemma 11, page 44.

Lemma 20. $\vdash A \vee C \to B \vee D$ and $\vdash A \wedge C \to B \wedge D$, provided that $\vdash A \to B$, $\vdash C \to D$, and $A \vee C \to B \vee D$ is a wff.

Lemma 21. $\vdash A \to C$ if $\vdash A \to B$ and $\vdash A \wedge B \to C$.

Here are two examples that illustrate the power of Lemma 17.

Example 1. Show that $\vdash \forall tA \vee \forall tB \to \forall t(A \vee B)$ if this is a wff.

Solution. Now $\vdash A \to A \vee B$ by AS 2; thus $\vdash \forall tA \to \forall t(A \vee B)$ by Lemma 17. Similarly, $\vdash B \to A \vee B$; thus $\vdash \forall tB \to \forall t(A \vee B)$ by Lemma 17. Applying Lemma 20 to these results yields

$$\vdash \forall tA \vee \forall tB \to \forall t(A \vee B) \vee \forall t(A \vee B)$$

By AS 1,

$$\vdash \forall t(A \vee B) \vee \forall t(A \vee B) \to \forall t(A \vee B)$$

So, by Lemma 3, $\vdash \forall tA \to \forall tB \to \forall t(A \vee B)$.

Example 2. Show that $\vdash \forall t(A \wedge B) \to A \wedge \forall tB$, assuming this is a wff.

Solution. Now, $\vdash A \wedge B \to B$; thus, by Lemma 17, $\vdash \forall t(A \wedge B) \to \forall tB$. Let s be a pure individual that is not bound in $\forall t(A \wedge B)$; by AS 4,

$$\vdash \forall t(A \wedge B) \to A \wedge \mathrm{S}_t^s[B]$$

since $\mathrm{S}_t^s[A \wedge B] = A \wedge \mathrm{S}_t^s[B]$. But $\vdash A \wedge \mathrm{S}_t^s[B] \to A$; so $\vdash \forall t(A \wedge B) \to A$. Applying Lemma 21 to these results yields

$$\vdash \forall t(A \wedge B) \wedge \forall t(A \wedge B) \to A \wedge \forall tB$$

Clearly, $\vdash \forall t(A \wedge B) \to \forall t(A \wedge B) \wedge \forall t(A \wedge B)$; so, by Lemma 3, $\vdash \forall t(A \wedge B) \to A \wedge \forall tB$.

We now present an important rule of inference which involves $\exists$. The label "I$\exists$L" stands for *introduce $\exists$ to the left.*

I$\exists$L. $\vdash \exists tA \to B$ provided that $\vdash A \to B$, t is free in A, and t does not occur in B.

Dem. We are given that $\vdash A \to B$:

Thus	$\vdash \neg B \to \neg A$	(by Lemma 1)
So	$\vdash \neg B \to \forall t(\neg A)$	(by I$\forall$R)
Therefore	$\vdash \neg\forall t(\neg A) \to \neg(\neg B)$	(by Lemma 1)
i.e.	$\vdash \exists tA \to \neg(\neg B)$	
But	$\vdash \neg(\neg B) \to B$	(by Lemma 12)
Thus	$\vdash \exists tA \to B$	(by Lemma 3)

Here is an interesting fact.

Lemma 22. $\vdash \forall tA \to \exists tA$ provided that $\forall tA \in \mathsf{W}$.

Dem. Let s be an individual that does not occur in A.

Now	$\vdash \forall t(\neg A) \to \mathrm{S}_t^s[\neg A]$	(by AS 4)
Thus	$\vdash \neg\mathrm{S}_t^s[\neg A] \to \neg\forall t(\neg A)$	(by Lemma 1)
i.e.	$\vdash \neg(\neg\mathrm{S}_t^s[A]) \to \exists tA$	
But	$\vdash \mathrm{S}_t^s[A] \to \neg(\neg\mathrm{S}_t^s[A])$	(by Lemma 12)
So	$\vdash \mathrm{S}_t^s[A] \to \exists tA$	(by Lemma 3)
Also	$\vdash \forall tA \to \mathrm{S}_t^s[A]$	(by AS 4)
Hence	$\vdash \forall tA \to \exists tA$	(by Lemma 3)

Here are some more useful facts about $\prod$ that carry over from Part I.

Lemma 23. $A \to B \equiv \neg B \to \neg A$ for any wff $A \to B$.

Lemma 24. $\vdash A \to B \dot{\to} \neg B \to \neg A$ if $A \vee B \in \mathsf{W}$.

Lemma 25. If $\vdash C \dot{\to} A \to B$, then $\vdash C \dot{\to} \neg B \to \neg A$.

Lemma 26. $\vdash A \to B \dot{\to} C \wedge A \to C \wedge B$ if $A \vee (B \vee C) \in \mathsf{W}$.

Exercises

1. Prove that $\vdash \exists tA \to \exists tB$ if $\vdash A \to B$ and if t is free in both A and B.
2. Prove that $\vdash \exists tA$ provided that $\vdash A$ and t is free in A.
3. Prove that $\vdash \exists tA$ if $\vdash A$ and if t is free in A.
4. Show that the conjecture: "$\vdash A$ if $\vdash \exists tA$" is false.
5. Let $A \vee B$ be a wff and let t be free in both A and B. Show that

$$\forall t(A \vee B) \to \forall tA \vee \forall tB$$

   is not provable.
6. Prove that $\forall t(A \vee B) \equiv A \vee \forall tB$ given that $A \vee B$ is a wff, t is free in B, and t does not occur in A.
7. Prove that $\forall t(A \wedge B) \equiv A \wedge \forall tB$ given that $A \wedge B$ is a wff, t is free in B, and t does not occur in A.
8. Find the fallacy in the following argument. "We shall show that $\vdash \mathsf{S}^s_t[A]$ if $\vdash A$, t is free in A, and s is any individual. Since $\vdash A$ and t is free in A, it follows that $\vdash \forall tA$ (see the Principle of Generalization). By AS 4, $\vdash \forall tA \to \mathsf{S}^s_t[A]$; thus, by Modus Ponens, $\vdash \mathsf{S}^s_t[A]$."
9. Prove that $\vdash A \to \exists tB$ if this is a wff and $\vdash A \to B$.
10. Prove that $\exists t(A \wedge B) \equiv A \wedge \exists tB$ and $\exists t(A \vee B) \equiv A \vee \exists tB$ given that $A \vee B$ is a wff, t is free in B, and t does not occur in A.
11. Given that $\vdash \exists tA$, does it follow that there is an individual, say s, such that $\vdash \mathsf{S}^s_t[A]$? Justify your answer.

9.6. A Fact about the Interchange Transform

The purpose of this section is to prove the following fact.

Theorem 1. Let $A_1, \ldots, A_n$ be any proof and let s and t be any pure individuals; then $\mathbf{I}^s_t[A_1], \ldots, \mathbf{I}^s_t[A_n]$ is a proof.

Dem. Suppose that $\mathrm{I}_t^s[A_1], \ldots, \mathrm{I}_t^s[A_n]$ is *not* a proof; then there is a smallest natural number k such that $\mathrm{I}_t^s[A_1], \ldots, \mathrm{I}_t^s[A_k]$ is not a proof. Then $\mathrm{I}_t^s[A_1], \ldots, \mathrm{I}_t^s[A_{k-1}]$ is a proof. The wffs A_k and $\mathrm{I}_t^s[A_k]$ are especially significant here. Accordingly, we now concentrate our attention on the provable wff A_k and the proof $A_1, \ldots, A_k$. There are three possibilities.

Case 1. A_k is a member of an axiom scheme. We claim that $\mathrm{I}_t^s[A_k]$ is a member of the same axiom scheme as A_k. This is easy to verify for the first three schemes (see page 159), so we shall consider only the fourth axiom scheme. Let the wff $\forall uA \rightarrow \mathrm{S}_v^u[A]$ be any member of the fourth axiom scheme, and let $B = \mathrm{I}_t^s[\forall uA \rightarrow \mathrm{S}_v^u A]$. Then:

(i) $B = \forall u \mathrm{I}_t^s[A] \rightarrow \mathrm{I}_t^s[\mathrm{S}_u^v[A]] = \forall u \mathrm{I}_t^s[A] \rightarrow \mathrm{S}_u^v[\mathrm{I}_t^s[A]]$ if s, t, u, v are all different.

(ii) $B = \forall u \mathrm{I}_t^s[A] \rightarrow \mathrm{I}_t^s[\mathrm{S}_u^s[A]] = \forall u \mathrm{I}_t^s[A] \rightarrow \mathrm{S}_u^t[\mathrm{I}_t^s[A]]$ if $v = s$ and s, t, and u are all different. Note that s is not bound in A and u is free in A.

(iii) $B = \forall s \mathrm{I}_t^s[A] \rightarrow \mathrm{I}_t^s[\mathrm{S}_t^v[A]] = \forall s \mathrm{I}_t^s[A] \rightarrow \mathrm{S}_s^v[\mathrm{I}_t^s[A]]$ if $u = t$ and s, t, and v are all different.

(iv) $B = \forall s \mathrm{I}_t^s[A] \rightarrow \mathrm{I}_t^s[\mathrm{S}_t^v[A]] = \forall s \mathrm{I}_t^s[A] \rightarrow \mathrm{S}_s^t[\mathrm{I}_t^s[A]]$ if $v = s$ and $u = t$.

This proves that $\mathrm{I}_t^s[A_k]$ is a member of an axiom scheme if A_k is a member of an axiom scheme. We conclude that A_k is *not* a member of an axiom scheme.

Case 2. A_k is preceded in the given proof by A_i and $A_i \rightarrow A_k$. Then $\mathrm{I}_t^s[A_k]$ is preceded in the sequence $\mathrm{I}_t^s[A_1], \ldots, \mathrm{I}_t^s[A_k]$ by $\mathrm{I}_t^s[A_i]$ and $\mathrm{I}_t^s[A_i] \rightarrow \mathrm{I}_t^s[A_k]$. This is impossible; we conclude that A_k is not preceded in the given proof by wffs of the form A_i and $A_i \rightarrow A_k$.

Case 3. A_k has the form $B \rightarrow \forall uC$, and A_k is preceded in the given proof by $B \rightarrow C$ (so u is free in C and does not occur in B). Now $\mathrm{I}_t^s[A_k] = \mathrm{I}_t^s[B] \rightarrow \mathrm{I}_t^s[\forall uC]$. Essentially there are just two cases.

(i) $u \neq t$ and $u \neq s$. Here $\mathrm{I}_t^s[A_k] = \mathrm{I}_t^s[B] \rightarrow \forall u \mathrm{I}_t^s[C]$. But $\mathrm{I}_t^s[B \rightarrow C] = \mathrm{I}_t^s[B] \rightarrow \mathrm{I}_t^s[C]$ (again, u is free in $\mathrm{I}_t^s[C]$ and does not occur in $\mathrm{I}_t^s[B]$). This is impossible.

(ii) $u = s$. Here $\mathrm{I}_t^s[A_k] = \mathrm{I}_t^s[B] \rightarrow \forall t \mathrm{I}_t^s[C]$. Again, $\mathrm{I}_t^s[A_k]$ is preceded in the sequence of I_t^s-images by $\mathrm{I}_t^s[B] \rightarrow \mathrm{I}_t^s[C]$. This is impossible. We conclude that $\mathrm{I}_t^s[A_1], \ldots, \mathrm{I}_t^s[A_k]$ is a proof. This establishes Theorem 1.

Corollary 1. $\vdash \mathrm{I}_t^s[A]$ provided that $\vdash A$ and s and t are pure individuals.

Corollary 2. If $\vdash A$ and if t is a pure individual that is not in A, then there is a proof of A in which t does not occur.

Dem. Let π be a proof of A. Since only a finite number of symbols occur in π, there is a pure individual, say s, that is not in π. Apply I_t^s to each term of π. The resulting sequence of wffs is a proof, by Theorem 1; this sequence is a proof of $\mathrm{I}_t^s[A]$, which is A since neither s nor t occurs in A. By construction, t does not occur in this proof of A.

We mention that the substitution transforms S_t^s do not possess the property of Theorem 1; i.e., applying S_t^s to each term of a proof does not necessarily yield a proof. Nevertheless, applying S_t^s to a provable wff yields a provable wff. We shall prove this later (see Theorem 2, Section 10.2). To illustrate our point, consider the proof

$$\forall xFx \rightarrow Fy, \quad \forall xFx \rightarrow \forall yFy$$

Applying S_x^y to each term of this proof, we obtain the sequence

$$\forall xFx \rightarrow Fy, \quad \forall xFx \rightarrow \forall xFx$$

This is a sequence of provable wffs, but it is not a proof.

Exercises

1. Find a member of an axiom scheme, say A, such that $\mathrm{S}_x^y[A]$ is *not* in an axiom scheme.
2. Find a wff, say A, such that x and y are both bound in A, $\mathrm{S}_x^y[A] \neq A$, and $\mathrm{S}_x^y[A] \neq \mathrm{I}_x^y[A]$.
3. Show that $\vdash \mathrm{S}_t^s[A]$ given that $\vdash A$, t is free in A, and s is not free in A.
4. Show that $\vdash \mathrm{S}_t^s[A]$, given that $\vdash A$, and both s and t are free in A.
5. Show that $\vdash \mathrm{S}_t^s[A]$, given that $\vdash A$ and t is not in A.
6. Show that $\vdash \mathrm{S}_t^s[A]$, given that $\vdash A$, t is bound in A, and s is not bound in A.

10

Substitution Theorems

10.1. Subwffs, Components, and Wff-Builders

As for the propositional calculus, we now introduce the notions of *subwff*, *component*, and *wff-builder*.

Definition of Subwff. We say that B is a *subwff* of a wff A provided that B is a wff and there are expressions ϕ and θ such that $A = \phi B \theta$.

So, a subwff of a wff A is a block of symbols contained in A which is itself a wff. For example, each of $(\exists y(Gyy))$, $(\forall y(\neg(Gyy)))$, $(\neg(Gyy))$, (Gyy) is a subwff of $(\exists y(Gyy))$. We point out that "$(\exists y(Gyy))$" is an abbreviation for $(\neg(\forall y(\neg(Gyy))))$. Note that (Gy) is *not* a subwff of $(\exists y(Gyy))$ even if (Gy) is a wff.

Definition of Component. Each atomic wff, say A, has exactly one component, namely A itself; the components of $\neg B$ are $\neg B$ and each component of B; the components of $C \vee D$ are $C \vee D$, each component of C, and each component of D; the components of $\forall tE$ are $\forall tE$ and each component of E.

For example, the components of $(\exists y(Gyy))$ are the following wffs: $\exists yGyy$, $\forall y(\neg Gyy)$, $\neg Gyy$, Gyy.

We now show that our definition of *component* is proper.

Lemma 1. The definition of *component* associates a unique set of wffs with each wff.

Dem. Apply the Construction Theorem for Wffs.

We want to prove that our notions of *subwff* and *component* are the same; i.e., for any wffs A and B, B is a subwff of A iff B is a component of A. We shall do this in two steps.

Lemma 2. Each component of a wff A is a subwff of A.

Dem. We continue the argument of the proof of Lemma 2, page 50.

4. Let $E \in S$ and let t be free in E; we shall show that $\forall tE \in S$. The components of $\forall tE$ are $\forall tE$ and each component of E. By definition, $\forall tE$ is a subwff of $\forall tE$; by assumption, each component of E is a subwff of E. But each subwff of E is a subwff of $\forall tE$; thus, each component of $\forall tE$ is a subwff of $\forall tE$. Thus $\forall tE \in S$.

This completes the proof of Lemma 2.

Lemma 3. Each subwff of a wff A is a component of A.

Dem. We continue the argument given in the proof of Lemma 3, page 50.

4. Let $E \in S$ and let t be free in E. As in part 1 of the proof of Lemma 3, page 50, by considering mates of parentheses we can prove that $\forall tE \in S$.

This completes the proof of Lemma 3.

Corollary 1. Let A and B be any wff; then B is a component of A iff B is a subwff of A.

We turn our attention from a subwff of a wff $\phi B\theta$ to the accompanying expressions ϕ and θ.

Definition of Wff-Builder. We say that $[\phi, \theta]$ is a *wff-builder* if there is a wff A such that $\phi A\theta$ is a wff.

For example, $[(\forall y, \ )]$ is a wff-builder since $(\forall y(Fy))$ is a wff.

Here, there is no analog to Lemma 4, page 51; we point out that Lemma 4 is false for $\prod$. Moreover, the contribution of Lemma 4 to the proof of the Substitution Theorem for Wffs (page 52) is built into the present form of the Substitution Theorem for Wffs (see page 175) as an assumption of the theorem.

Exercises

1. List the subwffs of $\forall xFx \rightarrow Fy$.
2. List the subwffs of $\exists x(Fx \land \neg Gxy)$.
3. List the subwffs of $\forall xGxx \leftrightarrow \forall yGyy$.
4. List the components of $\forall xFx \rightarrow Fy$.
5. List the components of $\exists x(Fx \land \neg Gxy)$.
6. List the components of $\forall xGxx \leftrightarrow \forall yGyy$.
7. Show that $[(\neg,\ )]$ is a wff-builder.
8. Show that $[((Fx) \lor,\ )]$ is a wff-builder.
9. Show that $[((Fx) \lor (\neg,\ ))]$ is a wff-builder.
10. Is $[(,\ )]$ a wff-builder?

10.2. Substitution Theorem for Wffs

Here is the important Substitution Theorem for Wffs. The present version is slightly more complicated than in the case of the propositional calculus because of a possible conflict between free and bound individuals.

Substitution Theorem for Wffs. Let $A \equiv B$ and let $[\phi, \theta]$ be a wff-builder such that both $\phi A\theta$ and $\phi B\theta$ are wffs. Then $\phi A\theta \equiv \phi B\theta$.

A more convenient formulation of this theorem is obtained by adopting the following notation. Let C be a wff; by "C'" we denote any wff obtained from C by replacing a subwff of C by an equivalent wff. Notice here that C' must be a wff; i.e., we rule out a substitution when the resulting expression is not a wff.

We now formulate our substitution theorem in terms of the many-valued operation $'$.

Substitution Theorem for Wffs. For each wff C and for each corresponding wff C', $C \equiv C'$.

Dem. Let $S = \{C \in \mathsf{W} \mid C \equiv C' \text{ for each } C'\}$; we shall apply the Fundamental Theorem about Wffs to prove that $S = \mathsf{W}$. For the first three steps of our argument, see page 52ff. Here is the final step.

4. Let $E \in S$ and let t be free in E. We shall show that $\forall tE \in S$.

Let $C = \forall tE$, let A be any subwff of C, and let $A \equiv B$, where the corresponding C' is a wff. There is no problem in showing that $C \equiv C'$ in case $A = C$. Accordingly, take A to be a subwff of E; so there are expressions ϕ and θ such that $E = \phi A\theta$. Since C' is a wff, we see that $\phi B\theta$ is a wff; thus $\phi A\theta \equiv \phi B\theta$ by assumption. Notice that t is free in $\phi B\theta$ since $C' = \forall t\phi B\theta$ is a wff. By Corollary 1, page 167,

$$\forall t\phi A\theta \equiv \forall t\phi B\theta$$

i.e., $C \equiv C'$. This proves that $\forall tE \in S$. We conclude that $S = \mathsf{W}$. This completes our proof of the Substitution Theorem for Wffs.

Our Substitution Theorem for Wffs has the following useful corollary.

Substitution Theorem for Provable Wffs. $\vdash \phi B\theta$ provided that:

1. $\vdash \phi A\theta$.
2. $A \equiv B$.
3. $\phi B\theta$ is a wff.

We now apply these results.

Lemma 1. Let $[\phi, \theta]$ be a wff-builder and let $\phi\forall tA\theta$ and $\phi\forall s\mathsf{S}_t^s[A]\theta$ be wffs, where s is a pure individual that does not occur in A. Then

$$\phi\forall tA\theta \equiv \phi\forall s\mathsf{S}_t^s[A]\theta \tag{1}$$

$$\vdash\phi\forall tA\theta \text{ iff } \vdash\phi\forall s\mathsf{S}_t^s[A]\theta \tag{2}$$

Dem. By Lemma 2, Section 9.4, $\forall tA \equiv \forall s\mathsf{S}_t^s[A]$.

Up to equivalence, Lemma 1 allows us to replace "$\forall s\mathsf{S}_t^s$" by "$\forall t$," provided that s is not in the scope of the quantifier involved.

Next, we want to extend Corollary 1, page 163, as far as possible; i.e., we shall show that $\forall tA \equiv \forall s\mathsf{S}_t^s[A]$ whenever $\forall tA \rightarrow \forall s\mathsf{S}_t^s[A]$ is a wff.

Lemma 2. $\vdash \forall tA \rightarrow \forall s\mathsf{S}_t^s[A]$ if this is a wff.

Dem. There is no problem if $s = t$; it is clear also that s is not free in A (if s is free in A, then $\forall tA \rightarrow \forall s\mathsf{S}_t^s[A]$ is not a wff). The case in which s does not occur in A has already been treated (see Corollary 1, page 163). It remains only to consider the case in which s is bound in A. Consider the provable wff

$$\vdash \forall tA \rightarrow \forall tA \tag{3}$$

We shall operate on the RHS of (3) in three steps, which will yield the required provable wff.

Step 1. Substitute u for t throughout the RHS of (3), where u is a pure individual that does not occur in A. By Lemma 1,

$$\vdash \forall tA \rightarrow \forall u\mathrm{S}_t^u[A] \tag{4}$$

Step 2. Substitute t for s throughout the RHS of (4). This can be achieved by replacing each subwff of the RHS of (4) that possesses the form "$\forall sC$" by $\forall t\mathrm{S}_s^t[C]$. Since t does not occur in $\forall u\mathrm{S}_t^u[A]$, Lemma 1 guarantees that the result is a provable wff. This step yields

$$\vdash \forall tA \rightarrow \forall u\mathrm{I}_s^t[\mathrm{S}_t^u[A]] \tag{5}$$

Step 3. Substitute s for u throughout the RHS of (5). By construction, s does not occur in $\mathrm{S}_s^t[\mathrm{S}_t^u[A]]$. Thus, by Lemma 1,

$$\vdash \forall tA \rightarrow \forall s\mathrm{S}_u^s[\mathrm{I}_s^t[\mathrm{S}_t^u[A]]] \tag{6}$$

The effect of the three substitutions is to interchange s and t throughout the RHS of (3); i.e., the RHS of (6) is $\forall s\mathrm{S}_t^s[A]$; therefore

$$\vdash \forall tA \rightarrow \forall s\mathrm{S}_t^s[A] \tag{7}$$

Here is our extension of Corollary 1, page 163.

Theorem 1. $\forall tA \equiv \forall s\mathrm{S}_t^s[A]$ provided that $\forall tA \rightarrow \forall s\mathrm{S}_t^s[A]$ is a wff.

Dem. There is nothing to prove if $s = t$; also, if $s \neq t$, s is not free in A since $\forall tA \rightarrow \forall s\mathrm{S}_t^s[A]$ is a wff. We must show that $\vdash \forall s\mathrm{S}_t^s[A] \rightarrow \forall tA$ provided that t is free in A and s is bound in A (there is no problem if s does not occur in A). By Lemma 2,

$$\vdash \forall s\mathrm{S}_t^s[A] \rightarrow \forall t\mathrm{S}_t^s[\mathrm{S}_t^s[A]]$$

since this is a wff; i.e., $\vdash \forall s\mathrm{S}_t^s[A] \rightarrow \forall tA$. We conclude that $\forall tA \equiv \forall s\mathrm{S}_t^s[A]$ provided that $\forall tA \rightarrow \forall s\mathrm{S}_t^s[A]$ is a wff.

This result permits us to ease the restrictions on Lemma 1.

Lemma 3. Let $[\phi, \theta]$ be a wff-builder and let $\phi\forall tA\theta$ and $\phi\forall s\mathsf{S}_t^s[A]\theta$ be wffs, where s is not free in A. Then (1) and (2) hold:

$$\phi\forall tA\theta \equiv \phi\forall s\mathsf{S}_t^s[A]\theta \tag{1}$$

$$\vdash\phi\forall tA\theta \text{ iff } \vdash\phi\forall s\mathsf{S}_t^s[A]\theta \tag{2}$$

Dem. Since s is not free in A, $\forall tA \to \forall s\mathsf{S}_t^s[A]$ is a wff. We obtain (1) from the Substitution Theorem for Wffs and Theorem 1; we obtain (2) from the Substitution Theorem for Provable Wffs and Theorem 1.

Here are some examples that illustrate the power of our substitution theorems.

Example 1. Show that $\forall s\forall tA \equiv \forall t\forall sA$ provided that $\forall s\forall tA \to \forall t\forall sA$ is a wff.

Solution. Let A be any wff in which s and t are free, $s \neq t$. Let u and v be pure individuals that do not occur in A, $u \neq v$. By AS 4

$$\vdash \forall s\forall tA \to \forall t\mathsf{S}_s^u[A] \qquad \text{and} \qquad \vdash \forall t\mathsf{S}_s^u[A] \to \mathsf{S}_t^v[\mathsf{S}_s^u[A]]$$

Thus, $\vdash \forall s\forall tA \to \mathsf{S}_t^v[[\mathsf{S}_s^u A]]$; by Lemma 15, page 166, we obtain first

$$\vdash \forall s\forall tA \to \forall u\mathsf{S}_s^u[\mathsf{S}_t^v[A]]$$

and then

$$\vdash \forall s\forall tA \to \forall v\forall u\mathsf{S}_t^v[\mathsf{S}_s^u[A]] \tag{8}$$

Since u, v, and t are all different, the RHS of (8) is $\forall v\mathsf{S}_t^v[\forall u\mathsf{S}_s^u[A]]$; thus

$$\vdash \forall s\forall tA \to \forall v\mathsf{S}_t^v[\forall u\mathsf{S}_s^u[A]] \tag{9}$$

We now simplify the RHS of (9). By Lemma 3, "$\forall v\mathsf{S}_t^v$" simplifies to "$\forall t$"; i.e.,

$$\vdash \forall s\forall tA \to \forall t\forall u\mathsf{S}_s^u[A] \tag{10}$$

Again, by Lemma 3, "$\forall u\mathsf{S}_s^u$" simplifies to "$\forall s$"; i.e.,

$$\vdash \forall s\forall tA \to \forall t\forall sA \tag{11}$$

Of course, this also proves that $\vdash \forall t\forall sA \to \forall s\forall tA$; we conclude that $\forall s\forall tA \equiv \forall t\forall sA$.

Example 2. Show that $\vdash \forall tA \wedge \forall tB \to \forall t(A \wedge B)$ if this is a wff.

Solution. Let s be a pure individual that does not occur in $A \wedge B$. By AS 4,

$$\vdash \forall tA \to \mathsf{S}_t^s[A] \qquad \text{and} \qquad \vdash \forall tB \to \mathsf{S}_t^s[B]$$

By Lemma 21, page 167,

$$\vdash \forall tA \wedge \forall tB \rightarrow \mathsf{S}_t^s[A] \wedge \mathsf{S}_t^s[B]$$

By Lemma 15, page 166,

$$\vdash \forall tA \wedge \forall tB \rightarrow \forall s\mathsf{S}_t^s[A \wedge B]$$

By Lemma 3,

$$\vdash \forall tA \wedge \forall tB \rightarrow \forall t(A \wedge B)$$

Example 3. Show that $\forall tA \wedge \forall tB \equiv \forall t(A \wedge B)$ provided that t is free in both A and B and provided that $A \wedge B$ is a wff.

Solution. Now, $\vdash A \wedge B \rightarrow A$ and $\vdash A \wedge B \rightarrow B$; by Lemma 17, page 166,

$$\vdash \forall t(A \wedge B) \rightarrow \forall tA \qquad \text{and} \qquad \vdash \forall t(A \wedge B) \rightarrow \forall tB$$

Thus, by Lemma 21, page 167,

$$\vdash \forall t(A \wedge B) \wedge \forall t(A \wedge B) \rightarrow \forall tA \wedge \forall tB$$

So, by the Substitution Theorem for Provable Wffs,

$$\vdash \forall t(A \wedge B) \rightarrow \forall tA \wedge \forall tB$$

We conclude from Example 2 that $\forall tA \wedge \forall tB \equiv \forall t(A \wedge B)$.

In Section 9.6 (see Corollary 1, page 171) we proved that $\vdash \mathbf{I}_t^s[A]$ if $\vdash A$, provided that s and t are pure individuals. We commented at the time that the S_t^s transforms also preserve provability. With the help of Lemma 3, we can now prove this statement.

Theorem 2. $\vdash \mathsf{S}_t^s[A]$ provided that $\vdash A$, s and t are individuals, and t is pure.

Dem. If t is not in A, then

$$\mathsf{S}_t^s[A] = \begin{cases} \mathbf{I}_t^s[A] & \text{if } s \text{ is bound in } A \\ A & \text{otherwise} \end{cases}$$

By assumption, $\vdash A$; by Corollary 1, page 171, $\vdash \mathbf{I}_t^s[A]$. We now assume that t is in A. There are four cases as follows.

Case 1. s is not bound in A and t is free in A. By Lemma 16, page 166, $\vdash \forall tA$; thus, by Lemma 18, page 167, $\vdash \mathsf{S}_t^s[A]$.

Case 2. s is not bound and t is bound. Here $\mathbf{S}^s_t[A] = A$; but $\vdash A$.

Case 3. s is bound and t is free. Here $\mathbf{S}^s_t[A] = \mathbf{I}^s_t[A]$; but $\vdash \mathbf{I}^s_t[A]$ (note that s is pure, since only pure individuals are quantified in a wff).

Case 4. s and t are both bound in A. The impact of $\mathbf{S}^s_t$ on A is as follows.

(a) Each s that is not within the scope of "$\forall t$" is replaced by t.
(b) Each s within the scope of "$\forall t$" is left as is.
(c) Each t within the scope of "$\forall s$" is replaced by s.
(d) Each t that is not within the scope of "$\forall s$" is left as is.

This can be achieved by transforming each subwff of A with the form "$\forall sC$" that is not within the scope of "$\forall t$," as follows.

(e) Replace "$\forall sC$" by "$\forall t\mathbf{S}^t_s[C]$."

The impact of $\mathbf{S}^t_s$ on C is to interchange s and t throughout C. Thus (e) achieves (a) and (c) directly; of course, (b) and (d) require no action. We must show that a provable wff is obtained by carrying out (e). By Lemma 3, $\vdash\phi\forall sC\theta$ iff $\vdash\phi\forall t\mathbf{S}^t_s[C]\theta$; therefore, each application of (e) yields a provable wff. We conclude that $\vdash\mathbf{S}^s_t[A]$. This completes our proof of Theorem 2.

We present some more examples that illustrate our methods.

Lemma 4. $\vdash \forall t(A \to B) \wedge \forall tA \to \forall tB$ if this is a wff.

Dem. Let s be a pure individual that is not in $A \to B$. Then

	$\vdash \forall s\mathbf{S}^s_t[A \to B] \dot{\to} A \to B$	(AS 4)
	$\vdash A \dot{\to} \forall s\mathbf{S}^s_t[A \to B] \to B$	(Lemma 7, page 164)
Also	$\vdash \forall s\mathbf{S}^s_t[A] \to A$	(AS 4)
Thus	$\vdash \forall s\mathbf{S}^s_t[A] \dot{\to} \forall s\mathbf{S}^s_t[A \to B] \to B$	($\to$ is transitive)
	$\vdash \forall s\mathbf{S}^s_t[A \to B] \wedge \forall s\mathbf{S}^s_t[A] \to B$	(Lemma 7, page 164)
	$\vdash \forall s\mathbf{S}^s_t[A \to B] \wedge \forall s\mathbf{S}^s_t[A] \to \forall tB$	(I$\forall$R)
	$\vdash \forall t(A \to B) \wedge \forall tA \to \forall tB$	[Theorem 1, page 177, and Substitution Theorem for Provable Wffs (applied twice)]

Corollary 1. $\vdash \forall t(A \to B) \dot{\to} \forall tA \to \forall tB$ if this is a wff.

Dem. Apply Lemma 7, page 164.

Lemma 5. $\vdash \forall t(A \rightarrow B) \wedge \exists tA \rightarrow \exists tB$ if this is a wff.

Dem.

$\vdash A \rightarrow B \mathrel{\dot{\rightarrow}} \neg B \rightarrow \neg A$	(Lemma 24, page 169)
$\vdash \forall t(A \rightarrow B) \rightarrow \forall t(\neg B \rightarrow \neg B)$	(Lemma 17, page 166)
$\vdash \forall t(\neg B \rightarrow \neg A) \mathrel{\dot{\rightarrow}} \forall t(\neg B) \rightarrow \forall t(\neg A)$	(Corollary 1)
$\vdash \forall t(A \rightarrow B) \mathrel{\dot{\rightarrow}} \forall t(\neg B) \rightarrow \forall t(\neg B)$	($\rightarrow$ is transitive)
$\vdash \forall t(A \rightarrow B) \mathrel{\dot{\rightarrow}} \neg\forall t(\neg A) \rightarrow \neg\forall t(\neg B)$	(Lemma 25, page 169)
$\vdash \forall t(A \rightarrow B) \wedge \exists tA \rightarrow \exists tB$	(Lemma 7, page 164)

Corollary 2. $\vdash \forall t(A \rightarrow B) \mathrel{\dot{\rightarrow}} \exists tA \rightarrow \exists tB$ if this is a wff.

Exercises

1. Prove that $\exists tA \equiv \exists s\mathbf{S}^s_t[A]$, provided that $\exists tA \rightarrow \exists s\mathbf{S}^s_t[A]$ is a wff.
2. Let $[\phi, \theta]$ be a wff-builder, and let $\phi\exists tA\theta$ and $\phi\exists s\mathbf{S}^s_t[A]\theta$ be wffs where s is not free in A. Prove that:
   (a) $\phi\exists tA\theta \equiv \phi\exists s\mathbf{S}^s_t[A]\theta$.
   (b) $\vdash \phi\exists tA\theta$ iff $\vdash \phi\exists t\mathbf{S}^s_t[A]\theta$.
3. Prove that $\vdash \forall t(A \wedge B) \rightarrow A \wedge \forall tB$, provided this is a wff.
4. Prove that $\vdash A \wedge \forall tB \rightarrow \forall t(A \wedge B)$, provided this is a wff.
5. Prove that $\forall t(A \wedge B) \equiv A \wedge \forall tB$, provided $A \wedge \forall tB$ is a wff.
6. Prove that $\vdash A \vee \forall tB \rightarrow \forall t(A \vee B)$, provided this is a wff.
7. Prove that $\forall t(A \vee B) \equiv A \vee \forall tB$, provided $A \vee \forall tB$ is a wff.
8. Prove that $\forall t(A \rightarrow B) \equiv A \rightarrow \forall tB$, provided $A \rightarrow \forall tB$ is a wff.
9. (a) Prove Corollary 1.
   (b) Prove Corollary 2.
10. Prove that $\vdash \forall t(A \leftrightarrow B) \rightarrow (\forall tA \leftrightarrow \forall tB)$, provided this is a wff.
11. Prove that $\vdash \forall t(A \rightarrow B) \rightarrow (\exists tA \rightarrow \exists tB)$, provided this is a wff.
12. Prove that $\vdash \exists s\forall tA \rightarrow \forall t\exists sA$, provided this is a wff.
13. Prove that $\neg\forall tA \equiv \exists t(\neg A)$, provided $\forall tA$ is a wff.
14. Prove that $\neg\exists tA \equiv \forall t(\neg A)$, provided $\forall tA$ is a wff.

15. Let $\forall t \forall s A$ be a wff; prove that $\forall t \forall s A \equiv \forall s \forall t \mathbf{I}_t^s[A]$. *Hint:* Apply Theorem 1 to $\forall t \forall s A$.

16. Is $A \equiv \mathbf{I}_t^s[A]$ for each wff A, where $s \neq t$?

17. Prove that $\forall t A \equiv \forall s \mathbf{I}_t^s[A]$ provided that $\forall t A \rightarrow \forall s \mathbf{I}_t^s[A]$ is a wff.

18. Prove that $\vdash \forall t(A \rightarrow B) \wedge \exists t(A \wedge C) \rightarrow \exists t(B \wedge C)$ provided that this is a wff. *Hint:* Use Lemma 26, page 169.

11

Duality

11.1. Normal Form

In this section we shall present an algorithm that yields a wff equivalent to the negation of a given wff, say A. The first step is to put A into a certain standard form, which we call *normal* form. This involves expressing A in terms of $\neg$, $\vee$, $\wedge$, $\forall$, and $\exists$; moreover, we require that each instance of $\neg$ is prefixed to an atomic wff. All this is subject to the requirement that the resulting wff be equivalent to A.

We begin the task of putting A into normal form by expressing all occurrences of $\rightarrow$ and $\leftrightarrow$ in terms of $\neg$, $\vee$, and $\wedge$. Next, we must *recognize* each $\wedge$-like subwff of A and each $\exists$-like subwff of A. A wff is said to be $\wedge$*-like* if it has the form "$\neg(C \vee D)$," where C and D are wffs; a wff is $\exists$*-like* if it has the form "$\neg(\forall tE)$." To *recognize* an $\wedge$-like subwff of A, say $\neg(C \vee D)$, we replace this subwff by $\neg C \wedge \neg D$, an equivalent wff. Similarly, to *recognize* an $\exists$-like subwff of A, say $\neg(\forall tE)$, we replace this subwff by $\exists t(\neg E)$, an equivalent wff. We call these operations *recognizing and* and *recognizing there exists* in the given wff.

Recall the following facts:

(a) $\neg(\neg D) \equiv D$ for any wff D.
(b) $\neg(D \vee E) \equiv \neg D \wedge \neg E$ for any wffs D and E.
(c) $\neg(D \wedge E) \equiv \neg D \vee \neg E$ for any wffs D and E.
(d) $\neg(\exists tD) \equiv \exists t(\neg D)$ for any wff D, where t is free in D.
(e) $\neg(\exists tD) \equiv \forall t(\neg D)$ for any wff D, where t is free in D.

These equivalences and the Substitution Theorem for Wffs allow us to obtain a *name* of a wff equivalent, to A, for which each instance of $\neg$ is prefixed to an atomic wff.

We now illustrate this procedure for putting a wff into normal form.

Example 1. Put $E = \neg(\forall xFx) \rightarrow \neg(\forall xFx)$ into normal form.

Solution. First, we eliminate $\rightarrow$:

$$E = \neg(\neg(\forall xFx)) \vee \neg(\forall xFx)$$

Next, we recognize each $\wedge$-like and each $\exists$-like subwff of the preceding wff:

$$E \equiv \neg(\exists x(\neg Fx)) \vee \exists x(\neg Fx)$$

Finally, we apply (a)–(e) repeatedly, to ensure that each instance of $\neg$ is prefixed to an atomic wff:

$$E \equiv \forall x(\neg(\neg(Fx))) \vee \exists x(\neg Fx) \equiv \forall xFx \vee \exists x(\neg Fx)$$

We have put E into normal form.

Here is our algorithm for obtaining a wff equivalent to the negation of a given wff.

Algorithm for Negation. Let A be any wff; then $\neg A \equiv C$, where C is obtained from A in three steps, as follows.

Step 1. Put A into normal form, say B.

Step 2. Interchange $\vee$ and $\wedge$, and interchange $\forall$ and $\exists$, throughout B.

Step 3. If an atomic wff is prefixed by $\neg$, delete this $\neg$; if an atomic wff is not prefixed by $\neg$, then insert $\neg$.

We shall justify this algorithm in Section 11.3 (see Lemma 3, page 192). Meanwhile, we illustrate our algorithm.

Example 2. Find a wff equivalent to $\neg E$, where E is the wff of Example 1.

Solution. From Example 1, Step 1 of the algorithm yields B, where

$$B = \forall xFx \vee \exists x(\neg Fx)$$

Step 2, applied to B, yields

$$\exists xFx \wedge \forall x(\neg Fx) \tag{1}$$

Step 3, applied to (1), yields $\exists x(\neg Fx) \wedge \forall xFx$. By our algorithm,

$$\neg(\neg\forall xFx \rightarrow \neg\forall xFx) \equiv \exists x(\neg Fx) \wedge \forall xFx$$

Exercises

Put each of the following wffs into normal form.

1. $\exists yGyy \leftrightarrow \forall xFx$.
2. $\exists yGyy \rightarrow \forall x(Fx \wedge Gxx)$.
3. $\neg\exists y(Fy \vee Gxx) \leftrightarrow Fx \wedge \neg Gxx$.
4. $\forall z(\neg(Fz \vee Gzz)) \rightarrow \neg(Fy \vee Gyy)$.

Use the Algorithm for Negation to obtain a wff equivalent to $\neg E$, where E is:

5. The wff of Exercise 1.
6. The wff of Exercise 2.
7. The wff of Exercise 3.
8. The wff of Exercise 4.
9. What is the normal form of an atomic wff $(P\alpha)$?
10. If A is atomic, what is the normal form of $\neg A$?
11. Let C^* and D^* be the normal forms of wffs C and D, respectively; what is the normal form of $C \vee D$ (given that $C \vee D$ is a wff)?
12. Let E^* be the normal form of a wff E in which t is free; what is the normal form of $\forall tE$?
13. Let B^* be the normal form of a wff B; what is the normal form of $\neg(\neg B)$?
14. Let A and B be the normal form of $\neg C$ and $\neg D$, respectively; what is the normal form of $\neg(C \vee D)$?
15. Let D be the normal form of a wff $\neg E$ in which t is free; what is the normal form of $\neg(\forall tE)$?

11.2. Normal Transforms

We shall now formalize the notion of *normal form* introduced in Section 11.1. The idea is to associate with each wff, say A, a unique wff which is in normal form (in the sense of Section 11.1) and is equivalent to A. Notice that the wff associated with any wff A is obtained from A in a constructive manner by applying a specified algorithm, namely the definition of the syntactical transform N. Here is our definition of N. Throughout this section we shall rely on Theorem 1, page 144, to characterize specific syntactical transforms.

Definition of N:

1. $\mathsf{N}A = A$ for each atomic wff A.

$$2.\quad \mathsf{N}[\neg A] = \begin{cases} \neg A & \text{if } A \text{ is atomic} \\ \mathsf{N}B & \text{if } A = \neg B \\ \mathsf{N}[\neg C] \wedge \mathsf{N}[\neg D] & \text{if } A = C \vee D \\ \exists t\mathsf{N}[\neg E] & \text{if } A = \forall tE \end{cases}$$

3. $\mathsf{N}[C \vee D] = \mathsf{N}C \vee \mathsf{N}D$ for each wff $C \vee D$.
4. $\mathsf{N}[\forall tE] = \forall t\mathsf{N}E$ for each wff $\forall tE$.

For example,

$$\begin{aligned} &\mathsf{N}[\neg(\neg\forall xFx \vee \forall x\forall y(\neg Gxy))] \\ &\quad = \mathsf{N}[\neg(\neg\forall xFx)] \wedge \mathsf{N}[\neg\forall x\forall y(\neg Gxy)] && \text{(by 2)} \\ &\quad = \mathsf{N}[\forall xFx] \wedge \exists x\mathsf{N}[\neg\forall y(\neg Gxy)] && \text{[by 2 (twice)]} \\ &\quad = \forall x\mathsf{N}[Fx] \wedge \exists x\exists y\mathsf{N}[\neg(\neg Gxy)] && \text{(by 4 and 2)} \\ &\quad = \forall xFx \wedge \exists x\exists y\mathsf{N}[Gxy] && \text{(by 1 and 2)} \\ &\quad = \forall xFx \wedge \exists x\exists yGxy && \text{(by 1)} \end{aligned}$$

Our goal in this chapter is to establish the Principle of Duality (see page 193). To this purpose we shall need four syntactical transforms N, M, R, and D. We have already defined N; we now present definitions of M, R, and D.

Definition of M:

1. $\mathsf{M}A = \neg A$ for each atomic wff A.

$$2.\quad \mathsf{M}[\neg A] = \begin{cases} A & \text{if } A \text{ is atomic} \\ \mathsf{M}B & \text{if } A = \neg B \\ \mathsf{M}[\neg C] \vee \mathsf{M}[\neg D] & \text{if } A = C \vee D \\ \forall t\mathsf{M}[\neg E] & \text{if } A = \forall tE \end{cases}$$

3. $\mathsf{M}[C \vee D] = \mathsf{M}C \wedge \mathsf{M}D$ for each wff $C \vee D$.
4. $\mathsf{M}[\forall tE] = \exists t\mathsf{M}E$ for each wff $\forall tE$.

Definition of R:

1. $\mathsf{R}A = \neg A$ for each atomic wff A.

$$2.\quad \mathsf{R}[\neg A] = \begin{cases} A & \text{if } A \text{ is atomic} \\ \mathsf{R}B & \text{if } A = \neg B \\ \mathsf{R}[\neg C] \wedge \mathsf{R}[\neg D] & \text{if } A = C \vee D \\ \exists t\mathsf{R}[\neg E] & \text{if } A = \forall tE \end{cases}$$

3. $\mathsf{R}[C \vee D] = \mathsf{R}C \vee \mathsf{R}D$ for each wff $C \vee D$.
4. $\mathsf{R}[\forall tE] = \forall t\mathsf{R}E$ for each wff $\forall tE$.

Definition of D:

1. $\mathsf{D}A = A$ for each atomic wff A.
2. $$\mathsf{D}[\neg A] = \begin{cases} \neg A & \text{if } A \text{ is atomic} \\ \mathsf{D}B & \text{if } A = \neg B \\ \mathsf{D}[\neg C] \wedge \mathsf{D}[\neg D] & \text{if } A = C \vee D \\ \forall t\mathsf{D}[\neg E] & \text{if } A = \forall tE \end{cases}$$
3. $\mathsf{D}[C \vee D] = \mathsf{D}C \wedge \mathsf{D}D$ for each wff $C \vee D$.
4. $\mathsf{D}[\forall tE] = \exists t\mathsf{D}E$ for each wff $\forall tE$.

As we have mentioned, the syntactical transform N is designed to put any wff A into normal form in the sense of Section 11.1. Notice in particular that N has the following impact on A:

(a) N recognizes each $\wedge$-like subwff of A.
(b) N recognizes each $\exists$-like subwff of A.
(c) N eliminates $\neg\neg$ wherever this appears in A.

The rationale underlying the definitions of the transforms M, R, and D is easily explained. Now, M is designed to first transform a wff into normal form, then to interchange $\vee$ and $\wedge$ throughout, and interchange $\forall$ and $\exists$ throughout, and finally to attach $\neg$ to each atomic wff that is not prefixed by $\neg$, at the same time deleting $\neg$ from each atomic wff to which it is prefixed. This is achieved in four steps as follows:

(a) Write down the definition of N.
(b) Put $\wedge$ for the displayed $\vee$ in the RHS of 3; put $\vee$ for the displayed $\wedge$ in the RHS of the third line of 2.
(c) Put $\exists$ for the displayed $\forall$ in the RHS of 4; put $\forall$ for the displayed $\exists$ in the RHS of the fourth line of 2.
(d) Attach $\neg$ to the RHS of 1; delete $\neg$ from the RHS of the first line of 2.

Moving on to R, we point out that this transform is designed to first put a wff into normal form and next to reverse the impact of M on atomic wffs. Now, the operation of adjoining $\neg$ or deleting $\neg$ is reversed by repeating the operation. So our definition of R is built up in two steps:

(a) Write down the definition of N.
(b) Attach $\neg$ to the RHS of 1; delete $\neg$ from the RHS of the first line of 2.

Finally, we consider the transform D. This transform is designed to first put a wff into normal form, then to interchange $\vee$ and $\wedge$ throughout, and finally to interchange $\forall$ and $\exists$ throughout. Accordingly, our definition of D is built up in three steps:

(a) Write down the definition of N.
(b) Put $\wedge$ for the displayed $\vee$ in the RHS of 3; put $\vee$ for the displayed $\wedge$ in the RHS of the third line of 2.
(c) Put $\exists$ for the displayed $\forall$ in the RHS of 4; put $\forall$ for the displayed $\exists$ in the RHS of the fourth line of 2.

It is important to recognize that the transforms N, M, R, and D share a common pattern. Notice that for each of these transforms, say T, TA is either A or $\neg A$ if A is atomic, T$[C \vee D]$ is either T$C \vee$ TD or T$C \wedge$ TD, and T$[\forall tE]$ is either $\forall t$TE or $\exists t$TE. Moreover, the definition of T$[\neg A]$ can be reconstructed from the remaining three parts of the definition of T. For example, if A is atomic, then T$[\neg A] = A$ whenever T$A = \neg A$, whereas T$[\neg A] = \neg A$ whenever T$A = A$.

Table 1 displays the pattern that runs through our definitions of N, M, R, and D.

Table 1

	N	M	R	D
1. A atomic	A	$\neg A$	$\neg A$	A
2. (a) $\neg A$, A atomic	$\neg A$	A	A	$\neg A$
(b) $\neg\neg B$	NB	MB	RB	DB
(c) $\neg(C \vee D)$	N$[\neg C]$ $\wedge$ N$[\neg D]$	M$[\neg C]$ $\vee$ M$[\neg D]$	R$[\neg C]$ $\wedge$ R$[\neg D]$	D$[\neg C]$ $\vee$ D$[\neg D]$
(d) $\neg\forall tE$	$\exists t$N$[\neg E]$	$\forall t$M$[\neg E]$	$\exists t$R$[\neg E]$	$\forall t$D$[\neg E]$
3. $C \vee D$	N$C \vee$ ND	M$C \wedge$ MD	R$C \vee$ RD	D$C \wedge$ DD
4. $\forall tE$	$\forall t$NE	$\exists t$ME	$\forall t$RE	$\exists t$DE

In summary, each of the transforms N, M, R, and D is characterized by three parameters, which we denote by n, d, and Q; let T be any of these transforms; then T$A = nA$ for A atomic, T$[C \vee D] =$ T$C\, d\,$TD, and T$[\forall tE] = Qt$TE. So n is either $\neg$ or is blank, d is either $\vee$ or $\wedge$, and Q is either $\forall$ or $\exists$.

In the spirit of this observation, we now present the notion of a *normal* transform.

Definition. A syntactical transform T is said to be *normal* if it possesses parameters n, d, and Q such that:

1. $\mathsf{T}A = nA$ for each atomic wff A.

2. $$\mathsf{T}[\neg A] = \begin{cases} mA & \text{if } A \text{ is atomic} \\ \mathsf{T}B & \text{if } A = \neg B \\ \mathsf{T}[\neg C]\; c\; \mathsf{T}[\neg D] & \text{if } A = C \vee D \\ qt\mathsf{T}[\neg E] & \text{if } A = \forall tE \end{cases}$$

3. $\mathsf{T}[C \vee D] = \mathsf{T}C\; d\; \mathsf{T}D$ for each wff $C \vee D$.
4. $\mathsf{T}[\forall tE] = Qt\mathsf{T}E$ for each wff $\forall tE$.

Here n is $\neg$ or is blank, m is $\neg$ or is blank, and $m \neq n$; $\{d, c\} = \{\vee, \wedge\}$, and $\{Q, q\} = \{\forall, \exists\}$.

Clearly, there are exactly eight normal transforms; we are primarily interested in the normal transforms N, M, R, and D whose parameters are displayed in Table 2.

Table 2

T	N	M	R	D
n		$\neg$	$\neg$	
d	$\vee$	$\wedge$	$\vee$	$\wedge$
Q	$\forall$	$\exists$	$\forall$	$\exists$

The following lemmas are easy to verify; throughout, T is a normal transform with parameters n, d, and Q.

Lemma 1. For each wff A, $\mathsf{T}A$ and A involve the same bound individuals and the same free individuals.

Lemma 2. $\mathsf{T}[nA] = A$ for each atomic wff A.

Lemma 3. $\mathsf{T}[mA] = \neg A$ for each atomic wff A.

Lemma 4. $\mathsf{T}[A\; d\; B] = \mathsf{T}A \vee \mathsf{T}B$ for each wff $A\; d\; B$.

Lemma 5. $\mathsf{T}[A\; c\; B] = \mathsf{T}A \wedge \mathsf{T}B$ for each wff $A\; c\; B$.

Lemma 6. $\mathsf{T}[A \wedge B] = \mathsf{T}A\; c\; \mathsf{T}B$ for each wff $A \wedge B$.

Lemma 7. $\mathsf{T}[QtE] = \forall t\mathsf{T}E$ for each wff QtE.

Lemma 8. $\mathsf{T}[qtE] = \exists t\mathsf{T}E$ for each wff qtE.

Lemma 9. $\mathsf{T}[\exists tE] = qt\mathsf{T}E$ for each wff $\exists tE$.

We want to show that normal transforms form a group under composition. Composition of maps is associative; so it is only necessary to verify the following:

1. The product of any two normal transforms is a normal transform.
2. N is the identity.
3. Each normal transform has an inverse, namely itself.

This is achieved by establishing Theorem 1, which follows. Here "$\mathsf{T}_1\mathsf{T}_2$" denotes the syntactical transform that associates $\mathsf{T}_1[\mathsf{T}_2 A]$ with A, for each wff A. Throughout the following, "$n = \text{bl}$" means that n is blank.

Theorem 1. Let T_1 and T_2 be normal transforms with parameters n_1, d_1, Q_1 and n_2, d_2, Q_2, respectively. Then $\mathsf{T}_1\mathsf{T}_2$ is a normal transform and its parameters are n, d, and Q, where

$$n = \begin{cases} n_1 & \text{if } n_2 = \text{bl} \\ m_1 & \text{if } n_2 = \neg \end{cases} \qquad d = \begin{cases} d_1 & \text{if } d_2 = \vee \\ c_1 & \text{if } d_2 = \wedge \end{cases}$$

$$Q = \begin{cases} Q_1 & \text{if } Q_2 = \forall \\ q_1 & \text{if } Q_2 = \exists \end{cases}$$

Dem. Follow the pattern of the proof of Theorem 1, page 66. The third part of that proof will now involve a fourth case:

$$(\mathsf{T}_1\mathsf{T}_2)[\neg\forall tE] = \mathsf{T}_1[q_2 t\mathsf{T}_2[\neg E]] = \begin{cases} Q_1 t(\mathsf{T}_1\mathsf{T}_2)[\neg E] & \text{if } Q_2 = \exists \\ q_1 t(\mathsf{T}_1\mathsf{T}_2)[\neg E] & \text{if } Q_2 = \forall \end{cases}$$

Note our use of Lemma 9 in the latter case.

With the aid of Theorem 1 we obtain the following facts.

Lemma 10. $\mathsf{TT} = \mathsf{N}$ for each normal transform T.

Lemma 11. $\mathsf{TN} = \mathsf{T}$ for each normal transform T.

Thus, normal transforms form a group under composition; N is the group identity, and each group element is its own inverse. As we have shown on page 67, it follows from the latter statement that our group is Abelian. Thus, we have the following result.

Lemma 12. $\mathsf{T}_1\mathsf{T}_2 = \mathsf{T}_2\mathsf{T}_1$ for any normal transforms T_1 and T_2.

The transforms N, M, R, and D form a subgroup of the group of normal transforms. Indeed, from Theorem 1 we obtain the multiplication table—Table 3—for this subgroup. In particular, note that MR = D. Without using Table 3, we can compute DR as follows: DR = (MR)R = M(RR) = MN = M, so DR = M.

Table 3

	N	M	R	D
N	N	M	R	D
M	M	N	D	R
R	R	D	N	M
D	D	R	M	N

Exercises

1. Show that $\mathsf{N}[A \wedge B] = \mathsf{N}A \wedge \mathsf{N}B$ for any wff $A \wedge B$.
2. Show that $\mathsf{N}[A \rightarrow B] = \mathsf{N}[\neg A] \vee \mathsf{N}B$ for any wff $A \rightarrow B$.
3. Compute $\mathsf{N}[\forall yFy \rightarrow \exists xGxx]$.
4. Compute $\mathsf{M}[\exists xFx \leftrightarrow \forall yGyy]$.
5. Compute $\mathsf{R}[\forall x(\neg Fx) \rightarrow \forall xFx]$.
6. Compute $\mathsf{D}[\exists x(Fx \vee Gxx) \rightarrow \forall y(Fy \wedge Gyy)]$.
7. Prove Lemma 1.
8. Prove Lemma 2.
9. Prove Lemma 3.
10. Prove Lemma 4.
11. Prove Lemma 5.
12. Prove Lemma 6.
13. Prove Lemma 7.
14. Prove Lemma 8.
15. Prove Lemma 9.
16. Complete the proof of Theorem 1.
17. Prove Lemma 10.
18. Prove Lemma 11.

11.3. Duality

We intend to prove that if $\mathsf{T} \in \{\mathsf{N}, \mathsf{M}, \mathsf{R}, \mathsf{D}\}$, then $\mathsf{T}A \equiv \mathsf{T}B$ iff $A \equiv B$. First, we require the following fact.

Lemma 1. Let T be any normal transform; then $\mathsf{T}[\neg A] \equiv \neg\mathsf{T}A$ for each wff A.

Dem. Let T be a normal transform with parameters n, d, and Q. The idea is to use the Fundamental Theorem about Wffs to prove that $\mathsf{T}[\neg A] \equiv \neg\mathsf{T}A$ for each wff A. To this purpose, let

$$S = \{A \in \mathsf{W} \mid \mathsf{T}[\neg A] \equiv \neg\mathsf{T}A\}.$$

Considering our proof of Lemma 1, page 69, we see that S meets the first three requirements of the Fundamental Theorem about Wffs. We now show that S also satisfies the fourth requirement.

4. Given that $E \in S$ and that t is free in E, we must show that $\forall tE \in S$. Now,

$$\mathsf{T}[\neg\forall tE] = qt\mathsf{T}[\neg E] \equiv qt(\neg\mathsf{T}E)$$

since $E \in S$, and

$$\neg\mathsf{T}[\forall tE] = \neg Qt\mathsf{T}E \equiv qt(\neg\mathsf{T}E)$$

Thus $\mathsf{T}[\neg\forall tE] \equiv \neg\mathsf{T}[\forall tE]$; so $\forall tE \in S$. We conclude that $S = \mathsf{W}$; this establishes Lemma 1.

Next, we exhibit some facts about N and M that will help us verify the comment that precedes Lemma 1.

Lemma 2. $\mathsf{N}A \equiv A$ for each wff A.

Dem. Apply the Fundamental Theorem about Wffs.

Lemma 3. $\mathsf{M}A \equiv \neg A$ for each wff A.

Dem. Apply the Fundamental Theorem about Wffs.

This result verifies our algorithm on page 184. We can now establish the following facts.

Lemma 4. $\mathsf{N}A \equiv \mathsf{N}B$ if $A \equiv B$.

Dem. We are given that $A \equiv B$; by Lemma 2, $A \equiv \mathsf{N}A$ and $B \equiv \mathsf{N}B$. Since $\equiv$ is not transitive, it would now be premature to conclude that $\mathsf{N}A \equiv \mathsf{N}B$. However, $A \rightarrow B$ is a wff, by assumption; thus, by Lemma 1, page 189, $\mathsf{N}A \rightarrow \mathsf{N}B$ is a wff. Therefore, by Lemma 1, page 163, $\mathsf{N}A \equiv \mathsf{N}B$.

Lemma 5. $\mathsf{M}A \equiv \mathsf{M}B$ if $A \equiv B$.

Dem. Use Lemma 3 and apply the preceding argument.

We need certain facts about R in order to prove that $\mathsf{R}A \equiv \mathsf{R}B$ if $A \equiv B$.

Lemma 6. $\mathsf{R}[A \rightarrow B] \equiv \mathsf{R}A \rightarrow \mathsf{R}B$ provided that $A \rightarrow B$ is a wff.

Dem. See the proof of Lemma 6, page 70.

Lemma 7. $\vdash \mathsf{R}A$ if $\vdash A$.

Dem. Use the Fundamental Theorem about Provable Wffs and Lemma 6; follow the pattern of the proof of Lemma 7, page 70.

Corollary 1. $\vdash \mathsf{R}A$ iff $\vdash A$.

Dem. See the proof of Corollary 1, page 70.

We are now in a position to prove the following result.

Lemma 8. $\mathsf{R}A \equiv \mathsf{R}B$ if $A \equiv B$.

Dem. See the proof of Lemma 8, page 70.

Since both M and R preserve equivalent wffs, and since $\mathsf{D} = \mathsf{MR}$, we can show algebraically that D also preserves equivalent wffs.

Principle of Duality. $\mathsf{D}A \equiv \mathsf{D}B$ if $A \equiv B$.

We can establish the converse of Lemmas 4, 5, and 8 and the converse of the Principle of Duality by applying Lemma 10, page 190, Lemma 2 page 192, and Lemma 1, page 163 (the fact that $\equiv$ is almost transitive).

Corollary 2. Let $\mathsf{T} \in \{\mathsf{N}, \mathsf{M}, \mathsf{R}, \mathsf{D}\}$; then $\mathsf{T}A \equiv \mathsf{T}B$ iff $A \equiv B$.

Dem. See the proof of Corollary 2, page 71.

Here is a useful property of D that follows from the preceding results.

Lemma 9. $\vdash A$ iff $\vdash \neg \mathsf{D}A$.

Dem. See the proof of Lemma 9, page 71.

Corollary 3. $\vdash \mathsf{D}[\neg A]$ if $\vdash A$.

Dem. See the proof of Corollary 3, page 71.

Corollary 4. $\vdash \mathsf{D}A$ iff $\vdash \neg A$.

Dem. See the proof of Corollary 4, page 71.

We can now establish the following fact.

Lemma 10. $\vdash A \rightarrow B$ iff $\vdash \mathsf{D}B \rightarrow \mathsf{D}A$.

Dem. See the proof of Lemma 10, page 71.

We showed earlier that the transforms N, M, R, and D form a subgroup of the group of all normal transforms. There are precisely eight normal transforms; let the remaining four normal transforms be called A, B, C, and E, and let these normal transforms have the parameters displayed in Table 4. Since $\mathsf{T}A \equiv \mathsf{T}B$ if $A \equiv B$, for $\mathsf{T} \in \{\mathsf{N}, \mathsf{M}, \mathsf{R}, \mathsf{D}\}$, we might conjecture that each normal transform has this property. This is a mistake; in fact, N, M, R, and D are the only normal transforms that preserve equivalent wffs. Indeed, for each $\mathsf{T} \in \{\mathsf{A}, \mathsf{B}, \mathsf{C}, \mathsf{E}\}$ we can produce a pair of equivalent wffs whose T-images are not equivalent. Alternatively, we can use a little group theory to the same purpose. Observe that

$$G = \{\mathsf{T} \mid \mathsf{T} \text{ is normal and } \mathsf{T}A \equiv \mathsf{T}B \text{ whenever } A \equiv B\}$$

is a subgroup of the group of all normal transforms. Certainly,

$$\{\mathsf{N}, \mathsf{M}, \mathsf{R}, \mathsf{D}\} \subset G$$

Thus, by Lagrange's Theorem, either $G = \{\mathsf{N}, \mathsf{M}, \mathsf{R}, \mathsf{D}\}$ or G is the set of all normal transforms. We shall show that $\mathsf{A} \notin G$. Now,

$$Fx \vee \neg Fx \equiv \forall y Gy \rightarrow Gz$$

Table 4

T	A	B	C	E
n			$\neg$	$\neg$
d	$\wedge$	$\vee$	$\wedge$	$\vee$
Q	$\forall$	$\exists$	$\forall$	$\exists$

But

$$\mathsf{A}[Fx \vee \neg Fx] \not\equiv \mathsf{A}[\forall y Gy \to Gz]$$

So $\mathsf{A} \notin G$; we conclude that $G = \{\mathsf{N}, \mathsf{M}, \mathsf{R}, \mathsf{D}\}$.

Here are some examples.

Example 1. Show that $\vdash \exists t(A \wedge B) \to \exists tA \wedge \exists tB$, if this is a wff.

Solution. By Example 1, page 167, $\vdash \forall t\mathsf{D}A \vee \forall t\mathsf{D}B \to \forall t(\mathsf{D}A \vee \mathsf{D}B)$; so, by Lemma 10,

$$\vdash \mathsf{D}[\forall t(\mathsf{D}A \vee \mathsf{D}B)] \to \mathsf{D}[\forall t\mathsf{D}A \vee \forall t\mathsf{D}B]$$

i.e.,

$$\vdash \exists t(\mathsf{DD}A \wedge \mathsf{DD}B) \to \exists t\mathsf{DD}A \wedge \exists t\mathsf{DD}B$$

By the Substitution Theorem for Provable Wffs, since $\mathsf{DD}A \equiv A$ and $\mathsf{DD}B \equiv B$,

$$\vdash \exists t(A \wedge B) \to \exists tA \wedge \exists tB$$

Example 2. Show that $\exists t(A \vee B) \equiv A \vee \exists tB$ if $A \vee B$ is a wff, t is free in B, and t does not occur in A.

Solution. By Exercise 7, page 169,

$$\forall t(\mathsf{D}A \wedge \mathsf{D}B) \equiv \mathsf{D}A \wedge \forall t\mathsf{D}B$$

Thus, by the Principle of Duality,

$$\mathsf{D}[\forall t(\mathsf{D}A \wedge \mathsf{D}B)] \equiv \mathsf{D}[\mathsf{D}A \wedge \forall t\mathsf{D}B]$$

i.e.,

$$\exists t(\mathsf{DD}A \vee \mathsf{DD}B) \equiv \mathsf{DD}A \vee \exists t\mathsf{DD}B$$

So, by the Substitution Theorem for Wffs,

$$\exists t(A \vee B) \equiv A \vee \exists tB$$

Exercises

1. Prove Lemma 2.
2. Prove Lemma 3.
3. Prove Lemma 5.

4. Prove Lemma 6.

5. Prove Lemma 7.

6. Prove Corollary 1.

7. Prove Lemma 8.

8. Prove the Principle of Duality.

9. Prove Corollary 2.

10. Prove Lemma 9.

11. Prove Corollary 3.

12. Prove Corollary 4.

13. Prove Lemma 10.

14. Show that $\mathsf{A}[Fx \vee \neg Fx] \not\equiv \mathsf{A}[\forall yGy \rightarrow Gz]$. (See Table 4.)

15. Prove that $\exists t(A \wedge B) \equiv A \wedge \exists tB$, provided that $\exists t(A \wedge B) \rightarrow A \wedge \exists tB$ is a wff.

16. Use Lemma 10 of this section and Lemma 17, page 166, to prove that

$$\vdash \exists tA \rightarrow \exists tB$$

given that $\vdash A \rightarrow B$ and t is free in both A and B.

11.4. Prenex Normal Form

Let A be a wff, which may or may not involve quantifiers; here we shall establish the existence of a wff equivalent to A that possesses a very special form called *prenex normal form*. Our technique is based upon the Substitution Theorem for Wffs, Theorem 1, page 177, and six results previously obtained, which we now gather together for convenience.

Let C and D be any wffs and let t be any individual such that $C \vee \forall tD \rightarrow \forall t(C \vee D)$ is a wff; then

$$C \wedge \forall tA \equiv \forall t(C \wedge D) \quad \text{(Exercise 7, page 169)} \quad (1)$$

$$C \vee \forall tD \equiv \forall t(C \vee D) \quad \text{(Exercise 6, page 169)} \quad (2)$$

$$C \wedge \exists tD \equiv \exists t(C \wedge D) \quad \text{(Exercise 15, above)} \quad (3)$$

$$C \vee \exists tD \equiv \exists t(C \vee D) \quad \text{(Example 2, page 195)} \quad (4)$$

$$\neg\forall tD \equiv \exists t(\neg D) \quad (5)$$

$$\neg\exists tD \equiv \forall t(\neg D) \quad (6)$$

We shall illustrate our technique in a moment; first, we recall (see page 140) what it means for a wff to be in prenex normal form.

Definition. A wff, say B, is said to be in *prenex normal form* if there is a natural number n, $n \geq 1$, n quantifiers $Q_1, \ldots, Q_n$ (where each Q_i is $\forall$ or $\exists$), and n distinct individuals $t_1, \ldots, t_n$, together with a quantifier-free wff M such that $B = Q_1 t_1 \cdots Q_n t_n M$. Here, $Q_1 t_1 \cdots Q_n t_n$ is called the *prefix* of B and M is called the *matrix* of B.

We shall prove that each wff is equivalent to some wff in prenex normal form. Let us illustrate the idea.

Example 1. Show that $A = (\forall y Fy \vee Fx) \vee \neg(\forall w Gww \vee \neg\forall y Fy)$ is equivalent to a wff in prenex normal form.

Solution. By Theorem 1, page 177, $\forall y Fy \equiv \forall z Fz$; so by the Substitution Theorem for Wffs

$$\begin{aligned}
A &\equiv (\forall y Fy \vee Fx) \vee \neg(\forall w Gww \vee \neg\forall z Fz) & & \\
&\equiv \forall y(Fy \vee Fx) \vee \exists w(\neg Gww \wedge \forall z Fz) & & \text{[by (2) and (5)]} \\
&\equiv \forall y(Fy \vee Fx) \vee \exists w \forall z(\neg Gww \wedge Fz) & & \text{[by (1)]} \\
&\equiv \forall y((Fy \vee Fx) \vee \exists w \forall z(\neg Gww \wedge Fz)) & & \text{[by (2)]} \\
&\equiv \forall y \exists w((Fy \vee Fx) \vee \forall z(\neg Gww \wedge Fz)) & & \text{[by (4)]} \\
&\equiv \forall y \exists w \forall z((Fy \vee Fx) \vee (\neg Gww \wedge Fz)) & & \text{[by (2)]}
\end{aligned}$$

Notice our use of the following algorithm in solving Example 1.

Algorithm for Prenex Normal Form (PNF). A wff in PNF equivalent to a given wff A, where A involves a quantifier, is obtained from A as follows:

1. Apply Theorem 1, page 177, repeatedly until a wff is obtained in which no individual is quantified more than once.
2. Ensure that $\neg$ appears attached only to quantifier-free wffs; this requires applying (5) and (6) above, as well as the De Morgan laws.

(3). Apply (1)–(4) until all quantifiers appear together at the LHS of the resulting wff.

This algorithm certainly supports the conjecture that each wff is equivalent to some wff in prenex normal form, but does not in itself establish this statement. Here is a rigorous demonstration; we shall need the following facts.

Lemma 1. If E is in PNF and t is free in E, then $\forall tE$ and $\exists tE$ are both in PNF.

Dem. Obvious.

Lemma 2. $\neg(Q_1t_1 \cdots Q_nt_nM) \equiv q_1t_1 \cdots q_nt_n(\neg M)$, where $q_i = \forall$ if $Q_i = \exists$, and $q_i = \exists$ if $Q_i = \forall$, $i = 1, \ldots, n$.

Dem. Mathematical induction on n.

Lemma 3. Given any wff in PNF, say B, there is a wff in PNF, say C, such that $\neg B \equiv C$; furthermore, t is free in C iff t is free in B, whenever t is an individual.

Dem. Consider Lemma 2.

Lemma 4. Let $A \vee B$ be a wff, where A and B are in PNF. Then there is a wff E such that $A \vee B \equiv E$ and E is in PNF; furthermore, t is free in E iff t is free in $A \vee B$, whenever t is an individual.

Dem. By assumption there are natural numbers r and n such that

$$A = Q_1s_1 \cdots Q_rs_rM \qquad \text{and} \qquad B = Q_1't_1 \cdots Q_n't_nM'$$

where M and M' are quantifier-free. In view of Theorem 1, page 177, and the Substitution Theorem for Wffs, we may assume that no individual is bound in both A and B. Applying (2) and (4) as necessary, we obtain

$$\begin{aligned} A \vee B &\equiv Q_1't_1(A \vee Q_2't_2 \cdots Q_n't_nM') \\ &\equiv Q_1't_1Q_2't_2(A \vee Q_3't_3 \cdots Q_n't_nM') \\ &\ \vdots \\ &\equiv Q_1't_1 \cdots Q_n't_n(A \vee M') \end{aligned}$$

Similarly, $A \vee M' \equiv Q_1s_1 \cdots Q_rs_r(M \vee M')$, so, by the Substitution Theorem for Wffs, $A \vee B \equiv Q_1't_1 \cdots Q_n't_nQ_1s_1 \cdots Q_rs_r(M \vee M')$.

Lemma 5. Let A be any quantifier-free wff. There is a wff in PNF, say E, such that $A \equiv E$; furthermore, t is free in E iff t is free in A, whenever t is a pure individual.

Dem. Let s be a pure individual that does not occur in A. Let $E = \forall s(A \wedge (Fs \vee Fs))$; clearly E is in PNF and t is free in E iff t is free in A, whenever t is an individual. We shall show that $A \equiv E$. By Lemma 11, page 166, $A \wedge B \equiv A$ if $\vdash B$, provided that $A \wedge B$ is a wff.

Now, $\vdash Fs \vee \neg Fs$; so $A \equiv A \wedge (Fs \vee \neg Fs)$, in particular $\vdash A \rightarrow A \wedge (Fs \vee \neg Fs)$, so $\vdash A \rightarrow \forall s(A \wedge (Fs \vee \neg Fs))$, i.e., $\vdash A \rightarrow E$. Also, $\vdash \forall s(A \wedge (Fs \vee \neg Fs)) \rightarrow A \wedge (Ft \vee \neg Ft)$, where t is an individual that does not occur in $A \vee Fs$. But $\vdash A \wedge (Ft \vee \neg Ft) \rightarrow A$, so $\vdash \forall s(A \wedge (Fs \vee \neg Fs)) \rightarrow A$, i.e., $\vdash E \rightarrow A$; so $E \equiv A$.

We are now ready to establish our main theorem.

Fundamental Theorem about Prenex Normal Form. Corresponding to each wff A there is a wff E such that $A \equiv E$, E is in PNF, and each individual t is free in E iff t is free in A.

Dem. We shall apply our Fundamental Theorem about Wffs. Let $S = \{A \mid$ there is a wff E such that $A \equiv E$, E is in PNF, and each individual t is free in E iff t is free in $A\}$.

1. By Lemma 5, $A \in S$ if A is atomic.
2. $B \in S$. We must show that $\neg B \in S$. By assumption, there is a wff E in PNF such that $B \equiv E$ and each individual t is free in E iff t is free in B. By the Substitution Theorem for Wffs, $\neg B \equiv \neg E$; by Lemma 3, there is a wff C in PNF such that $\neg E \equiv C$ and t is free in C iff t is free in E. We conclude that $\neg B \equiv C$ as required. Thus, $\neg B \in S$.
3. $C \in S$ and $D \in S$. We must show that $C \vee D \in S$. By assumption there are wffs C' and D' in PNF such that $C \equiv C'$ and $D \equiv D'$, C' involves the same free individuals as C, and D' involves the same free individuals as D. Now, $C \vee D \equiv C' \vee D' \equiv E$, as required, by Lemma 4. So $C \vee D \in S$.
4. $B \in S$ and s is free in B. We must show that $\forall sB \in S$. By assumption, there is a wff E in PNF that involves the same free individuals as B, such that $B \equiv E$. Thus, by the Substitution Theorem for Wffs, $\forall sB \equiv \forall sE$. By Lemma 1, $\forall sE$ is in PNF and involves the same free individuals as $\forall sB$. So $\forall sB \in S$. Thus $S = \mathsf{W}$.

By the *length* of the prefix of a wff in PNF we mean the natural number n such that $Q_1t_1 \cdots Q_nt_n$ is the prefix of the wff. In this way a natural number is associated with each wff in PNF, i.e., the length of its prefix. This turns out to be useful.

Exercises

Find a wff in prenex normal form equivalent to each of the following wffs.

1. $Fy \rightarrow \forall xGxy$.
2. $\neg(\forall xFx \vee \exists yFy)$.

3. $\forall xFx \wedge \forall x\exists yGxy \rightarrow \forall xGxx$.

4. $\exists yGyy \vee \forall yGyy \leftrightarrow \exists yFy$.

5. Fz.

6. $\neg Fz$.

7. $Fz \wedge \neg Fz$.

8. $Fz \vee \neg Fz$.

9. Prove Lemma 2.

10. Prove Lemma 3.

11. Show that corresponding to each wff A there is a wff E in prenex normal form such that $A \equiv E$, but the free individuals of E are not identical with the free individuals of A.

12

Deducibility and Completeness

12.1. Deducibility

We begin our discussion of deducibility by generalizing the notion of a *true* wff. Recall that a wff A is true provided that μA is a true swff of $\sum$ for each semantical system $\sum$ matched to the given predicate calculus and for each $\sum$-interpreter μ (see page 155). We shall generalize this concept by weakening the requirement that μA is a true swff of $\sum$ for *each* semantical system $\sum$ matched to the given predicate calculus. Instead, we shall require that μA is a true swff of $\sum$ for *certain* semantical systems $\sum$ matched to the given predicate calculus, say $\prod$.

Now, our purpose in this section is to characterize the notion that a wff A is *deducible* from a set of wffs K. Accordingly, we choose the family of semantical systems involved in the preceding paragraph to be the models of K (see page 157). This brings us to the concept of a K-*true* wff.

Definition. A wff A is said to be K-*true* provided that μA is a true swff of $\sum$ whenever $\sum$ is a model of K under μ.

This means that A is K-true iff for each semantical system $\sum$ and for each $\sum$-interpreter μ, either: (a) There is a wff $B \in \mathsf{K}$ such that μB is *not* a true swff of $\sum$; or (b) μA is a true swff of $\sum$.

If A is K-true, we also say that A is a *consequence* of K.

For example, Fy is $\{\forall x Fx\}$-true; each wff is $\{Fx, \neg Fx\}$-true. The latter example illustrates the fact that A is K-true if K has no models. Clearly, $\{Fx, \neg Fx\}$ does not have a model.

We point out that each semantical system $\sum$ matched to $\prod$ is a model of the empty set under any $\sum$-interpreter. Therefore, our definition of K-true wff reduces to the definition of *true* wff in case K is the empty set; i.e., A is true iff A is $\varnothing$-true.

Just as we defined the notion of a provable wff in order to characterize internally (i.e., syntactically) the concept of a true wff, we shall now introduce the notion of deducibility in order to characterize K-true wffs within $\prod$.

Definition. Let K be a nonempty set of wffs and let B be a wff. We say that B is *deducible* from K (in symbols, $\mathsf{K} \vdash B$) if there is a nonempty, finite set of wffs $\{C_1, \ldots, C_n\}$, where each C_i is equivalent to a member of K, and a wff D equivalent to B, such that $\vdash C_1 \wedge \cdots \wedge C_n \rightarrow D$. We say that B is *deducible* from $\varnothing$ (in symbols, $\varnothing \vdash B$) if $\vdash B$.

Here are some examples.

Example 1. Show that $\{\forall xFx\} \vdash Fx$.

Solution. $\forall yFy \equiv \forall xFx$ and $Fx \equiv Fx$; also, $\vdash \forall yFy \rightarrow Fx$. Therefore, $\{\forall xFx\} \vdash Fx$.

Example 2. Show that $\{Fx, \neg Fx\} \vdash \forall xFx$.

Solution. $\forall yFy \equiv \forall xFx$ and $\vdash Fx \wedge \neg Fx \rightarrow \forall yFy$. Therefore, $\{Fx, \neg Fx\} \vdash \forall xFx$.

Example 3. Show that $\{\forall x(A \rightarrow B), \exists xA\} \vdash \exists xB$ provided that $A \rightarrow B$ is a wff and x is free in both A and B.

Solution. By Lemma 5, page 181,

$$\vdash \forall x(A \rightarrow B) \wedge \exists xA \rightarrow \exists xB$$

so

$$\{\forall x(A \rightarrow B), \exists xA\} \vdash \exists xB$$

Later (see Theorem 1, page 224), we shall prove that our formal definition of *deducibility* has captured the intended idea, i.e., that $\mathsf{K} \vdash A$ iff A is K-true, where K is any set of wffs and A is any wff. Bear in mind that when we assert $\mathsf{K} \vdash A$ we are thinking syntactically, i.e., in terms of a predicate calculus. On the other hand, when we assert that A is K-true we are thinking semantically, i.e., in terms of semantical systems and truth-values.

Our first task is to simplify the characterization of deducibility provided by the definition. To this purpose we now introduce the *-operation (read "star operation"), a many-valued operation on wffs. Let A be any wff; then A^* denotes any wff obtained from A by replacing some (or none) of its bound individuals by pure individuals not in A.

For example,

$$[\forall xFx]^* = \forall yFy \quad \text{and} \quad [\forall xFx]^* = \forall zFz$$
$$[\forall x\exists yGxy]^* = \forall w\exists zGwz \quad \text{and} \quad [\forall x\exists yGxy]^* = \forall z\exists wGzw$$

We mention that each instance of the star operation can be expressed by a sequence of substitution transforms. For example, let A^* be obtained from A by replacing $x_1, \ldots, x_n$ by $y_1, \ldots, y_n$, respectively, where the x's are distinct individuals bound in A and the y's are distinct pure individuals that do not occur in A. Then

$$A^* = \mathrm{S}^{x_n}_{y_n}[\cdots[\mathrm{S}^{x_1}_{y_1}[A]]\cdots]$$

since the y's do not occur in A. Also, $A^* = \mathrm{I}^{x_n}_{y_n}[\cdots[\mathrm{I}^{x_1}_{y_1}[A]]\cdots]$.

We now present our result.

Theorem 1. Let K be a nonempty set of wffs and let B be a wff. Then $\mathsf{K} \vdash B$ iff there is a nonempty, finite subset of K, say $\{A_1, \ldots, A_n\}$, and wffs $A_1^*, \ldots, A_n^*$ and B^* such that $\vdash A_1^* \wedge \cdots \wedge A_n^* \to B^*$.

Dem. 1. Let $\mathsf{K} \vdash B$. Then there is a finite number of wffs $C_1, \ldots, C_n$, where $n \geq 1$, and there is a wff D such that each C_i is equivalent to a member of K, say A_i, $D \equiv B$, and

$$\vdash C_1 \wedge \cdots \wedge C_n \to D \tag{1}$$

By the Substitution Theorem for Provable Wffs (see page 176),

$$\vdash A_1 \wedge \cdots \wedge A_n \to B \tag{2}$$

provided that this is a wff. If (2) is not a wff, then some individual t is free in one of its disjuncts and is bound in another disjunct of (2). The idea is to replace each bound occurrence of t by s, where s is a pure individual that does not occur in (2), leaving each free occurrence of t as is. The "replacing" is to be carried out on the RH sides of the equivalences: $C_1 \equiv A_1, \ldots, C_n \equiv A_n, D \equiv B$. We use the star operation to denote the resulting RH sides. By Lemma 3, page 178,

$$C_1 \equiv A_1^*, \ldots, \quad C_n \equiv A_n^*, \quad D \equiv B^*$$

From (1), by the Substitution Theorem for Provable Wffs,

$$\vdash A_1^* \wedge \cdots \wedge A_n^* \to B^*$$

and this is a wff.

2. Let $\vdash A_1^* \wedge \cdots \wedge A_n^* \to B^*$, where $\{A_1, \ldots, A_n\} \subset \mathsf{K}$ and $n \geq 1$. Clearly $B^* \equiv B$ and $A_i^* \equiv A_i$ for $i = 1, \ldots, n$; so $\mathsf{K} \vdash B$ by definition. This establishes Theorem 1.

Note. It is important to observe that the star operation does not affect the free individuals of the wffs involved, i.e., for each wff E, the individuals that are free in E^* are precisely the individuals that are free in E.

Here are some facts about *deducibility*; throughout, K is any set of wffs and A and B are any wffs.

Fact 1. $\mathsf{K} \vdash A \wedge B$ if $\mathsf{K} \vdash A$, $\mathsf{K} \vdash B$, and $A \wedge B \in \mathsf{W}$.

Dem. If $\mathsf{K} = \varnothing$, then $\vdash A$ and $\vdash B$; so $\vdash A \wedge B$ by Lemma 4, page 165. Assume that $K \neq \varnothing$. By Theorem 1, there are nonempty, finite subsets of K, say $\{A_1, \ldots, A_m\}$ and $\{B_1, \ldots, B_n\}$, such that

$$\vdash A_1^* \wedge \cdots \wedge A_m^* \to A^* \qquad \text{and} \qquad \vdash B_1^* \wedge \cdots \wedge B_n^* \to B^* \tag{3}$$

We wish to show that

$$(A_1^* \wedge \cdots \wedge A_m^*) \wedge (B_1^* \wedge \cdots \wedge B_n^*) \to A^* \wedge B^* \tag{4}$$

is provable. If (4) is a wff, then it is provable by Lemma 20, page 167. If (4) is not a wff, it is only because of a conflict between free and bound individuals, i.e., one or more individuals are free in $A_1^* \wedge \cdots \wedge A_m^* \to A^*$ and are bound in $B_1^* \wedge \cdots \wedge B_n^* \to B^*$ or vice versa. This can be avoided by carrying out the star operations suitably in the first place. Accordingly, (4) is provable. But $\{A_1, \ldots, A_m, B_1, \ldots, B_n\} \subset K$; thus, by Theorem 1, $\mathsf{K} \vdash A \wedge B$.

Fact 2. $\mathsf{K} \vdash B$ if $\mathsf{K} \vdash A$ and $A \equiv B$.

Dem. Use Theorem 1 and the Substitution Theorem for Provable Wffs.

Fact 3. $\mathsf{K} \vdash B$ if $\mathsf{K} \vdash A$ and $\vdash A \to B$.

Dem. If $\mathsf{K} = \varnothing$, then $\vdash A$ and $\vdash A \to B$; so $\vdash B$. Thus $\varnothing \vdash B$. Assume that $\mathsf{K} \neq \varnothing$. By Theorem 1, there is a nonempty, finite subset of K, say $\{A_1, \ldots, A_n\}$, such that

$$\vdash A_1^* \wedge \cdots \wedge A_n^* \to A^*$$

Now $\vdash A \to B$; thus, by the Substitution Theorem for Provable Wffs, $\vdash A^* \to B^*$. We can assume that the star operations have been chosen so that

$$A_1^* \wedge \cdots \wedge A_n^* \to B^* \tag{5}$$

is a wff. By the transitivity of $\to$, it follows that (5) is provable. We conclude from Theorem 1 that $\mathsf{K} \vdash B$.

Fact 4. $\mathsf{K} \vdash A$ if $A \in \mathsf{K}$.

Fact 5. $\mathsf{K}_1 \vdash A$ if $\mathsf{K} \vdash A$ and $\mathsf{K} \subset \mathsf{K}_1$.

Fact 6. $\mathsf{K} \vdash A$ if $\vdash A$.

Fact 7. $\mathsf{K} \vdash A \vee B$ if $\mathsf{K} \vdash A$ and $A \vee B \in \mathsf{W}$.

Fact 8. $\mathsf{K} \vdash B$ if $\mathsf{K} \vdash A$ and $\mathsf{K} \vdash A \to B$.

Our goal is to demonstrate that $\mathsf{K} \vdash A$ iff A is K-true. Whereas it is quite easy to show that A is K-true if $\mathsf{K} \vdash A$, it is quite another matter to show that $\mathsf{K} \vdash A$ if A is K-true (we do this in Section 12.6).

Lemma 1. A is K-true if $\mathsf{K} \vdash A$.

Dem. See Lemma 1, page 89.

Here is the Deduction Theorem.

Deduction Theorem. Let $A \to B$ be a wff; then $\mathsf{K} \cup \{A\} \vdash B$ iff $\mathsf{K} \vdash A \to B$.

Dem. See the demonstration of the Deduction Theorem for the propositional calculus, page 89; use Theorem 1.

It is interesting to formulate *deducibility* in a manner analogous to the definition of *provability*. First, we introduce the notion of a K-proof where K is a given set of wffs.

Definition. Let K be any set of wffs. A finite sequence of wffs is called a K-*proof* if each of its terms, say E, satisfies one or more of the following conditions:

(a) $E \in \mathsf{K}$.
(b) E is a member of one of the four axiom schemes on page 159.
(c) There is a wff D such that both D and $D \to E$ precede E in the sequence.
(d) E has the form $A \to \forall tB$ and is preceded in the sequence by $A \to B$, where t is not free in any wff of K.

Certainly each finite sequence of wffs is a W-proof. If K is a set each of whose members is contained in one of the four axiom schemes, then each K-proof is a proof.

To illustrate our definition we point out that

$$\forall xFx, \quad \forall xFx \to Fy, \quad Fy$$

is a $\{\forall xFx\}$-proof, and

$$\forall x(Fz \to Gx), \quad \forall x(Fz \to Gx) \mathbin{\dot\to} Fz \to Gy, \quad Fz \to Gy, \quad Fz \to \forall yGy$$

is a $\{\forall x(Fz \to Gx)\}$-proof.

The following criterion gives a sufficient condition for deducibility.

Criterion for Deducibility. Let K be any set of wffs and let A be any wff; then $\mathsf{K} \vdash A$ if there is a K-proof whose last term is A.

Dem. See the demonstration of the Criterion for Deducibility, page 90. Continuing the argument given there, we assume that $\mathsf{K} \neq \varnothing$ and that $B_1, \ldots, B_m$ is a K-proof with last term A. We shall show that $\mathsf{K} \vdash A$. Let $A_1, \ldots, A_n$ be the terms of this K-proof that are in K. We claim that

$$\vdash A_1^* \wedge \cdots \wedge A_n^* \to B_j^* \qquad (\text{i.e., } \mathsf{K} \vdash B_j) \tag{6}$$

for $j = 1, \ldots, m$. Three of the four possible cases concerning B_i are considered on page 91; we now consider the remaining case.

Case 4. B_i meets condition (d). Then $B_i = C \to \forall tD$, where $C \to D$ precedes B_i in the given K-proof and t is not free in any wff of K. By assumption, $\vdash A_1^* \wedge \cdots \wedge A_n^* \mathbin{\dot\to} C^* \to D^*$; so

$$\vdash A_1^* \wedge \cdots \wedge A_n^* \wedge C^* \to D^* \tag{7}$$

We can carry out the star operations so that t is free in the RHS of (7) and t does not occur in the LHS of (7); by Lemma 15, page 166,

$$\vdash A_1^* \wedge \cdots \wedge A_n^* \wedge C^* \to \forall tD^*$$

Thus

$$\vdash A_1^* \wedge \cdots \wedge A_n^* \to B_i^*$$

This establishes (6); in particular, $\vdash A_1^* \wedge \cdots \wedge A_n^* \rightarrow A^*$, so $\mathsf{K} \vdash A$. We have proven the Criterion for Deducibility.

From the preceding example, $\{\forall x(Fz \rightarrow Gx)\} \vdash Fz \rightarrow \forall yGy$.

We point out that the key step in proving the Criterion for Deducibility is the following fact.

Lemma 2. Let $B_1, \ldots, B_m$ be a K-proof and let $A_1, \ldots, A_n$ be the members of K that are terms of the given K-proof. Then

$$\vdash A_1^* \wedge \cdots \wedge A_n^* \rightarrow B_m^*$$

Dem. See the demonstration of the Criterion for Deducibility.

The Criterion for Deducibility provides a standard for mathematical proofs. Each term of a K-proof, where K is the postulate set of a mathematical theory, may be regarded as a step in a mathematical proof; the theorem being proven is the final term of the K-proof. Since K-proofs are usually very long, it is convenient to relax the conditions of the above criterion; the aim is to reduce the number of steps in a mathematical proof.

Theorem 2. Let K be a set of wffs and let A be any wff. Then $\mathsf{K} \vdash A$ if there is a finite sequence of wffs, with last term A, such that each of its terms, say E, satisfies one or more of the following conditions:

1. $E^* \in \mathsf{K}$ for some choice of the star operation.
2. $\vdash E$.
3. E is equivalent to a preceding term of the sequence.
4. There is a wff D such that both D and $D \rightarrow E$ precede E in the sequence.
5. E has the form $C \rightarrow \forall tD$ and is preceded in the sequence by $C \rightarrow D$, where t is not free in any wff of K.
6. E has the form $\forall tD$ and is preceded by D, where t is not free in any wff of K.

Dem. Let $C_1, \ldots, C_m$ be a sequence with last term A that meets the above conditions. This sequence can be extended to a K-proof; so $\mathsf{K} \vdash A$.

We now illustrate Theorem 2. Let $\mathsf{K} = \{\forall x(Fx \wedge Gx)\}$; then the following sequence satisfies the conditions of Theorem 2:

$$\forall x(Fx \wedge Gx),\ \ \forall x(Fx \wedge Gx) \rightarrow Fy \wedge Gy,\ \ Fy \wedge Gy,\ \ Fy \wedge Gy \rightarrow Fy,$$
$$Fy,\ \ \forall yFy,\ \ Fy \wedge Gy \rightarrow Gy,\ \ Gy,\ \ \forall yGy,$$
$$\forall yFy \rightarrow \forall yGy \rightarrow \forall yFy \wedge \forall yGy,\ \ \forall yGy \rightarrow \forall yFy \wedge \forall yGy,$$
$$\forall yFy \wedge \forall yGy,\ \ \forall xFx \wedge \forall xGx.$$

Thus, by Theorem 2, $\{\forall x(Fx \wedge Gx)\} \vdash \forall xFx \wedge \forall xGx$.

By including more rules of inference in the conditions of our theorem, we can shorten the sequence involved; this allows us to reduce the number of steps required to demonstrate that $\mathsf{K} \vdash A$, given K and A (see Theorem 2, page 92).

Exercises

1. Show that $\exists xFx$ is $\{\forall xFx\}$-true.
2. Show that $Fy \vee Gxx$ is $\{\forall xFx\}$-true.
3. Show that $Fy \wedge Gxx$ is $\{\forall xFx, \forall xGxx\}$-true.
4. Show that $\exists z(Fz \wedge Gzz)$ is $\{\forall xFx, \forall yGyy\}$-true.
5. Prove that $\{\forall xFx\} \vdash \exists xFx$.
6. Prove that $\{\forall xFx\} \vdash Fy \vee Gxx$.
7. Prove that $\{\forall xFx, \forall xGxx\} \vdash Fy \wedge Gxx$.
8. Prove that $\{\forall xFx, \forall yGyy\} \vdash \exists z(Fz \wedge Gzz)$.
9. Let E be any wff; prove that the individuals that are free in any E^* are precisely the individuals that are free in E.
10. Let $\sum$ be any semantical system matched to $\prod$ and let μ be any $\sum$-interpreter. Prove that for each wff A, μA is true for $\sum$ iff μA^* is true for $\sum$ for each choice of the star operation.
11. Prove Fact 2.
12. (a) Prove Fact 4.
    (b) Prove Fact 5.
    (c) Prove Fact 6.
    (d) Prove Fact 7.
13. Prove Fact 8. *Hint:* Use Lemma 21, page 167.
14. Prove Lemma 1.
15. Prove the Deduction Theorem.
16. (a) Show that each finite sequence of wffs is a W-proof.
    (b) Let K be any set of wffs; is each proof a K-proof?

(c) Show that each K-proof is a proof given that K contains only members of axiom schemes.
(d) Show that each $\varnothing$-proof is a proof and that each proof is a $\varnothing$-proof.

17. Let K be a set of provable wffs; is each K-proof necessarily a proof?

18. Prove Theorem 2.

19. (a) Use Theorem 2 to show that $\{\forall x(Fy \vee Gx)\} \vdash Fy \vee \forall xGx$.
(b) Use Theorem 2 to show that $\{Fy \vee \forall xGx\} \vdash \forall x(Fy \vee Gx)$.
(c) Prove that $\forall x(Fy \vee Gx) \equiv Fy \vee \forall xGx$.

20. Show that $\{\forall x(A \rightarrow B), \exists x(A \wedge C)\} \vdash \exists x(B \wedge C)$ provided that $A \vee (B \vee C)$ is a wff and x is free in A, B, and C. *Hint:* Use Exercise 18, page 182.

12.2. Consistent Sets and Contradictory Sets

Let K be any set of wffs and consider the set of all wffs that are deducible from K; we shall denote this set by "C[K]."

Definition. $\mathsf{C}[\mathsf{K}] = \{A \in \mathsf{W} \mid \mathsf{K} \vdash A\}$.

For example, $\mathsf{C}[\varnothing] = \{A \in \mathsf{W} \mid \vdash A\}$, the set of all provable wffs. Of course, $\mathsf{C}[\mathsf{W}] = \mathsf{W}$. If K is a set of wffs such that $\mathsf{K} \vdash B$ and $\mathsf{K} \vdash \neg B$, for some wff B, then $\mathsf{K} \vdash B \wedge \neg B$; it follows that $\mathsf{K} \vdash A$ for each wff A. In this case, $\mathsf{C}[\mathsf{K}] = \mathsf{W}$, the set of all wffs.

We now define the terms *contradictory* and *consistent.*

Definition. We say that K is *contradictory* if $\mathsf{C}[\mathsf{K}] = \mathsf{W}$.

Definition. We say that K is *consistent* if K is not contradictory.

Thus, K is consistent iff some wff is *not* deducible from K. For example, $\{Fx, \neg Fx\}$ is contradictory; indeed, $\{Fx \wedge \neg Fx\}$ is contradictory. Notice that $\varnothing$, the empty set, is consistent. Let A be any atomic wff; then both $\{A\}$ and $\{\neg A\}$ are consistent.

Our definition of contradictory sets is not easily applied to the job of determining whether a given set of wffs is contradictory. Fortunately, there is a simple criterion that achieves this purpose.

Criterion for Contradictory Sets. K is contradictory iff there is a wff B such that $\mathsf{K} \vdash B \wedge \neg B$.

Dem. See the proof of the Criterion for Contradictory Sets, page 95. Use the star operation.

Lemma 1. Each superset of a contradictory set is contradictory.

Lemma 2. Each subset of a consistent set is consistent.

Here is a useful fact.

Theorem 1. An infinite set of wffs is consistent iff each of its finite subsets is consistent.

Dem. See the proof of Lemma 4, page 97. Use the star operation.

Notice that $\mathsf{K} \cup \{A\}$ is contradictory if $\mathsf{K} \vdash \neg A$ (apply our Criterion for Contradictory Sets). Therefore, if $\mathsf{K} \cup \{A\}$ is consistent, then $\neg A \notin \mathsf{C}[\mathsf{K}]$. The converse of this statement is also correct.

Lemma 3. $\mathsf{K} \cup \{A\}$ is consistent if $\neg A \notin \mathsf{C}[\mathsf{K}]$.

Dem. See the proof of Lemma 1, page 96; use the star operation.

In view of the comment that precedes Lemma 3, we obtain the following corollary to Lemma 3.

Corollary 1. $\mathsf{K} \cup \{A\}$ is consistent iff $\neg A \notin \mathsf{C}[\mathsf{K}]$.

Our next observation is vital.

Lemma 4. K is consistent if K has a model.

Dem. See the proof of Lemma 3, page 96; use the star operation.

Notice that our next lemma has no analog in the propositional calculus. We shall need this lemma in our proof of the Strong Completeness Theorem.

Lemma 5. $\mathsf{K} \cup \{\mathsf{S}_t^s[D]\}$ is consistent if

1. K is consistent.
2. $\exists tD \in \mathsf{K}$.
3. s is a pure individual that does not occur in any wff of K.

Dem. Assume that $\mathsf{K} \cup \{\mathsf{S}_t^s[D]\}$ is contradictory. Then there is a finite subset of K, say $\{A_1, \ldots, A_n\}$, such that

$$\vdash A_1^* \wedge \cdots \wedge A_n^* \wedge [\mathsf{S}_t^s[D]]^* \rightarrow Fx \wedge \neg Fx$$

where individuals different from s are used in carrying out the star operation of replacing bound individuals (so as to avoid a conflict between free and bound individuals). Now, $[S_t^s[D]]^* = S_t^s[D^*]$; so

$$\vdash S_t^s[D^*] \dot{\rightarrow} A_1^* \wedge \cdots \wedge A_n^* \rightarrow Fx \wedge \neg Fx$$

Here, s is free in the LHS and does not occur in the RHS; so

$$\vdash \exists s S_t^s[D^*] \dot{\rightarrow} A_1^* \wedge \cdots \wedge A_n^* \rightarrow Fx \wedge \neg Fx$$

By Theorem 1, page 177, and the Substitution Theorem for Provable Wffs,

$$\vdash \exists t D^* \dot{\rightarrow} A_1^* \wedge \cdots \wedge A_n^* \rightarrow Fx \wedge \neg Fx$$

Thus

$$\vdash \exists t D^* \wedge A_1^* \wedge \cdots \wedge A_n^* \rightarrow Fx \wedge \neg Fx$$

This means that $\mathsf{K} \vdash Fx \wedge \neg Fx$, since $\exists tD \in \mathsf{K}$ and $\exists tD^* \equiv \exists tD$. So K is contradictory. This contradiction proves that $\mathsf{K} \cup \{S_t^s[D]\}$ is consistent.

We shall need the following theorem in our proof of the Strong Completeness Theorem.

Theorem 2. Let K be a consistent set of wffs in a predicate calculus $\prod$, and let $\prod'$ be a predicate calculus obtained from $\prod$ by adjoining more pure individuals to the individuals of $\prod$. Then K is consistent in $\prod'$.

Dem. Assume that K is contradictory in $\prod'$. Then

$$\mathsf{K} \vdash Fx \wedge \neg Fx \qquad (\text{in } \textstyle\prod')$$

Therefore, there are members of K, say $A_1, \ldots, A_n$ such that

$$\vdash A_1^* \wedge \cdots \wedge A_n^* \rightarrow Fx \wedge \neg Fx \qquad (\text{in } \textstyle\prod') \tag{1}$$

Let π be a proof of (1) in $\prod'$. If each individual that occurs in π belongs to $\prod$, then (1) is provable in $\prod$; indeed π is a proof of (1) in $\prod$. Assume that π involves individuals of $\prod'$ that are not in $\prod$, say individuals $s_1, \ldots, s_m$; certainly, the s's are pure. Choose pure individuals of $\prod$, say $t_1, \ldots, t_m$, that do not occur in π, and apply the interchange transforms $I_{s_1}^{s_1}, \ldots, I_{t_m}^{s_m}$ to π; i.e., apply $I_{t_1}^{s_1}$ to π, then apply $I_{t_2}^{s_2}$ to the resulting sequence, and so on. By Theorem 1, page 169, each interchange transform yields a proof when applied to a proof. So we obtain a proof of (1) which involves only individuals of $\prod$. Thus (1) is provable in $\prod$. But this means that K is contradictory in $\prod$. This contradiction proves that K is consistent in $\prod'$.

Exercises

1. Show that C[W] = W.
2. Show that C[K] = W if K = $\{Fx, \neg Fx\}$.
3. Prove that $\varnothing$ is consistent.
4. (a) Prove that $\{A\}$ is consistent if A is atomic.
   (b) Prove that $\{\neg A\}$ is consistent if A is atomic.
5. Prove Lemma 1.
6. Prove Lemma 2.
7. Prove Theorem 1.
8. Prove Lemma 3.
9. Prove Lemma 4.

12.3. Strong Completeness Theorem

The *Strong Completeness Theorem* is the bridge connecting mathematical logic and modern algebra. Modern algebra is the study of classes of relational systems, more generally, semantical systems; mathematical logic is the study of the pure theory of deduction. The Strong Completeness Theorem shows that with certain restrictions the two disciplines are aspects of the same thing.

Some comments on the general picture may be helpful at this point. First, let us consider the general method of abstract algebra. A family, or class, of semantical systems is specified by first describing the sort of system in the family, i.e., by stating the number and type of each relation symbol involved (so that any two systems in the family are alike in this respect). Next, the postulates for the family are listed; these are statements which are true for each system in the family. In short, any two systems in the family agree as to the number and type of their relation symbols, and satisfy given statements—the postulates for the family. The idea is to find more statements which are true for each system in the family; any statement with this property is said to be a *theorem* of the theory. This is achieved by considering any system in the family, without announcing precisely which system is being considered, and showing that the given statement is true for it; to this purpose, we may use only the postulates and theorems of the theory that have been obtained earlier. Since the only assumption made about the semantical system considered is that it is a member of the family, we can conclude that the given statement is true for each system of the family. Here, semantical systems are uppermost in our thoughts and play a key role in the technique for proving that a statement is a theorem of the theory.

In a pure theory of deduction, on the other hand, semantical systems play no role at all. Attention is concentrated on the postulates which characterize the family under investigation, and the abstract theory of deduction is used to characterize its consequences, i.e., the statements which are deducible from the postulates. So, the theorems of the algebraic theory are established by applying the theory of deduction to the postulates of the theory. There is no involvement with semantical systems; instead, the theory of deduction plays the key role. Notice that mathematical logic and modern algebra have a common goal—to establish the theorems of a theory. We hasten to mention that these disciplines have other, more far-reaching goals.

In this chapter we shall demonstrate that the theory of deduction presented in Section 12.1 achieves our goal. Some insight into the problem is obtained by considering the following three statements; throughout, K is any set of wffs and A is any wff.

I. $\vdash A$ iff A is true.
II. K $\vdash A$ iff A is K-true.
III. K is consistent iff K has a model.

Each of these statements is correct. They are known collectively as *Gödel's Completeness Theorem* and were demonstrated by Gödel (1930) for countable languages. The extension to uncountable languages was established by Henkin (1949). Statements II and III are equivalent in the sense that either can be deduced from the other (see Lemmas 1 and 2). They are generalizations of I inasmuch as I can be deduced from either II or III. Statement III is known as the *Strong Completeness Theorem*.

We begin by establishing the connection between II and III.

Lemma 1. If III is true, so is II.

Dem. See the proof of Theorem 2, page 102.

Lemma 2. If II is true, so is III.

Dem. 1. Assume that K is consistent. We must show that K has a model. If K does not have a model, then each wff is K-true. By II, K $\vdash A$ for each wff A; so K is contradictory. This contradiction proves that K has a model.

2. Assume that K has a model. Then K is consistent by Lemma 4, page 210. We conclude that III is correct if II is correct.

Next, we consider the connection between I and II.

Lemma 3. If II is true, so is I.

Dem. Now, II asserts that for any set of wffs K and for any wff A,

$$\mathsf{K} \vdash A \qquad \text{iff} \qquad A \text{ is } \mathsf{K}\text{-true}$$

With the empty set for K, this yields: For each wff A,

$$\varnothing \vdash A \qquad \text{iff} \qquad A \text{ is } \varnothing\text{-true}$$

But $\varnothing \vdash A$ iff $\vdash A$; and A is $\varnothing$-true iff A is true. Thus $\vdash A$ iff A is true.

Corollary 1. If III is true, so is I.

The goal of this chapter is to establish statement III, the Strong Completeness Theorem. We shall need two more concepts: the notion of a *maximal-consistent* set of wffs and the notion of an $\exists$-*complete* set of wff. We shall consider these ideas in the next two sections.

Exercises

1. Prove Corollary 1 directly, i.e., without using Lemma 3.
2. Assuming I and that $\mathsf{K} \vdash A$, prove that A is K-true.
3. Let K be a finite set of wffs and let A be K-true. Assuming I, prove that $\mathsf{K} \vdash A$.

12.4. Maximal-Consistent Sets

Here we shall present the concept of a maximal-consistent set of wffs; we need this notion to prove the Strong Completeness Theorem.

We have seen that each superset of a contradictory set of wffs is contradictory; on the other hand, a contradictory set may possess consistent subsets (certainly, the empty set is consistent). The situation is reversed for a consistent set of wffs, say K. Each subset of K is necessarily consistent; but K certainly possesses contradictory supersets. The question that interests us is this: Does K possess a proper superset that is consistent? If not, then we say that K *is* maximal-consistent.

Definition. A set of wffs, say K, is said to be *maximal-consistent* if:

1. K is consistent.
2. Each proper superset of K is contradictory.

To illustrate this concept, we present a method of constructing a semantical system $\sum$ from a given predicate calculus $\prod$. Let $\sum$ be the semantical system such that:

1. $R\alpha \in \text{dom} \sum$ iff $(R\alpha)$ is an atomic wff of $\prod$.
2. $\sum$ associates "true" with each member of its domain.

Now that we have defined $\sum$, we go about our business of constructing a maximal-consistent set. Let ι be the identity interpreter and form the set of wffs

$$\mathsf{K} = \{A \in \mathsf{W} \mid \iota A \text{ is true for } \textstyle\sum\}$$

Then K is maximal-consistent. To verify this claim, consider the argument on page 98ff.

Here is a criterion for maximal-consistent sets.

Criterion for Maximal-Consistent Sets. A consistent set of wffs, say K, is maximal-consistent iff $\mathsf{K} \cup \{B\}$ is contradictory for each wff B not in K.

Dem. 1. Assume that K is maximal-consistent. If $B \notin \mathsf{K}$, then $\mathsf{K} \cup \{B\}$ is a proper superset of K, so $\mathsf{K} \cup \{B\}$ is contradictory.

2. Assume that $\mathsf{K} \cup \{B\}$ is contradictory for each wff B not in K. Let L be any proper superset of K; then there is a wff A such that $A \in \mathsf{L}$ and $A \notin \mathsf{K}$. By assumption, $\mathsf{K} \cup \{A\}$ is contradictory; so L is contradictory. We conclude that K is maximal-consistent.

Here are some useful facts about maximal-consistent sets.

Lemma 1. Let K be maximal-consistent and let $\mathsf{K} \vdash A$; then $A \in \mathsf{K}$.

Dem. See the proof of Lemma 1, page 99.

Lemma 2. Let K be maximal-consistent and let $B \notin \mathsf{K}$; then $\neg B \in \mathsf{K}$.

Dem. See the proof of Lemma 2, page 99.

Corollary 1. Let K be maximal-consistent and let B be any wff; then either $B \in \mathsf{K}$ or else $\neg B \in \mathsf{K}$.

Dem. Apply Lemma 2.

Lemma 3. Let K be maximal-consistent, let $A \in \mathsf{K}$, and let $\vdash A \rightarrow B$; then $B \in \mathsf{K}$.

Dem. By assumption, $\mathsf{K} \vdash B$; so, by Lemma 1, $B \in \mathsf{K}$.

Our proof of the Strong Completeness Theorem relies on the following fact.

Theorem 1. Each consistent set possesses a maximal-consistent superset.

Dem. See the proof of Theorem 1, page 101.

In part 4(b) of the proof of Lemma 1, page 219, we shall require the fact that for any maximal-consistent set K such that $\forall tE \in \mathsf{K}$, $\mathsf{S}_t^s[E] \in \mathsf{K}$ for each individual s. In the following lemmas we build up a proof of this fact.

Lemma 4. Let K be maximal-consistent and let $\forall tE \in \mathsf{K}$; then $\mathsf{S}_t^s[E] \in \mathsf{K}$ provided that s is not bound in E.

Dem. Since s is not bound in E, $\vdash \forall tE \rightarrow \mathsf{S}_t^s[E]$; so $\mathsf{S}_t^s[E] \in \mathsf{K}$ by Lemma 3.

Lemma 5. Let K be maximal-consistent and let $\forall tE \in \mathsf{K}$; then $E \in \mathsf{K}$.

Dem. Let s be a pure individual that is not in E. By Lemma 2, page 163, $\vdash \forall tE \rightarrow \forall s\mathsf{S}_t^s[E]$. By Lemma 3, $\forall s\mathsf{S}_t^s[E] \in \mathsf{K}$. Thus, by Lemma 4, $\mathsf{S}_s^t[\mathsf{S}_t^s[E]] \in \mathsf{K}$; i.e., $E \in \mathsf{K}$.

Lemma 6. Let K be maximal-consistent and let $\forall tE \in \mathsf{K}$; then $\mathsf{S}_t^s[E] \in \mathsf{K}$ for each individual s.

Dem. By Lemma 4, $\mathsf{S}_t^s[E] \in \mathsf{K}$ if s is not bound in E. Accordingly, we now consider the case in which s is bound in E; thus, s is a pure individual. By Lemma 2, page 176, $\vdash \forall tE \rightarrow \forall s\mathsf{S}_t^s[E]$. So, by Lemma 3,

$\forall s \mathsf{S}_t^s[E] \in \mathsf{K}$. Therefore, by Lemma 5, $\mathsf{S}_t^s[E] \in \mathsf{K}$. This completes our proof of Lemma 6.

For the propositional calculus, a model of a maximal-consistent set K can be constructed from K itself, in a simple and direct fashion (see page 100). It is not so easy, however, to construct a model of a maximal-consistent set in the case of a predicate calculus.

The direct procedure for constructing a model involves assigning truth-values to atomic wffs so that an atomic wff A is true, under the assignment, if $A \in \mathsf{K}$; whereas A is false, under the assignment, if $\neg A \in \mathsf{K}$. Each assignment can be extended in a natural way to a valuation under which each wff receives a truth-value. So, each assignment yields an interpretation. This procedure works for the propositional calculus; to see why it fails for a predicate calculus, consider the following example. Let K be a maximal-consistent superset of

$$\{Ft \mid t \text{ is an individual}\} \cup \{\exists x(\neg Fx)\}$$

In the semantical system $\sum$ yielded by the direct procedure, Ft is interpreted as a true swff of $\sum$ for each individual t. Accordingly, the swff $\exists x(\neg Fx)$ is false for $\sum$. Thus, $\sum$ is not a model of K. In more direct terms, using a valuation in place of a semantical system, we make an assignment under which each atomic wff Ft is true; thus, the wff $\exists x(\neg Fx)$ is false under the ensuing valuation.

To overcome this difficulty, we need to increase the stock of pure individuals available in our predicate calculus; i.e., we must extend the given predicate calculus to another predicate calculus which possesses the same predicates but has more pure individuals. Here is the notion of an *extension* of a predicate calculus. We used this idea in Theorem 2, page 211.

Definition. A predicate calculus $\prod'$ is said to be an *extension* of a predicate calculus $\prod$ provided that:

1. $\prod$ is a subset of $\prod'$.
2. $\prod$ and $\prod'$ have the same predicates, and each predicate of $\prod$ has the same type as in $\prod'$.

Thus $\prod'$ is obtained by adjoining more pure individuals to $\prod$.

The idea is to use an extension of $\prod$, say $\prod'$, to construct a semantical system $\sum$, which we regard as an interpretation, via a $\sum$-interpreter. of $\prod$. This is achieved as follows. Define $\sum$ so that:

(a) $\alpha \in \text{dom} \sum$ iff (α) is an atomic wff of $\prod'$.
(b) Truth-values can be assigned as required.

Using truth-tables, we can compute the truth-value of each swff of $\sum$; but each wff of $\prod$ goes over, under the identity interpreter, into a swff of $\sum$. In this sense, the extension $\prod'$ allows us to interpret $\prod$.

The distinction between a predicate calculus and a semantical system rests on the fact that the latter includes truth-values. We may regard a semantical system as a predicate calculus plus an assignment of truth-values. A given predicate calculus can be interpreted in many ways. First, we are free to assign truth-values to its atomic wffs as we wish. Second, by increasing the stock of individuals to suit our purpose (this is the role of an extension) we vary our method of interpreting the given predicate calculus. We need this flexibility when we face the problem of constructing a model for a given set of wffs.

Essentially, a model is an interpretation of a language. For the propositional calculus, an interpretation is obtained once an assignment of truth-values to its atomic wffs is made. For a predicate calculus an interpretation requires not only an *assignment*, but also an *extension* of the given predicate calculus.

Exercises

1. Prove Lemma 1.
2. Prove Lemma 2.
3. Prove Corollary 1.
4. Prove Theorem 1.
5. Prove that $\mathsf{C}[\mathsf{K}] = \mathsf{K}$ if K is maximal-consistent.
6. Show that the conjecture

$$\text{“ } \mathsf{K} \text{ is maximal-consistent iff } \mathsf{C}[\mathsf{K}] = \mathsf{K}\text{”}$$

   is false.
7. Let $\sum$ be a semantical system matched to $\prod$ and let μ be a $\sum$-interpreter.
   (a) Show that $\{A \in \mathsf{W} \mid \mu A \text{ is true for } \sum\}$ is maximal-consistent.
   (b) Is $\{A \in \mathsf{W} \mid \mu(\neg A) \text{ is true for } \sum\}$ maximal-consistent? Is it consistent?

12.5. ∃-Complete Sets

Henkin's proof of the Strong Completeness Theorem requires one more notion, the notion of an ∃-complete set of wffs.

Definition. A set of wffs K is said to be ∃-*complete* if corresponding to each wff in K of the form "$\exists tA$" there is an individual s such that $S_t^s[A] \in K$.

For example, $\{\exists xFx, Fy\}$ is ∃-complete; $\{Fx\}$ is ∃-complete; the empty set is ∃-complete; $\{\forall yFy\}$ is ∃-complete; $\{\neg\forall yFy\}$ is ∃-complete. On the other hand, none of the following set is ∃-complete: $\{\exists yFy\}$, $\{\exists yFy, Fx, \neg\forall z(\neg Gzz)\}$, $\{\neg\forall y(\neg Fy), \neg(\neg Fx)\}$.

The notion of ∃-complete sets is important because of the fact that a set of wffs has a model if it is both ∃-complete and maximal-consistent.

Theorem 1. Let K be maximal-consistent and ∃-complete; then K has a model.

Our plan is to construct a semantical system $\sum$ from K, more precisely, from the atomic wffs of K. With an appropriate interpreter, $\sum$ turns out to be a model of K.

The first step is to construct $\sum$.

Construction of $\sum$. Let $\sum$ be the semantical system such that:

(a) $R\alpha \in \text{dom} \sum$ iff $(R\alpha)$ is an atomic wff of $\prod$.
(b) $\sum$ associates "true" with $R\alpha$ iff $(R\alpha) \in K$.

So, an atomic swff $(R\alpha)$ is true for $\sum$ if the corresponding wff $(R\alpha) \in K$; $(R\alpha)$ is false for $\sum$ if $\neg(R\alpha) \in K$. Remember that K is maximal-consistent, so either $(R\alpha) \in K$ or else $\neg(R\alpha) \in K$.

Let ι be the identity interpreter; we claim that $\sum$ is a model of K under ι. This claim is based on Lemma 1, which follows. Notice that this lemma asserts more than we actually need. We point out that the strengthened induction assumption allows us to carry through our argument.

Lemma 1. For each wff A, ιA is true for $\sum$ iff $A \in K$.

Dem. Let $S = \{A \in W \mid \iota A$ is true for $\sum$ iff $A \in K\}$. We shall show that $S = W$ by applying the Fundamental Theorem about Wffs.

1. Let A be any atomic wff; we shall show that $A \in S$. Take $A = (R\alpha)$, where R is a predicate and α is an R-string. By construction, $(R\alpha)$ is true for $\sum$ iff $(R\alpha) \in K$. So $A \in S$.

2. Let $B \in S$; we shall show that $\neg B \in S$.

(a) Assume that $\iota(\neg B)$ is true for $\sum$. Then ιB is false for $\sum$, so $B \notin K$. But K is maximal-consistent; so $\neg B \in K$.

(b) Assume that $\neg B \in \mathsf{K}$. Then $B \notin \mathsf{K}$, since K is consistent; so ιB is false for $\sum$. Therefore, $\iota(\neg B)$ is true for $\sum$.

3. Let $C, D \in S$; we shall show that $C \vee D \in S$.

(a) Assume that $\iota(C \vee D)$ is true for $\sum$. Then ιC is true for $\sum$ or ιD is true for $\sum$, or both; so $C \in \mathsf{K}$ or $D \in \mathsf{K}$. In either case, $C \vee D \in \mathsf{K}$ (by Lemma 3, page 216).

(b) Assume that $C \vee D \in \mathsf{K}$. If $\iota(C \vee D)$ is not true for $\sum$, then ιC is not true for $\sum$ and ιD is not true for $\sum$. Therefore, $C \notin \mathsf{K}$ and $D \notin \mathsf{K}$ (since $C, D \in S$). By Lemma 2, page 215, $\neg C \in \mathsf{K}$ and $\neg D \in \mathsf{K}$; so $\neg C \wedge \neg D \in \mathsf{K}$, and it follows that $\neg(C \vee D) \in \mathsf{K}$. Thus, K is contradictory. This contradiction proves that $\iota(C \vee D)$ is true for $\sum$.

4. Let $\mathsf{S}_t^s[E] \in S$ for each individual s, where E is a wff in which t is free; we shall show that $\forall tE \in S$.

(a) Assume that $\iota(\forall tE)$ is true for $\sum$. Then $\iota(\mathsf{S}_t^s[E])$ is true for $\sum$ whenever s is an individual. Thus, $\mathsf{S}_t^s[E] \in \mathsf{K}$ for each individual s. Now, either $\forall tE \in \mathsf{K}$ or $\exists t(\neg E) \in \mathsf{K}$. If the latter, then there is an individual a such that $\mathsf{S}_t^a[\neg E] \in \mathsf{K}$ (since K is $\exists$-complete). But $\mathsf{S}_t^a[E] \in \mathsf{K}$; so K is contradictory. This contradiction proves that $\forall tE \in \mathsf{K}$.

(b) Assume that $\forall tE \in \mathsf{K}$. By Lemma 6, page 216, $\mathsf{S}_t^s[E] \in \mathsf{K}$ for each individual s. Therefore, $\iota(\mathsf{I}_t^s[E])$ is true for $\sum$ whenever s is an individual. So $\iota(\forall tE)$ is true for $\sum$.

In view of the Fundamental Theorem about Wffs, this proves that $S = \mathsf{W}$, the set of all wffs, and completes our proof of Lemma 1.

It follows from Lemma 1 that $\sum$ is a model of K under ι. As we have already mentioned, our lemma asserts a little more; namely, that $A \in \mathsf{K}$ if A is a wff such that ιA is true for $\sum$. In short, K is the largest set of wffs for which $\sum$ is a model under ι.

Exercises

1. Let $\mathsf{K} = \{Ft \mid t \text{ is an individual}\} \cup \{\exists x(\neg Fx)\}$.
   (a) Show that K is not $\exists$-complete.
   (b) Show that K has a model.
   (c) Show that K is consistent.
   (d) Show that K is not maximal-consistent.

2. Let K be a set of wffs such that:
   (1) If $A \in \mathsf{K}$, then either A is atomic or else there is an atomic wff B such that A is $\neg B$.
   (2) If A is atomic, then $A \in \mathsf{K}$ or else $\neg A \in \mathsf{K}$ (but not both).
   (a) Is K $\exists$-complete?

(b) Show that K has a model.
(c) Show that K is consistent.
(d) Show that K is not maximal-consistent.
(e) Exhibit a superset of K that is maximal-consistent.
(f) Is the maximal-consistent set of part (e) also $\exists$-complete?

3. Let K be a nonempty set of wffs which contains only atomic wffs.

(a) Construct a model of K.
(b) Show that K is consistent.
(c) Show that K is not maximal-consistent.
(d) Exhibit a superset of K that is maximal-consistent.
(e) Is the maximal-consistent set of part (d) also $\exists$-complete?

4. Let $\sum$ be any semantical system such that $\alpha \in \operatorname{dom} \sum$ iff (α) is an atomic wff of $\prod$. Let $\mathsf{K} = \{A \in \mathsf{W} \mid \iota A \text{ is true for } \sum\}$, where ι is the identity interpreter.

(a) Show that K is maximal-consistent.
(b) Show that K is $\exists$-complete.

12.6. Proof of the Strong Completeness Theorem

In this section we shall prove that each consistent set of wffs possesses a superset that is both maximal-consistent and $\exists$-complete. More precisely, let $\prod$ be a predicate calculus and let K be a consistent set of wffs of $\prod$; we shall prove that there is an extension $\prod'$ of $\prod$ in which some superset K′ of K is both maximal-consistent and $\exists$-complete.

Of course, Theorem 1, page 219, applies to any predicate calculus. Therefore, K′ has a model, say $\sum$, obtained by following the construction given on page 219. But K is a subset of K′; so $\sum$ is also a model of K. If we can prove the claim that each consistent set can be extended to a maximal-consistent, $\exists$-complete set in some extension of the given predicate calculus, then we can conclude that each consistent set of wffs has a model. Previously (see Lemma 4, page 210), we proved that a set of wffs is consistent if it has a model. Therefore, we can prove the Strong Completeness Theorem by merely establishing our claim.

We now set to work. Let K be a consistent set of wffs in a predicate calculus $\prod$. We require a sequence of extensions of $\prod$, say

$$\textstyle\prod_0, \prod_1, \prod_2, \prod_3, \ldots$$

where $\prod_0 = \prod$, and for each $n = 0, 1, 2, \ldots, \prod_{n+1}$ is an extension of $\prod_n$. Remember that an extension involves merely increasing the stock of pure individuals. For simplicity, let us assume that $\prod$ has countably many predicates and individuals (therefore, countably many wffs). In this case, we require that each predicate calculus in our sequence of extensions is obtained from its predecessor by adjoining countably many new individuals. Thus $\prod_1$ is obtained from $\prod$ by adjoining countably many new individuals to the individuals of $\prod$, and for each $n \in N$, $\prod_{n+1}$ is obtained from $\prod_n$ by adjoining countably many new individuals to the individuals of $\prod_n$.

We need one more predicate calculus $\prod_\omega$, which is constructed from the sequence of predicate calculi $\prod_0, \prod_1, \prod_2, \ldots$. The predicates of $\prod_\omega$ are the predicates of $\prod_0$; the individuals of $\prod_\omega$ are the individuals of each $\prod_n$, $n \in N$, i.e., the set of individuals of $\prod_\omega$ is the union of the set of individuals of each $\prod_n$.

The union of a countable number of countable sets is a countable set; so the predicate calculus $\prod_\omega$ has a countable number of individuals. Since $\prod_\omega$ has the same predicates as $\prod$, which we assume forms a countable set, it follows that the set of all wffs of $\prod_\omega$ is countable. Therefore, the wffs of $\prod_\omega$ can be well-ordered; we choose one ordering of this set, which we shall call the standard ordering. Without our assumptions of countability, we still obtain a well-ordering of the wffs of $\prod_\omega$ by appealing to the well-ordering theorem, an equivalent of the Axiom of Choice.

We intend to extend K to a maximal-consistent, $\exists$-complete set of wffs in the predicate calculus $\prod_\omega$. Our technique for carrying out this extension involves two steps.

Step 1. Given a consistent set of wffs in a predicate calculus $\prod_n$, use the standard ordering of the wffs of $\prod_\omega$ to extend the given set to a maximal-consistent set of wffs in the predicate calculus $\prod_n$. We shall denote the resulting set of wffs by K_n.

Step 2. The purpose of this step is to treat K_n for $\exists$-completeness. Select the first wff of K_n, in the standard ordering, that has the form "$\exists tD$." Let a be the first individual in the list of new individuals of the predicate calculus $\prod_{n+1}$. By Lemma 5, page 210, $\mathsf{K}_n \cup \{\mathsf{S}_t^a[D]\}$ is consistent in $\prod_{n+1}$. Select the next wff of K_n, in the standard ordering, that has the form "$\exists uE$," and let b be the next individual in the list of new individuals of $\prod_{n+1}$. Form $\mathsf{K}_n \cup \{\mathsf{S}_t^a[D], \mathsf{S}_u^b[E]\}$, a consistent set in $\prod_{n+1}$. Repeat until each wff in K_n of the specified form has been treated. Denote the resulting set by K_n'; then K_n' is consistent in $\prod_{n+1}$.

Applying the twofold construction described in Steps 1 and 2 to the given set K yields first K_0, which is maximal-consistent in $\prod_0$, and then K_0', which is consistent in $\prod_1$. Next, we apply our twofold construction to K_0', obtaining in turn K_1 and K_1'; K_1 is maximal-consistent in $\prod_1$ and

K_1' is consistent in $\prod_2$. Continuing, we obtain the following sequence of supersets of K:

$$\mathsf{K}, \mathsf{K}_0, \mathsf{K}_0', \mathsf{K}_1, \mathsf{K}_1', \mathsf{K}_2, \mathsf{K}_2', \ldots, \mathsf{K}_n, \mathsf{K}_n', \ldots$$

These sets have the following properties, for each $n \in N$:

1. $\mathsf{K}_n \subset \mathsf{K}_{n+1}$.
2. K_n is maximal-consistent in $\prod_n$.
3. For each wff of the form "$\exists tD$" in K_n, there is an individual a of $\prod_{n+1}$ such that $\mathsf{S}_t^a[D] \in \mathsf{K}_{n+1}$.

We are now ready to produce a maximal-consistent, $\exists$-complete superset of K. Let $\mathsf{K}_\omega = \bigcup_{n \in N} \mathsf{K}_n$; i.e., $A \in \mathsf{K}_\omega$ iff there is a natural number n such that $A \in \mathsf{K}_n$. We claim that K_ω is maximal-consistent and $\exists$-complete.

Lemma 1. K_ω is maximal-consistent.

Dem. Assume that K_ω is contradictory. By Theorem 1, page 210, K_ω has a contradictory, finite subset, say $\{A_1, \ldots, A_n\}$. There are natural numbers $i_1, \ldots, i_n$ such that $A_1 \in \mathsf{K}_{i_1}, \ldots, A_n \in \mathsf{K}_{i_n}$; let $m = \max\{i_1, \ldots, i_n\}$; then $\{A_1, \ldots, A_n\} \subset \mathsf{K}_m$. It follows that K_m is contradictory; but K_m is consistent. This contradiction proves that K_ω is consistent. To prove that K_ω is maximal-consistent, let A be a wff of $\prod_\omega$ such that $A \notin \mathsf{K}_\omega$; we shall prove that $\mathsf{K}_\omega \cup \{A\}$ is contradictory. There is a natural number n such that A is a wff of $\prod_n$; by assumption, $A \notin \mathsf{K}_n$ (otherwise, $A \in \mathsf{K}_\omega$). Thus $\mathsf{K}_n \cup \{A\}$ is contradictory (recall that K_n is maximal-consistent in $\prod_n$). But $\mathsf{K}_n \cup \{A\}$ is a subset of $\mathsf{K}_\omega \cup \{A\}$; so $\mathsf{K}_\omega \cup \{A\}$ is contradictory. This proves that K_ω is maximal-consistent.

Lemma 2. K_ω is $\exists$-complete.

Dem. Let $\exists tD \in \mathsf{K}_\omega$; then $\exists tD \in \mathsf{K}_n$ for some $n \in N$. By property 3, there is an individual a of $\prod_{n+1}$ such that $\mathsf{S}_t^a[D] \in \mathsf{K}_{n+1}$. So $\mathsf{S}_t^a[D] \in \mathsf{K}_\omega$. This proves that K_ω is $\exists$-complete.

So K_ω is a maximal-consistent, $\exists$-complete superset of K. This establishes our claim that each consistent set can be extended to a maximal-consistent, $\exists$-complete set in $\prod_\omega$. As we have pointed out, this proves the Strong Completeness Theorem.

As for the propositional calculus, the Strong Completeness Theorem has the following corollaries. Here, K is any set of wffs.

Compactness Theorem. K has a model iff each finite subset of K has a model.

Dem. See page 102.

Theorem 1. Let A be any wff; then A is K-true iff $K \vdash A$.

Dem. See Theorem 2, page 102.

1. Let $\prod$ be a predicate calculus whose set of predicates has cardinal $\aleph$, where $\aleph \geq \aleph_0$, and whose set of pure individuals has cardinal $\aleph$.
   (a) Show that the set of wffs of $\prod$ has cardinal $\aleph$.
   (b) Let K be any consistent set of wffs of $\prod$. Use the method of the text to prove that K has a maximal-consistent, $\exists$-complete superset in some predicate calculus which is an extension of $\prod$.
2. Prove that K'_n is consistent in $\prod_{n+1}$ (see step 2, page 222).
3. Let K'_n be the set defined in step 2, page 222; is K'_n $\exists$-complete? Justify your answer.
4. Prove the Compactness Theorem.
5. Prove Theorem 1.
6. Let $\prod$ have $\aleph_0$ individuals and $\aleph_0$ predicates. Let K be any consistent set of wffs of $\prod$; prove that K has a model with $\aleph_0$ constants and $\aleph_0$ relation symbols.

PART III
Applications

13

Nonstandard Analysis

13.1. Extended Natural Number System

The natural number system is usually identified with the algebraic system $\langle N, +, \cdot, <, = \rangle$, where $N = \{1, 2, 3, \ldots\}$, $+$ and $\cdot$ represent the binary operations of addition and multiplication, $<$ denotes the *less than* relation, and $=$ is the equality relation on N. Now, any operation can be represented by a relation; e.g., addition is represented by the ternary relation S, where $(a, b, c) \in S$ iff $a + b = c$. In view of this observation, we can identify the natural number system with the semantical system $\mathscr{N}$ whose relation symbols are N, $+$, $\cdot$, $<$, and $=$; the constants of $\mathscr{N}$ are natural numbers; the placeholders of $\mathscr{N}$ are $x, y, z, \ldots$; the strings in the domain of $\mathscr{N}$ have the form

$$Na, \quad +abc, \quad \cdot abc, \quad <ab, \quad =ab$$

where a, b, and c are natural numbers. Truth-values are assigned in the obvious way; e.g., "$+abc$" is true for $\mathscr{N}$ if $a + b = c$.

In the language of $\mathscr{N}$ we have the following atomic swffs:

$$+123, \quad \cdot 236, \quad <45, \quad =55$$

Each of these is a true swff of $\mathscr{N}$. Here are some more true swffs of $\mathscr{N}$:

$\forall xyz(<xy \wedge <yz \rightarrow <xz)$	i.e., $<$ is transitive
$\forall xyz(+xyz \rightarrow +yxz)$	i.e., addition is commutative
$\forall xyz(+xyz \rightarrow <xz)$	i.e., $\forall xy(x < x + y)$

Moreover, equality is a substitutive equivalence relation; so the following swffs are true for $\mathscr{N}$:

$$\forall x(=xx), \quad \forall xy(=xy \rightarrow =yx), \quad \forall xyz(=xy \wedge =yz \rightarrow =xz)$$
$$\forall xu(Nx \wedge =xu \rightarrow Nu)$$
$$\forall xyzuvw(+xyz \wedge =xu \wedge =yv \wedge =zw \rightarrow +uvw)$$
$$\forall xyzuvw(\cdot xyz \wedge =xu \wedge =yv \wedge =zw \rightarrow \cdot uvw)$$
$$\forall xyuv(<xy \wedge =xu \wedge =yv \rightarrow <uv)$$
$$\forall xyuv(=xy \wedge =xu \wedge =yv \rightarrow =uv)$$

We want to extend $\mathscr{N}$ to a semantical system that includes infinite numbers. To this purpose we shall consider two predicate calculi $\prod$ and $\prod'$, where $\prod$ is the predicate calculus constructed from $\mathscr{N}$ as described below and $\prod'$ is the extension of $\prod$ obtained by adjoining the symbol "ω" to the individuals of $\prod$.

Construction of $\prod$. The predicates of $\prod$ are the relation symbols of $\mathscr{N}$; each predicate has the same type as the corresponding relation symbol. The individuals of $\prod$ are the constants of $\mathscr{N}$ and the placeholders of $\mathscr{N}$.

We point out that dom $\mathscr{N}$ can be regarded as a predicate calculus; $\prod$ is the extension of dom $\mathscr{N}$ obtained by adjoining the placeholders of $\mathscr{N}$ to the individuals of the predicate calculus dom $\mathscr{N}$.

Let ι be the identity interpreter from the predicate calculus dom $\mathscr{N}$ to the semantical system $\mathscr{N}$. We shall use ι to translate a wff of $\prod$, say A, into a swff of $\mathscr{N}$, provided that no placeholder of $\mathscr{N}$ is free in A.

Throughout the following, W is the set of all wffs of the predicate calculus $\prod$ constructed above.

Given that $A \in \mathsf{W}$, it is not necessarily the case that ιA is a swff of $\mathscr{N}$. However, ιA is a swff of $\mathscr{N}$ for each $A \in \mathsf{W}$ such that no placeholder of $\mathscr{N}$ is free in A (remember that bound individuals are ignored by each interpreter). Thus, if the statement "ιA is true for $\mathscr{N}$" is true, then ιA is a swff of $\mathscr{N}$; so, no placeholder of $\mathscr{N}$ is free in A.

Bearing this in mind, we point out that

$$\mathsf{K}' = \{A \in \mathsf{W} \mid \iota A \text{ is true for } \mathscr{N}\}$$

is a postulate set for $\mathscr{N}$, although rather large. For convenience, we shall denote "$<st$" and "Nm" by writing "$s < t$" and "$m \in N$," respectively. Using the predicate calculus $\prod'$, we can now form a set of wffs such that the corresponding set of swffs asserts the existence of an infinitely large natural number:

$$\mathsf{K}'' = \{\omega \in N, \quad 1 < \omega, \quad 2 < \omega, \quad 3 < \omega, \ldots\}$$

Abraham Robinson's idea is to take $\mathsf{K} = \mathsf{K}' \cup \mathsf{K}''$ as a postulate set for the semantical system we seek, namely an elementary extension of $\mathscr{N}$ that includes infinite numbers.

First we shall show that K has a model. In view of the compactness theorem, it is enough to show that each finite subset of K has a model. In fact, we shall show that $\mathscr{N}$ itself is a model of each finite subset of K. Let K_0 be any finite subset of K, and let $A \in \mathsf{K}_0$. Either ιA is true for $\mathscr{N}$ or A is in K''. Since K_0 is finite, there are only finitely many wffs in K_0 that have the form "$n < \omega$," say

$$n_1 < \omega, \quad n_2 < \omega, \ldots, \quad n \quad < \omega$$

Let $j = \max\{n_1, n_2, \ldots, n_k\}$ and let λ be the $\mathscr{N}$-interpreter such that

$$\lambda t = \begin{cases} j + 1 & \text{if} \quad t = \omega \text{ or } t \text{ is a placeholder of } \mathscr{N} \\ \iota t & \text{otherwise} \end{cases}$$

where t is any individual or predicate of $\prod'$. Note that $\lambda A = \iota A$ if none of $\omega, x, y, z, \ldots$ is free in A; therefore, λA is true for $\mathscr{N}$ for each $A \in \mathsf{K}_0$ that does not have the form "$n < \omega$." If A has the form "$n < \omega$," where $A \in \mathsf{K}_0$, then λA is "$n < j + 1$," which is true for $\mathscr{N}$. This proves that $\mathscr{N}$ is a model of K_0 under λ; so, each finite subset of K has a model. We conclude from the compactness theorem that K has a model; of course, K may have many models.

We now choose one of the models of K. Call this semantical system $^*\mathscr{N}$; also, we shall denote the interpreter involved by $*$. To simplify our notation, for each $n \in N$, substitute n for each instance of $^*n$ in each string in the domain of $^*\mathscr{N}$; similarly, substitute ω for each instance of $^*\omega$ in the domain of $^*\mathscr{N}$. Clearly, the resulting semantical system is also a model of K. We may as well assume that this semantical system is the model of K that we chose initially; so $^*\mathscr{N}$ is a model of K under an interpreter $*$ such that $^*n = n$ for each $n \in N$, and $^*\omega = \omega$. The relation symbols of $^*\mathscr{N}$ are denoted by $^*N$, $^*+$, $^*\cdot$, $^*<$, and $^*=$. It is usually clear from the context whether we are discussing $\mathscr{N}$ or $^*\mathscr{N}$; accordingly, it is customary to suppress the stars in the relation symbols of $^*\mathscr{N}$, writing $+$, $\cdot$, $<$, and $=$ for the relation symbols of $^*\mathscr{N}$.

Note that $^*N$, the image of N under the interpreter $*$, is the set of constants of $^*\mathscr{N}$. Thus, each quantifier in a swff of $^*\mathscr{N}$ refers to $^*N$. By construction, ω and each natural number are in $^*N$; so $N \subset {}^*N$. We shall prove, in a moment, that $^*\mathscr{N}$ has many more constants. Now we shall prove that $^*N \neq N$.

Lemma 1. $^*N$ is a proper superset of N.

Dem. Here, we use the fact that

$$1 < \omega, \quad 2 < \omega, \quad 3 < \omega, \ldots$$

are postulates for $*\mathcal{N}$. Notice that our postulates for $*\mathcal{N}$ also include statements that assert that $*<$ is an order relation; i.e., $*<$ is transitive and obeys the trichotomy law. Therefore, each of the statements

$$\omega * = 1, \quad \omega * = 2, \quad \omega * = 3, \ldots$$

is false for $*\mathcal{N}$. Thus $\omega \in *N - N$; we conclude that $*N$ is a proper superset of N.

Statements involving operations can be expressed in terms of relations, albeit at some cost in readability. For example, if the term "$x + y$" appears in a statement, we insert "$+xyz \rightarrow$," replace "$x + y$" by z, and quantify over z. To be specific, consider the associative law

$$\forall xyz(x + (y + z) = (x + y) + z)$$

Expanding as indicated yields

$$\forall xyzuvwt(+yzu \wedge +xuv \wedge +xyw \wedge +wzt \rightarrow =vt)$$

Our concern is to simplify our statements (i.e., wffs or swffs) in the interest of readability. Therefore, we shall express our statements in terms of operations rather than relations, wherever this is appropriate.

With the same goal in mind, we shall freely utilize the usual mathematical conventions to represent wffs or swffs. For example, we shall represent "$=xy$" and "$>xy$" by writing "$x = y$" and "$x > y$," respectively. Statements involving the relations $\leq$ or $\geq$ shall be handled in a similar fashion.

We return, now, to our task of showing that postulating the existence of a number greater than each natural number, in conjunction with the postulates for $\mathcal{N}$, ensures the existence of infinitely many new numbers.

Lemma 2. $*N - N$ is an infinite set.

Dem. The wff represented by $\forall x(x < x + 1)$ is in K; so, the swff represented by $\forall x(x < x + 1)$ is true for $*\mathcal{N}$ (i.e., we have interpreted the preceding wff to obtain this swff of $*\mathcal{N}$; the interpreter involved is *). Note that we have suppressed two stars in writing this swff; for clarification, let us put them back. We obtain: $\forall x(x *< x *+ 1)$. So $\omega *< \omega *+ 1$. Now, $*<$ is transitive in $*\mathcal{N}$; also, for each $n \in N$, $n *< \omega$. Thus, for each $n \in N$, $n *< \omega *+ 1$. As before, this allows us to conclude that for each $n \in N$, the swff "$n *= \omega *+ 1$" is false for $*\mathcal{N}$. Recall that $*=$ is an equivalence relation on $^*\mathcal{N}$ (in fact, a substitutive equivalence relation); so $\omega \;^*\!\!+ 1$ is not in N. By the same argument, we see that $\omega \;^*\!\!+ n$ is not in N if $n \in N$. Moreover, if $m \neq n$, then "$\omega \;^*\!\!+ m \;^*\!\!= \omega \;^*\!\!+ n$" is false for $^*\mathcal{N}$. Therefore, the following constants are all distinct, and are all members of $^*N - N$.

$$\omega, \quad \omega \;^*\!\!+ 1, \quad \omega \;^*\!\!+ 2, \quad \omega \;^*\!\!+ 3, \ldots$$

Moreover, $\omega \;^*\!\!+ \omega$ is a constant of $^*\mathcal{N}$; this number is different from each of the numbers listed above, and is not in N. So, the above list can be extended to include each of the following constants:

$$\omega \;^*\!\!+ \omega, \quad \omega \;^*\!\!+ (\omega \;^*\!\!+ 1), \quad \omega \;^*\!\!+ (\omega \;^*\!\!+ 2), \ldots$$

This completes our discussion of Lemma 2.

Note our practice of referring to members of $^*N$ as *numbers*. We shall recognize two sorts of numbers, *infinite* numbers and *finite* numbers.

Definition. A member of $^*N$, say κ, is said to be *infinite* provided that $n \;^*\!\!< \kappa$ for each $n \in N$. A member of $^*N$ is said to be *finite* provided it is not infinite.

Let us show that the finite numbers are the members of N and the infinite numbers are the members of $^*N - N$.

Lemma 3. Each member of N is finite; each member of $^*N - N$ is infinite.

Dem. 1. Let $n \in N$. Then $n < n + 1$, so $n \;^*\!\!< n \;^*\!\!+ 1$. But $n \;^*\!\!+ 1$ *is* $n + 1$; i.e., $n \;^*\!\!< n + 1$. Since $^*\!\!<$ is an order relation on $^*N$, it follows that the statement $n + 1 \;^*\!\!< n$ is false for $^*\mathcal{N}$. Therefore, n is finite.

2. Let $\kappa \in {}^*N - N$. If κ is not infinite, then $\kappa \;^*\!\!\leq n$ for some $n \in N$. The swff

$$\forall m(m \leq n \leftrightarrow m = 1 \vee \cdots \vee m = n), \qquad m \in N$$

is true for $\mathcal{N}$. The corresponding wff is in K; so

$$\forall m(m \leq n \leftrightarrow m = 1 \vee \cdots \vee m = n), \qquad m \in {}^*N$$

is true for $^*\mathcal{N}$. In particular,

$$\kappa \leq n \leftrightarrow \kappa = 1 \vee \cdots \vee \kappa = n$$

is true for $^*\mathcal{N}$. So

$$\kappa = 1 \vee \cdots \vee \kappa = n$$

is true for $^*\mathcal{N}$. We conclude that $\kappa \in N$. This contradiction proves that κ is infinite.

By construction, for each atomic swff $R\alpha$ of $\mathcal{N}$, "$\alpha \in R$" is true for $\mathcal{N}$ iff "$\alpha \in {}^*R$" is true for $^*\mathcal{N}$. Therefore, we have the following result.

Theorem 1. $^*\mathcal{N}$ is an extension of $\mathcal{N}$.

Since K is a postulate set for $^*\mathcal{N}$, we see that each swff that is true for $\mathcal{N}$ is also true for $^*\mathcal{N}$ when interpreted in $^*\mathcal{N}$ by the interpreter *. To explore the ramifications of this statement, we consider the following situation. Let A be any wff of $\prod$ in which none of $x, y, z, \ldots$ is free; then ιA is the corresponding swff of $\mathcal{N}$, and $^*A$ is the corresponding swff of $^*\mathcal{N}$. Moreover, assume that A is such that $^*A$ is true for $^*\mathcal{N}$ and ιA is false for $\mathcal{N}$. Then $\iota(\neg A)$ is true for $\mathcal{N}$, so $^*(\neg A)$ is true for $^*\mathcal{N}$. This contradiction proves that ιA is true for $\mathcal{N}$ if $^*A$ is true for $^*\mathcal{N}$. This establishes the vital Transfer Theorem, which we now state.

Transfer Theorem. Let A be any wff of $\prod$ in which none of $x, y, z, \ldots$ is free. Then ιA is true for $\mathcal{N}$ iff $^*A$ is true for $^*\mathcal{N}$.

It follows that $^*\mathcal{N}$ is an elementary extension of $\mathcal{N}$.

We shall enrich the natural number system $\mathcal{N}$ by incorporating within this semantical system such concepts as the notion of *even* natural numbers, the notion of *odd* natural numbers, the concept of *prime* numbers, the concept of finite tuples whose terms are natural numbers, the operation of summing the terms of a finite tuple, and the concept of *sets* of natural numbers. To this purpose, we require additional relation symbols:

E (even numbers), O (odd numbers), P (primes), T (finite tuples), S (the sum of a finite tuple), $\mathscr{P}N$ (subsets of N)

Furthermore we need the natural apparatus for working with tuples and sets. For example, we need to include a relation symbol, say U, of type $\{3\}$ such that $Ua n\alpha$ is true for $\mathcal{N}$ iff a is the nth term of α. Also we need the usual predicate symbol $\in$ of type $\{2\}$ for membership in a set of numbers. Naturally we write "$x \in S$" in place of "$\in xS$."

For the sake of readability we shall generally follow the convention of writing "$\alpha \in R$" for "$R\alpha$," where R is a relation symbol and α is an R-string. This is not to be confused with the other usage of the $\in$ symbol.

Note that the constants of $\mathcal{N}$ now include natural numbers, sets of natural numbers, and finite tuples of natural numbers. Therefore, each quantifier appearing in a swff of $\mathcal{N}$ must be *relativized*, i.e., we must indicate to which sort of constant the quantifier refers. For example, Peano's *induction postulate* is

$$\forall y(y \in \mathscr{P}N \dot{\rightarrow} 1 \in y \wedge \forall x(x \in N \rightarrow (x \in y \rightarrow x + 1 \in y)) \rightarrow y = N) \quad (1)$$

The usual practice is that the sort of each quantifier is indicated typographically; so capital letters indicate quantification over $\mathscr{P}N$, lower case letters indicate quantification over N, and Greek letters indicate quantification over T. Thus (1) is abbreviated by

$$\forall S(1 \in S \wedge \forall x(x \in S \rightarrow x + 1 \in S) \rightarrow S = N) \quad (2)$$

As before, we obtain an elementary extension of $\mathscr{N}$ by applying the compactness theorem to a suitable postulate set. First, we form the predicate calculus dom $\mathscr{N}$; this predicate calculus is obtained directly from the semantical system $\mathscr{N}$ (the variables of dom $\mathscr{N}$ are the relation symbols and constants of the semantical system $\mathscr{N}$). Next, we extend the predicate calculus dom $\mathscr{N}$ to $\prod'$ by adjoining $\omega, x, y, z, \ldots$ to the individuals of dom $\mathscr{N}$. Let ι be the identity interpreter from the predicate calculus dom $\mathscr{N}$ to the semantical system $\mathscr{N}$, and let W be the set of all wffs of $\prod'$; then for each $A \in \mathsf{W}$ such that none of $\omega, x, y, z, \ldots$ is free in A, ιA is a swff of $\mathscr{N}$.

Here is our postulate set for an elementary extension of $\mathscr{N}$:

$$\mathsf{K} = \{A \in \mathsf{W} \mid \iota A \text{ is true for } \mathscr{N}\} \cup \{\omega \in N, 1 < \omega, 2 < \omega, 3 < \omega, \ldots\}$$

By the same argument as before, it is easy to prove that each finite subset of K has a model, namely $\mathscr{N}$ itself; therefore, by the compactness theorem, K has a model. We choose one of the models of K, which we shall call $^*\mathscr{N}$, and let the interpreter involved be denoted by * also. We can assume that we have chosen $^*\mathscr{N}$ so that $^*n$ is n for each $n \in N$, and that $^*\omega$ is ω. Since each constant of $\mathscr{N}$ is in $N \cup \mathscr{P}N \cup T$, it follows from our postulate set for $^*\mathscr{N}$ that each constant of $^*\mathscr{N}$ is in $^*N \cup {}^*[\mathscr{P}N] \cup {}^*T$. Each constant in $\mathscr{P}N$ is a subset of N; so each constant in $^*[\mathscr{P}N]$ may be assumed to be a subset of $^*N$. Similarly, we use K to prove that each constant in $^*T$ is an n-tuple, where $n \in {}^*N$, whose terms are in $^*N$. The point is that the constants of $^*\mathscr{N}$ are expressed in terms of the members of $^*N$. Each constant of $^*\mathscr{N}$ is either a member of $^*N$, a subset of $^*N$, or an n-tuple whose terms are in $^*N$.

We now present a paradox, which is intended to develop our insight into $^*\mathscr{N}$ by bringing out some of the subtleties involved. Since (2) is true for $\mathscr{N}$, it is true for $^*\mathscr{N}$ when interpreted in $^*\mathscr{N}$. Note that N is a subset of $^*N$ such that $1 \in N$ and $\forall x(x \in N \rightarrow x + 1 \in N)$. Therefore, from (2) interpreted in $^*\mathscr{N}$, $^*N = N$. This contradicts Lemma 1.

The fallacy in this argument is easy to spot. Certainly (2) is true for $*\mathcal{N}$ when interpreted in $*\mathcal{N}$. But (2) is an abbreviation for (1); so (1) is true for $*\mathcal{N}$ when interpreted in $*\mathcal{N}$. To interpret (1) in $*\mathcal{N}$, we must replace each relation symbol of (1) by the corresponding relation symbol of $*\mathcal{N}$, and we must replace each free individual of (1) by its image under the interpreter $*$. This yields

$$\forall y(y \in *[\mathscr{P}N] \dot{\to} 1 \in y \wedge \forall x(x \in *N \to (x \in y \to x + 1 \in y)) \to y = *N) \tag{3}$$

which is true for $*\mathcal{N}$. In particular,

$$N \in *[\mathscr{P}N] \dot{\to} 1 \in N \wedge \forall x(x \in *N \to (x \in N \to x + 1 \in N)) \to N = *N \tag{4}$$

is true for $*\mathcal{N}$. The fallacy rests on the assumption that $N \in *[\mathscr{P}N]$, from which (4) allows us to conclude that $N = *N$. In fact $N \notin *[\mathscr{P}N]$, and this is what our paradox actually demonstrates. We point out that $*[\mathscr{P}N]$ is a certain collection of subsets of $*N$, but not all; i.e., $*[\mathscr{P}N] \neq \mathscr{P}(*N)$. A subset of $*N$ that is a member of $*[\mathscr{P}N]$ is called an *internal* subset of $*N$; any other subset of $*N$ is said to be an *external* subset of $*N$. These concepts allow us to verbalize the interpretation of a statement such as (2) in $*\mathcal{N}$. Simply, interpret "subset" as "internal subset." Thus, to interpret (2) in $*\mathcal{N}$, we read: "For each internal subset of $*N$, say S, if $1 \in S$ and if $\forall x(x \in S \to x + 1 \in S)$, then $S = *N$." This is how (2) is interpreted in $*\mathcal{N}$ (of course, $+$ must be interpreted as $*+$).

Similarly, a tuple whose terms are in $*N$ is said to be *internal* if it is in $*T$; otherwise, it is called *external*. For convenience, we shall call the members of $*N$ *natural numbers*; indeed, we shall generally refer to a concept of $*\mathcal{N}$ by the same name as the corresponding concept of $\mathcal{N}$. We mention that each finite tuple $(a_1, \ldots, a_n)$ is in $*T$ provided that $n \in N$ and each a_i is a natural number (i.e., is in $*N$). In this sense, each finite tuple is internal. Some infinite tuples are internal also. For example, $(1, 2, 3, \ldots, \omega)$ is internal; so is the ω-tuple $(1, \ldots, 1)$. Here is an infinite tuple that is external. Let $(a_1, \ldots, a_\omega)$ be the ω-tuple such that

$$a_i = \begin{cases} 1 & \text{if} \quad i \in N \\ 2 & \text{if} \quad i \in *N - N \text{ and } i \leq \omega \end{cases}$$

The point, here, is that for each tuple α, $\{n \in N \mid a \text{ is the } n\text{th term of } \alpha\}$ is a subset of N; here, $a \in N$. Since this is true for $\mathcal{N}$, it is also true for $*\mathcal{N}$ when interpreted in $*\mathcal{N}$. Therefore, if $(a_1, \ldots, a_\omega)$ is internal, then both N and $*N - N$ are internal subsets of $*N$. We have just proved that N is not an internal subset of $*N$. Therefore, $(a_1, \ldots, a_\omega)$ is not internal.

Exercises

1. Exhibit the wff of $\prod$ represented by $\forall x(x < x + 1)$.

2. Simplify the wff $\forall xyzuv(+xzu \wedge +yzv \wedge <xy \rightarrow <uv)$.

3. What is meant by saying that an equivalence relation of a relational system is substitutive?

4. Prove Theorem 1.

5. Show that $^*\mathscr{N}$ is a model of any postulate set for $\mathscr{N}$ that contains only swffs of $\mathscr{N}$.

6. Show that we can regard $\mathscr{N}$ and $^*\mathscr{N}$ as *similar*; i.e., $\mathscr{N}$ and $^*\mathscr{N}$ have the same relation symbols.

7. Let A be any swff of $\mathscr{N}$ and let m be a natural number that is free in A. Exhibit a swff of $\mathscr{N}$ that asserts the existence of the set $\{n \in N \mid \mathrm{S}^n_m[A]\}$.

8. Include an appropriate relation symbol in $\mathscr{N}$, so that the existence of

$$\{n \in N \mid a \text{ is the } n\text{th term of } \alpha\}$$

can be characterized by a swff of $\mathscr{N}$; here $\alpha \in T$ and $a \in N$. Exhibit the swff that asserts the existence of this set.

9. Let $\exists tE$ be a swff of $\mathscr{N}$ such that $\exists t{}^*E$ is true for $^*\mathscr{N}$. Prove directly that there is a constant a of $\mathscr{N}$ such that $\mathrm{S}^a_t[^*E]$ is true for $^*\mathscr{N}$; do not use the Test for Elementary Extensions, page 125.

10. Find the fallacy in the following argument. "The sum of the terms of the ω-tuple $(1, \ldots, 1)$ is ω; so "$S1 \cdots 1\omega$" is true for $^*\mathscr{N}$. But $^*\mathscr{N}$ is an elementary extension of $\mathscr{N}$; thus, by Lemma 5, page 126, there is a constant of $\mathscr{N}$, say n, such that "$S1 \cdots 1n$" is true for $^*\mathscr{N}$. Clearly, "$S1 \cdots 1n$" is false for $^*\mathscr{N}$ since n is finite. Contradiction!"

13.2. Extended Real Number System

We have presented Robinson's method for obtaining an elementary extension of a number system in Section 13.1; there, we applied his method to the natural number system. Robinson's procedure, which allows us to extend the natural number system to a number system that includes infinite numbers, also allows us to obtain an elementary extension of the real number system that includes both infinitely large and infinitely small numbers.

It is essential to incorporate in the real number system all concepts that we wish to discuss, now or later. Accordingly, let $\mathscr{R}$ be a semantical system whose relation symbols include

$$R,\ N,\ \mathscr{P}R,\ \mathscr{P}N,\ T,\ F, f,\ +,\ \cdot,\ <,\ >,\ =, \ldots$$

and whose constants include each real number, each set of real numbers, each finite tuple whose terms are real numbers, each function, and so no. The placeholders of $\mathscr{R}$ are $x, y, z, \ldots$.

The domain of $\mathscr{R}$ includes strings such as

$$R-2,\ N3,\ T(1, 3),\ Ff, f24,\ <45,\ =77$$

Truth-values are assigned in the obvious way; e.g., "Ff" is true for $\mathscr{R}$ provided that f is a function (i.e., a map of a subset of R into R, where R is the set of all real numbers).

Since each concept of real numbers is a relation symbol of the semantical system $\mathscr{R}$, the language of $\mathscr{R}$ is extremely rich. For example, we include the *upper bound* concept and the *least upper bound* concept in $\mathscr{R}$ by way of relation symbols for these concepts; so the Completeness Theorem of the real number system is a swff of $\mathscr{R}$. Similarly, Peano's induction postulate is a swff of $\mathscr{R}$. More simply, the postulates for an ordered field are swffs of $\mathscr{R}$; indeed, the Archimedean property can be expressed by a swff of $\mathscr{R}$.

We want to establish the existence of an elementary extension of $\mathscr{R}$ that includes both infinitely large and infinitely small numbers. To this purpose, let $\prod$ be the predicate calculus formed from the objects of $\mathscr{R}$ and symbols $\omega, x, y, z, \ldots$ as follows. The predicates of $\prod$ are the relation symbols of $\mathscr{R}$; the individuals of $\prod$ are the constants of $\mathscr{R}$ and $\omega, x, y, z, \ldots$. Let ι be the identity interpreter from the predicate calculus dom $\mathscr{R}$ to the semantical system $\mathscr{R}$; then ιA is a swff of $\mathscr{R}$ provided that A is a wff and none of $\omega, x, y, z, \ldots$ is free in A.

Let W' be the set of all wffs of $\prod$ in which none of $\omega, x, y, z, \ldots$ is free. Then

$$\{A \in \mathsf{W}' \mid \iota A \text{ is true for } \mathscr{R}\}$$

can be regarded as a postulate set for $\mathscr{R}$, admittedly a trifle oversized. Robinson now adjoins a few more postulates which collectively assert the existence of an infinite number; let

$$\mathsf{K} = \{A \in \mathsf{W}' \mid \iota A \text{ is true for } \mathscr{R}\} \cup \{\omega \in N,\quad 1 < \omega,\quad 2 < \omega,\quad 3 < \omega, \ldots\}$$

This is Robinson's postulate set for an extension of $\mathscr{R}$ that includes infinitely large and infinitely small numbers.

It is clear (see page 229) that each finite subset of K has a model, namely $\mathscr{R}$ itself. Thus, by the compactness theorem, K has a model; of course, K may have many models. We now choose one of the models of K;

call this semantical system $^*\mathscr{R}$, and let the interpreter involved be denoted by * also. We can choose the semantical system $^*\mathscr{R}$ so that the image of each object of $\mathscr{R}$, under the interpreter *, is the object itself. For simplicity, we shall assume that $^*\omega$ is ω itself (i.e., we shall denote the image of ω in $^*\mathscr{R}$, whatever it is, by ω).

Since we shall sometimes want to refer to a relation of $\mathscr{R}$ and the corresponding relation of $^*\mathscr{R}$ at the same time, it is necessary to insert stars on names for relations of $^*\mathscr{R}$; e.g., we shall write $^*N$ for the set of natural numbers of $^*\mathscr{R}$ and we shall write $^*\mathscr{R}$ for the set of real numbers of $^*\mathscr{R}$. Bear in mind the distinction between a *relation symbol* and the corresponding *relation*; a relation is a set.

Now, the constants of $\mathscr{R}$ are real numbers, sets of real numbers, tuples of real numbers, and functions; taking the union of the relations R, $\mathscr{P}R$, T, and F yields the set of all constants of R, namely

$$R \cup \mathscr{P}R \cup T \cup F$$

The wff $\forall x(x \in R \vee x \in \mathscr{P}R \vee x \in T \vee x \in F)$ is in K; so the set of constants of $^*\mathscr{R}$ is

$$^*\mathscr{R} \cup {}^*[\mathscr{P}R] \cup {}^*T \cup {}^*F$$

(later, we may include additional sorts of constants in $\mathscr{R}$, and so in $^*\mathscr{R}$).

Let A be any atomic wff in W'; so ιA is a swff of $\mathscr{R}$. Then $A \in \mathsf{K}$ iff ιA is true for $\mathscr{R}$. Of course, $^*A$ is true for $^*\mathscr{R}$ iff $A \in \mathsf{K}$. We conclude that

$$\iota A \text{ is true for } \mathscr{R} \text{ iff } {}^*A \text{ is true for } {}^*\mathscr{R}$$

This establishes the following result.

Lemma 1. $^*\mathscr{R}$ is an extension of $\mathscr{R}$.

Let A be any wff in W'; then $A \in \mathsf{K}$ iff $\neg A \notin \mathsf{K}$. Of course, if ιA is true for $\mathscr{R}$, then $A \in \mathsf{K}$, so $^*A$ is true for $^*\mathscr{R}$. Assume, next, that $^*A$ is true for $^*\mathscr{R}$; we wish to prove that ιA is true for $\mathscr{R}$. If not, then $\neg\iota A$ is true for $\mathscr{R}$; so $\neg A \in \mathsf{K}$; thus $\neg{}^*A$ is true for $^*\mathscr{R}$. This contradiction proves that ιA is true for $\mathscr{R}$. We conclude that $^*\mathscr{R}$ is an elementary extension of $\mathscr{R}$. In other words, we have established the transfer theorem.

Transfer Theorem. Let $A \in \mathsf{W}'$; then ιA is true for $\mathscr{R}$ iff $^*A$ is true for $^*\mathscr{R}$.

We want to show that $^*\mathscr{R}$ is a proper extension of $\mathscr{R}$. From K, our postulate set for $^*\mathscr{R}$, we see that $^*\omega \in {}^*N$ and $n \mathbin{^*<} {}^*\omega$ for each $n \in N$. Deleting stars, as agreed, we have $\omega \in {}^*N$ and $n < \omega$ for each $n \in N$.

Now, $\forall x \exists n(x \in R \rightarrow n \in N \wedge x < n)$ is in K; thus, for each $x \in R$, the wff $x < n$ is in K, for some $n \in N$. So, $x < n$ and $n < \omega$ are both true for $^*\mathscr{R}$. But $<$ is transitive in $\mathscr{R}$; so $<$ (i.e., $^*<$) is transitive in $^*\mathscr{R}$. Thus, $x < \omega$ is true for $^*\mathscr{R}$, for each $x \in R$. It follows from the trichotomy law that $\omega \neq x$ for each $x \in R$. This proves that $^*R$ is a proper superset of R; so we have the following result.

Lemma 2. $^*\mathscr{R}$ is a proper extension of $\mathscr{R}$.

The Transfer Theorem is extremely useful since it allows us to utilize our knowledge of $\mathscr{R}$ in studying $^*\mathscr{R}$ (and vice versa). Bear in mind the restriction on A; A must be a wff in which none of $\omega, x, y, z, \ldots$ is free. This ensures that ιA is a swff of $\mathscr{R}$. Also, we must take care when verbalizing a statement $^*A$ in the language of $^*\mathscr{R}$. The fact is that the relations of $^*\mathscr{R}$ need not have quite the same significance as the corresponding relations of $\mathscr{R}$; we must bear in mind that the domain of each quantifier of $^*A$ is not necessarily the same as the domain of the corresponding quantifier of ιA. The failure to verbalize $^*A$ accurately sometimes results in a paradox.

Since each nonzero member of R has a multiplicative inverse, the transfer theorem allows us to conclude that each nonzero member of $^*R$ has a multiplicative inverse. In particular, ω has a multiplicative inverse, which is denoted by $1/\omega$. We have just proved that $x < \omega$ for each $x \in R$; using this and facts about inequalities in $\mathscr{R}$, we can show that $1/\omega < h$ for each $h \in R, h > 0$. Thus $|1/\omega| < h$ for each $h \in R, h > 0$. Any member of $^*R$ with this property is called an infinitesimal.

Definition. Let $\varepsilon \in {}^*R$; then ε is said to be an *infinitesimal* provided that $|\varepsilon| < h$ for each $h \in R, h > 0$. Let $\kappa \in {}^*R$; then κ is said to be *infinite* provided that $|\kappa| > h$ for each $h \in R$. Let $a \in {}^*R$; then a is said to be *finite* provided that a is not infinite.

Notice that 0 is an infinitesimal, $-\varepsilon$ is an infinitesimal if ε is an infinitesimal, $-\kappa$ is infinite if κ is infinite, and $-a$ is finite if a is finite. Each infinitesimal is finite, and each member of R is finite. The sum or product of two finite numbers is finite. The sum or product of two infinitesimals is an infinitesimal. The product of two infinite numbers is infinite; however, the sum of two infinite numbers is not necessarily infinite.

The idea that a number is *approximated* by another number (or that a point is *close* to another point) underlies calculus. For example, when we declare that a function f is continuous at a, $a \in R$, we have in mind the following:

$$f(x) \text{ approximates } f(a) \text{ for each number } x \text{ that approximates } a \qquad (1)$$

If restricted to $\mathscr{R}$, we can regard (1) as intuitive only, since there is no way that we can express the idea that one real number approximates another real number within the semantical system $\mathscr{R}$. However, using the resources of the extension $^*\mathscr{R}$, we can pin down this notion. Intuitively, we must admit that two numbers are close if they differ by an infinitesimal. Accordingly, Robinson introduced the following equivalence relation on $^*R$.

Definition. Let $a, b \in {}^*R$; then $a \simeq b$ (read "a is *infinitely close* to b") provided that $a - b$ is an infinitesimal.

Note that we are enriching the language of $^*\mathscr{R}$ by adjoining the relation symbol $\simeq$ to $^*\mathscr{R}$. It is convenient to enlarge $^*\mathscr{R}$ in this way; we point out that the type of $\simeq$ is $\{2\}$ and that we use the definition to assign truth-values to atomic swffs of the form $a \simeq b$. Of course, the language of $^*\mathscr{R}$ is automatically enriched by including another relation symbol, namely $\simeq$, among the objects of $^*\mathscr{R}$. We can now express (1) directly, as follows:

$$\forall x(x \simeq a \rightarrow f(x) \simeq f(a)), \qquad x \in {}^*R, \quad x \in \operatorname{dom} f \qquad (2)$$

This is a swff of our enriched $^*\mathscr{R}$. Note that we have suppressed the * on f.

One of the achievements of 19th century mathematicians was the discovery that there is a swff C of $\mathscr{R}$, which involves a function f, such that C is true for $\mathscr{R}$ iff f is continuous at a [in the sense of (2)]. For example,

$$\forall \varepsilon \exists \delta \forall x(|x - a| < \delta \rightarrow |f(x) - f(a)| < \varepsilon) \qquad (3)$$
$$\varepsilon, \delta > 0 \quad \text{and} \quad x \in \operatorname{dom} f$$

is true for $\mathscr{R}$ iff (2) is true for $^*\mathscr{R}$. In Section 13.6 we shall prove this statement. We emphasize that (3) does not express the idea contained in (2); instead, it is merely equivalent to (2) in the sense that (3) is true for $\mathscr{R}$ iff (2) is true for $^*\mathscr{R}$.

The notion of *internal* vs. *external*, which we introduced on page 234, applies to $^*\mathscr{R}$ as well as to $^*\mathscr{N}$. Of course, $^*[\mathscr{P}R]$ is a subset of $\mathscr{P}(^*R)$. The members of $^*[\mathscr{P}R]$ are called *internal* subsets of $^*R$; any other subset of $^*R$ is said to be *external*. Similarly, in $^*\mathscr{R}$ a tuple α is said to be *internal* provided that $\alpha \in {}^*T$; otherwise α is said to be *external*. This notion also applies to functions in $^*\mathscr{R}$. Thus, a map of a subset of $^*R$ into $^*R$, say f, is said to be *internal* if $f \in {}^*F$; otherwise, f is said to be an *external* function.

For example, $\{x \in {}^*R \mid 3 < x < 4\}$ is an internal subset of $^*R$; the ω-tuples $(1, \ldots, 1)$ and $(1, 2, \ldots, \omega)$ are both internal tuples; the functions $\{(x, x) \mid x \in {}^*R\}$ and $\sin \omega x$ are both internal. On the other hand, both R and $\{\varepsilon \in {}^*R \mid \varepsilon \simeq 0\}$ are external subsets of $^*R$. Since the domain of an internal function is an internal subset of $^*R$, it follows that a function is external if its domain is an external subset of $^*R$. For example, the function $\{(x, x^2) \mid x \in R\}$ is an external function.

Exercises

1. Prove that K, the postulate set on page 236, has a model.
2. Let A be any atomic wff in W′; prove that $^*A$ is true for $^*\mathscr{R}$ iff $A \in$ K.
3. Prove that each nonzero member of $^*R$ has a multiplicative inverse in $^*R$.
4. (a) Prove that $-\varepsilon$ is an infinitesimal iff ε is an infinitesimal.
   (b) Prove that $-\kappa$ is infinite iff κ is infinite.
   (c) Prove that $-a$ is finite iff a is finite.
5. (a) Prove that the sum of two infinitesimals is an infinitesimal.
   (b) Prove that the sum of two finite numbers is finite.
   (c) Exhibit two infinite numbers whose sum is not infinite.
6. (a) Prove that each infinitesimal is finite.
   (b) Prove that each member of R is finite.
   (c) Let $a \in R$ and let ε be an infinitesimal; prove that $a + \varepsilon$ is finite.
7. (a) Prove that the product of two infinitesimals is an infinitesimal.
   (b) Prove that the product of two finite numbers is finite.
   (c) Prove that the product of two infinite numbers is infinite.
8. Prove that $\simeq$ is an equivalence relation on $^*R$.
9. Let $a \in {}^*R$; show that a is an infinitesimal iff $a \simeq 0$.
10. Prove that the domain of an internal function is an internal subset of $^*R$.
11. Prove that f is an external function if $\{x \in {}^*R \mid f(x) = a\}$ is external; here $a \in {}^*R$.
12. Let f be the function such that $f(t) = 1$ if $t \in R$, and $f(t) = 0$ if $t \in {}^*R - R$. Show that f is external.
13. Let A be a true swff of $^*\mathscr{R}$ in which a occurs, where $a \in {}^*R - R$, and suppose that each symbol of A, except a, can be interpreted in $\mathscr{R}$.
    (a) Prove that there is a member of R, say q, such that $\mathsf{S}^q_a[A]$ is true for $\mathscr{R}$.
    (b) Prove that there is a member of R, say q, such that $\mathsf{S}^q_a[A]$ is true for $^*\mathscr{R}$.

14. Let $\forall x \exists y \iota A$, $x, y \in R$, be a swff of $\mathscr{R}$. Prove that the following are equivalent:

$$\forall x \exists y \iota A, \quad x, y \in R, \quad \text{is true for } \mathscr{R} \tag{4}$$

$$\forall x \exists y {}^*A, \quad x, y \in R, \quad \text{is true for } {}^*\mathscr{R} \tag{5}$$

$$\forall x \exists y {}^*A, \quad x, y \in {}^*R, \quad \text{is true for } {}^*\mathscr{R} \tag{6}$$

15. Let $Q_1x_1 \cdots Q_nx_n\iota M$, $x_1, \ldots, x_n \in R$, be a swff of $\mathscr{R}$ in prenex normal form. Prove that the following are equivalent:

$$Q_1x_1 \cdots Q_nx_n\iota M, \quad x_1, \ldots, x_n \in R, \quad \text{is true for } \mathscr{R} \tag{7}$$

$$Q_1x_1 \cdots Q_nx_n{}^*M, \quad x_1, \ldots, x_n \in R, \quad \text{is true for } {}^*\mathscr{R} \tag{8}$$

$$Q_1x_1 \cdots Q_nx_n{}^*M, \quad x_1, \ldots, x_n \in {}^*R, \quad \text{is true for } {}^*\mathscr{R} \tag{9}$$

13.3. Properties of $\simeq$

Following Robinson, we shall apply the terminology of the semantical system $\mathscr{R}$ to ${}^*\mathscr{R}$; for example, each member of ${}^*R$ is called a *real* number and each member of ${}^*N$ is called a *natural* number. Moreover, let P be the set of primes of $\mathscr{R}$ (more accurately, the relation that represents the set of primes); then each member of ${}^*P$ is said to be a *prime* in ${}^*\mathscr{R}$. Robinson uses the term *standard* to refer to the members of R, N, or P, etc., in the context of ${}^*\mathscr{R}$; i.e., a member of ${}^*R$ is said to be *standard* if it is a member of R.

A basic fact concerning finite real numbers is that each finite real number is linked to a unique standard number in the sense that each finite real number is infinitely close to a unique standard number. We shall prove this in a moment.

Lemma 1. If t is finite, there is a standard number b such that $-b \leq t \leq b$.

Dem. Since t is finite, there is a standard positive number h such that $|t| \leq h$. Thus $-h \leq t \leq h$.

Fundamental Theorem about Finite Numbers. Each finite number is infinitely close to a unique standard number.

Dem. Let t be any finite number and let

$$S = \{y \in R \mid y < t\}$$

By Lemma 1, there is a standard positive number b such that $-b < t < b$. Thus $-b \in S$ and b is an upper bound of S; so S is nonempty and is bounded above. By the Completeness Theorem for $\mathscr{R}$, S has a least upper bound, say a; of course, a is standard. We claim that $t \simeq a$. If not, then $t - a$ is not an infinitesimal; so there is a positive, standard number h such that $h < |t - a|$. There are two possibilities.

Case 1: $a < t$. Then $h < t - a$, so $h + a < t$. Thus $h + a \in S$; so a is not an upper bound of S.

Case 2: $t < a$. Then $h < a - t$, so $t < a - h$. Thus $a - h$ is an upper bound of S; so a is not the least upper bound of S. These contradictions prove that $t - a$ is an infinitesimal, i.e., $t \simeq a$. To prove uniqueness, suppose that t is infinitely close to two standard numbers, say a_1 and a_2. Since $\simeq$ is an equivalence relation on $^*R$, it follows that $a_1 \simeq a_2$; so $a_1 - a_2$ is an infinitesimal. But 0 is the only infinitesimal that is standard; therefore, $a_1 - a_2 = 0$, so $a_1 = a_2$. This completes our proof.

This result can be expressed as follows.

Corollary 1. Each finite number can be expressed uniquely as the sum of a standard number and an infinitesimal.

Our result proves that each interval of infinitesimal length contains at most one standard number. For example, the open interval $(5 - \varepsilon, 5 + \varepsilon)$, where ε is a positive infinitesimal, contains exactly one standard number, namely 5. The open interval $(\varepsilon, 2\varepsilon)$ contains no standard numbers. The closed interval $[\omega, \omega + 1]$, which has length one, also contains no standard numbers.

We now look at the arithmetic of infinitesimals.

Lemma 2. The sum of two infinitesimals is an infinitesimal.

Dem. Let $\varepsilon \simeq 0$ and $\delta \simeq 0$; we shall show that $|\varepsilon + \delta| < h$ for each positive standard h. By assumption, $|\varepsilon| < h/2$ and $|\delta| < h/2$. By the triangle inequality, which transfers from $\mathscr{R}$ to $^*\mathscr{R}$,

$$|\varepsilon + \delta| \leq |\varepsilon| + |\delta < h$$

Thus $\varepsilon + \delta \simeq 0$.

Lemma 3. The product of two infinitesimals is an infinitesimal.

Dem. This is left as an exercise.

We mention that a product of an infinitesimal and another number is not necessarily an infinitesimal. For example, let ε be a nonzero infinitesimal; then $\varepsilon(1/\varepsilon) = 1$, which is not an infinitesimal; on the other hand, we have just observed that the product of two infinitesimals is an infinitesimal. Something more can be established in this direction.

Lemma 4. The product of an infinitesimal and a finite number is an infinitesimal.

Dem. Let $\varepsilon \simeq 0$ and let t be any finite number. We shall show that $|\varepsilon t| < h$ for each positive standard h. By Lemma 1, there is a standard number b such that $|t| \leq b$. Now, h/b is a positive, standard number; so $|\varepsilon| < h/b$ by assumption. Thus

$$|\varepsilon t| = |\varepsilon|\,|t| < (h/b)b = h$$

Therefore, $\varepsilon t \simeq 0$.

The following facts are easy to establish.

Lemma 5. $1/\varepsilon$ is infinite if ε is a nonzero infinitesimal.

Lemma 6. a/ε is infinite if ε is a nonzero infinitesimal and a is not an infinitesimal.

Dem. Assume that a/ε is finite. By Lemma 4, $\varepsilon(a/\varepsilon)$ is an infinitesimal; i.e., a is an infinitesimal. This contradiction proves that a/ε is infinite.

For some purposes, we are content with a calculation that yields a number infinitely close to the required number. The following lemma allows us to simplify certain calculations by replacing terms by equivalent terms.

Lemma 7. Let a, a', b, and b' be finite numbers such that $a \simeq a'$ and $b \simeq b'$. Then

(a) $a + b \simeq a' + b'$.
(b) $ab \simeq a'b'$.
(c) $a/b \simeq a'/b'$ provided that b is not an infinitesimal.

Dem. This is left as an exercise.

Let us illustrate the arithmetic of infinitesimals.

Example 1. Find the standard number that is infinitely close to

$$\frac{(5+\varepsilon)^2 - 25}{\varepsilon}$$

where ε is a nonzero infinitesimal.

Solution. We cannot apply Lemma 7(c), because $\varepsilon \simeq 0$. Instead, we apply the usual laws of algebra, which are true for $*\mathscr{R}$; thus

$$\begin{aligned}\frac{(5+\varepsilon)^2 - 25}{\varepsilon} &= (25 + 10\varepsilon + \varepsilon^2 - 25)/\varepsilon \\ &= \varepsilon(10+\varepsilon)/\varepsilon \\ &= 10 + \varepsilon \\ &\simeq 10\end{aligned}$$

We conclude that for each nonzero infinitesimal

$$\frac{(5+\varepsilon)^2 - 25}{\varepsilon} \simeq 10$$

In our next example we utilize Lemma 7.

Example 2. Find the standard number that is infinitely close to

$$\frac{(5+\varepsilon)^2 - 10}{3+\varepsilon}$$

where $\varepsilon \simeq 0$.

Solution. This time we can apply Lemma 7. By (b), $(5+\varepsilon)^2 \simeq 25$; by (a), $(5+\varepsilon)^2 - 10 \simeq 25 - 10 = 15$. Therefore, by (c),

$$\frac{(5+\varepsilon)^2 - 10}{3+\varepsilon} \simeq \frac{15}{3} = 5$$

Thus 5 is the required standard number.

Let t be any finite number in $*R$; we have shown (see page 241) that there is a unique standard number a such that $a \simeq t$. Robinson calls this standard number the *standard part of* t, and denotes it by $^\circ t$.

Here is a fact about finite numbers and their standard parts that we shall find quite useful.

Lemma 8. Let x and y be finite numbers such that $x < y$; then $^\circ x \leq {}^\circ y$.

Dem. By the trichotomy law, either $°x \le °y$ or else $°y < °x$. Assume the latter, and let $x = °x + \varepsilon$ and $y = °y + \delta$, where ε and δ are infinitesimals. We are given that

$$°x + \varepsilon < °y + \delta$$

Thus

$$0 < °x - °y < \delta - \varepsilon$$

By Lemma 2, $\delta - \varepsilon \simeq 0$; thus $°x - °y \simeq 0$. But $°x - °y$ is in R; so $°x - °y = 0$, and $°x = °y$. This contradicts our assumption that $°y < °x$. We conclude that $°x \le °y$.

We shall use Lemma 8 in our proofs of the Maximum Value Theorem (page 256), the Intermediate Value Theorem (page 256), and Lemma 3 (page 250).

Here is an interesting fact.

Lemma 9. Let a be a standard number such that $|a| < 1$; then $a^\kappa \simeq 0$ for each infinite natural number κ.

Dem. It is enough to consider the case in which $0 < a < 1$. Then there is a positive standard number h such that $a = 1 - h$. Now,

$$\forall n(n \in N \rightarrow (1 - h)^n \le 1/(1 + nh))$$

is true for $\mathscr{R}$; therefore, this statement is true for $*\mathscr{R}$ when interpreted in $*\mathscr{R}$. So

$$\forall n(n \in *N \rightarrow (1 - h)^n \le 1/(1 + nh))$$

is true for $*\mathscr{R}$. Let κ be any infinite natural number; then

$$0 < (1 - h)^\kappa \le 1/(1 + \kappa h)$$

Since h is not an infinitesimal, κh is infinite; thus $1/(1 + \kappa h) \simeq 0$. We conclude that $(1 - h)^\kappa \simeq 0$; so $a^\kappa \simeq 0$.

Exercises

1. Prove Lemma 3.
2. Prove that ab is finite if a and b are finite.
3. Prove that ab is *not* an infinitesimal if neither a nor b is an infinitesimal.
4. Prove that $1/a$ is finite if a is *not* an infinitesimal.

5. Prove Lemma 5.
6. Prove Lemma 7.
7. Let $0 < a$ where a is standard. Prove that $0 < a^n$ for each $n \in {}^*N$.
8. Complete the proof of Lemma 9, i.e., for the case $-1 < a \leq 0$.

13.4. Paradoxes

Paradoxes are useful for sharpening our appreciation of the finer points of nonstandard analysis. In addition, some paradoxes are really theorems which establish that a certain object of ${}^*\mathscr{R}$ is external.

Paradox 1. It is well known that the real number system $\mathscr{R}$ is Archimedean; i.e.,

$$\forall xy\exists n(0 < x < y \rightarrow y < nx), \qquad x, y \in R \quad \text{and} \quad n \in N$$

Therefore, ${}^*\mathscr{R}$ is Archimedean. We obtain a paradox by observing that ${}^*\mathscr{R}$ in fact is non-Archimedean. To see this, let ε be any positive infinitesimal. Then $0 < \varepsilon < 1$; but for each $n \in N$, $n\varepsilon = \varepsilon + \varepsilon + \cdots + \varepsilon$ (n ε's). Now, the sum of a finite number of infinitesimals is an infinitesimal; thus $n\varepsilon < 1$ for each $n \in N$. Therefore, ${}^*\mathscr{R}$ is non-Archimedean.

Paradox 2. The completeness theorem for $\mathscr{R}$ asserts that each nonempty set of real numbers that is bounded above has a least upper bound. By including a relation symbol for the concept of an *upper bound* and a relation symbol for the concept of the *least upper bound* in our language, we can express the completeness theorem by a swff of $\mathscr{R}$. Accordingly, the completeness theorem is true for ${}^*\mathscr{R}$; i.e., each nonempty subset of ${}^*\mathscr{R}$ that is bounded above has a least upper bound. In ${}^*\mathscr{R}$, R is bounded above and is nonempty, where R is the set of all standard numbers. Therefore, R has a least upper bound; i.e., there is a smallest number $x \in {}^*R$ such that

$$\forall y(y \in R \rightarrow y \leq x)$$

is true for ${}^*\mathscr{R}$. Clearly x is infinite by definition; thus $x - 1$ is infinite, and it follows that $x - 1$ is an upper bound of R. Of course, $x - 1 < x$; so x is not the least upper bound of R. Contradiction!

Paradox 3. Here is a fact about tuples in $\mathscr{R}$. Let $n \in N$ and let $x_1, \ldots, x_n$ be any n numbers (i.e., each $x_i \in R$); then $(x_1, \ldots, x_n)$ is a tuple in T. Thus, corresponding to any list of n numbers, there is a tuple whose terms are precisely the numbers in the given list. By the transfer theorem, $^*\mathscr{R}$ has the same property; i.e., for each $n \in {}^*N$ and for any list of n numbers $x_1, \ldots, x_n$ (where each $x_i \in {}^*R$), there is a tuple $(x_1, \ldots, x_n)$ in $^*T$. Now, $\omega \in {}^*N$ and form the list

$$0, 0, 0, \ldots; \ldots, 1, 1, 1, \ldots, 1$$

which has ω terms; the ith term of this list is 0 if $i \in N$, and the ith term is 1 if $i \in {}^*N - N$ and $i \leq \omega$. Therefore, there is an ω-tuple α, $\alpha \in {}^*T$, such that

$$\alpha(i) = \begin{cases} 0 & \text{if } \ i \in N \\ 1 & \text{if } \ i \in {}^*N - N \ \text{ and } \ i \leq \omega \end{cases}$$

So $\{n \in {}^*N \mid \alpha(n) = 0\} = N$ is an internal subset of $^*R$. But we have already shown that N is an external subset of $^*R$. Contradiction!

We shall now resolve these paradoxes. Consider the first paradox. $^*\mathscr{R}$ is Archimedean in the sense that

$$\forall xy \exists n(0 < x < y \rightarrow y < nx), \qquad x, y \in R \quad \text{and} \quad n \in N$$

is true for $^*\mathscr{R}$ when interpreted in $^*\mathscr{R}$. So

$$\forall xy \exists n(0 < x < y \rightarrow y < nx), \qquad x, y \in {}^*R \quad \text{and} \quad n \in {}^*N$$

is true for $^*\mathscr{R}$. This does not assert that $^*\mathscr{R}$ is Archimedean; indeed, there is no contradiction with the fact that $^*\mathscr{R}$ is non-Archimedean.

The second paradox is resolved by pointing out that the given argument merely proves that R is an external subset of $^*R$. The completeness theorem, when interpreted in $^*\mathscr{R}$, refers to *internal* subsets of $^*R$; indeed, this theorem asserts that each nonempty, internal subset of $^*R$ that is bounded above has a least upper bound.

The fallacy in the third paradox is of a different sort. The fact which is being transferred from $\mathscr{R}$ to $^*\mathscr{R}$ cannot be expressed by a swff of $\mathscr{R}$. This fact requires a swff in prenex normal form whose prefix has the form "$\forall n x_1 x_2 \cdots x_n$." One quantifier in the prefix indicates the number of quantifiers which follow it; our quantifiers are allowed to refer to the constants of the semantical system involved, and to nothing else. So, the given fact cannot be expressed by a swff of $\mathscr{R}$ and therefore cannot be transferred to $^*\mathscr{R}$.

Exercises

1. Prove that $\{\varepsilon \in {}^*R \mid \varepsilon \simeq 0\}$ is an external subset of $^*R$. *Hint:* Use the Completeness Theorem.

2. Prove that $\{\kappa \in {}^*R \mid \kappa \text{ is infinite}\}$ is an external subset of $^*R$.

3. Resolve the following paradox. "Each standard number in [0, 1] possesses a decimal expansion, which is a map of N into the digits $\{0, 1, 2, 3, 4, 5, 6, 7, 8, 9\}$. Moreover, each map of N into the digits is the decimal expansion of some number in [0, 1]. Transferring to $^*\mathscr{R}$, this means that each map of $^*N$ into $\{0, 1, 2, 3, 4, 5, 6, 7, 8, 9\}$ is the decimal expansion of some member of $^*[0, 1]$. Let d be the map such that

$$d_i = \begin{cases} 0 & \text{if } i \in N \\ 9 & \text{if } i \in {}^*N - N \end{cases}$$

Then d is the decimal expansion of

$$x = .000\cdots;\ \cdots 999\cdots$$

Clearly, $x < 1/10^n$ for each $n \in N$; so $0 < x < h$ for each $h > 0$, $h \in R$. Therefore, x is an infinitesimal. On the other hand, $x + 1/10^\omega > x$, so the decimal expansion of $x + 1/10^\omega$ has a nonzero digit to the left of the semicolon; this means that $x + 1/10^\omega$ is *not* an infinitesimal. Certainly, $1/10^\omega$ is an infinitesimal; moreover, the sum of two infinitesimals is necessarily an infinitesimal. Thus, $x + 1/10^\omega$ is an infinitesimal. Contradiction!"

13.5. The Limit Concept

A sequence in $\mathscr{R}$ is a map of N into R; so each sequence is a function. We require a relation symbol to represent the concept of a sequence in $\mathscr{R}$; of course, each specific sequence is a constant of $\mathscr{R}$.

Now, each sequence (a_n) is also a relation of $\mathscr{R}$; so (a_n) grows to $^*(a_n)$, a sequence in $^*\mathscr{R}$. Since each sequence in $\mathscr{R}$ is a map of N into R, it follows that each sequence in $^*\mathscr{R}$ is a map of $^*N$ into $^*R$. Thus $^*(a_n)$ is a map of $^*N$ into $^*R$. Moreover, the maps (a_n) and $^*(a_n)$ have the same image for each $i \in N$. Therefore, $^*(a_n)$ is a superset of (a_n).

We are especially interested in the values of $^*(a_n)$ at infinite natural numbers; i.e., the numbers a_κ, where κ is an infinite natural number. Indeed, we shall prove that (a_n) converges to L, where $L \in R$, iff

$$a_\kappa \simeq L \quad \text{for each infinite natural number } \kappa \tag{1}$$

Recall that (a_n) converges to L, where $L \in R$, iff

$$\forall h \exists q \forall n (n > q \rightarrow |a_n - L| < h), \qquad h > 0, \quad q, n \in N \tag{2}$$

Now, (2) is a swff of $\mathscr{R}$; (1) is a swff of $^*\mathscr{R}$, enriched by including the concept of the infinite natural numbers.

Our claim is that (1) is a criterion for convergence.

Theorem. For each sequence (a_n), (2) is true for $\mathscr{R}$ iff (1) is true for $^*\mathscr{R}$.

Dem. 1. Assume that (2) is true for $\mathscr{R}$. Choose $h > 0$; by (2) there is a natural number q, $q \in N$, such that

$$\forall n(n > q \to |a_n - L| < h), \qquad n \in N \tag{3}$$

is true for $\mathscr{R}$. By the transfer theorem, (3) is true for $^*\mathscr{R}$ when interpreted in $^*\mathscr{R}$; i.e.,

$$\forall n(n > q \to |a_n - L| < h), \qquad n \in {}^*N \tag{4}$$

is true for $^*\mathscr{R}$. Choose any infinite natural number κ; then $\kappa > q$, since $q \in N$. So, from (4), $|a_\kappa - L| < h$ is true for $^*\mathscr{R}$; here, h is any positive standard number. Therefore $|a_\kappa - L|$ is an infinitesimal; thus $a_\kappa \simeq L$. This proves that (1) is true for $^*\mathscr{R}$ if (2) is true for $\mathscr{R}$.

2. Assume that (1) is true for $^*\mathscr{R}$. Since ω is an infinite natural number, any number greater than ω is also infinite. Thus

$$\forall n(n > \omega \to a_n \simeq L), \qquad n \in {}^*N \tag{5}$$

is true for $^*\mathscr{R}$. Therefore,

$$\exists q \forall n(n > q \to a_n \simeq L), \qquad q, n \in {}^*N \tag{6}$$

is true for $^*\mathscr{R}$. Choose any positive standard number h; if $a_n \simeq L$, then $|a_n - L| < h$. Thus, from (6),

$$\exists q \forall n(n > q \to |a_n - L| < h), \qquad q, n \in {}^*N \tag{7}$$

is true for $^*\mathscr{R}$. By the transfer theorem,

$$\exists q \forall n(n > q \to |a_n - L| < h), \qquad q, n \in N \tag{8}$$

is true for $\mathscr{R}$; here, h is any positive standard number. Thus

$$\forall h \exists q \forall n(n > q \to |a_n - L| < h), \qquad h > 0, \quad q, n \in N \tag{9}$$

is true for $\mathscr{R}$. This proves that (2) is true for $\mathscr{R}$ if (1) is true for $^*\mathscr{R}$.

Example 1. By (1), $\lim(1/n) = 0$ since $1/\kappa \simeq 0$ for each infinite natural number κ.

Example 2. Use (1) to prove that $\lim(a_n + b_n) = \lim(a_n) + \lim(b_n)$, where (a_n) and (b_n) both converge.

Solution. Let $\lim(a_n) = L$, $\lim(b_n) = M$, and let κ be any infinite natural number. By Lemma 7, page 243,

$$a_\kappa + b_\kappa \simeq L + M$$

Thus, by (1),

$$\lim(a_n + b_n) = L + M$$

i.e.,

$$\lim(a_n + b_n) = \lim(a_n) + \lim(b_n)$$

In Example 2, we have used the fact that a convergent sequence converges to a unique number. With (1), it is easy to prove this.

Lemma 1. If (a_n) converges, then (a_n) converges to a unique number in R.

Dem. Since (a_n) converges, there is a standard number L such that $a_\kappa \simeq L$ for each infinite natural number κ. Assume that (a_n) converges to a standard number L', where $L' \neq L$; then $a_\kappa \simeq L'$ for each infinite natural number. Thus $L \simeq L'$ and it follows that $L = L'$.

Throughout, we are following the usual convention of writing "$\lim(a_n) = L$" if (a_n) converges to L.

The following well-known facts about convergent sequences are easy to verify by applying (1). We mention that if $\lim(a_n) = L$, then ${}^\circ a_\kappa = L$ for each infinite natural number.

Lemma 2. Let $\lim(a_n) = L$ and let $c \in R$; then $\lim(ca_n) = cL$.

Dem. By assumption, for each infinite natural number κ, $a_\kappa \simeq L$; so $a_\kappa - L \simeq 0$. By Lemma 4, page 243, $c(a_\kappa - L) \simeq 0$; thus $ca_\kappa \simeq cL$ for each infinite natural number κ. By (1), $\lim(ca_n) = cL$.

Lemma 3. Let $\lim(a_n) = L$ and $\lim(b_n) = L'$, where $\forall n(a_n \leq b_n)$. Then $L \leq L'$.

Dem. By the transfer theorem, $a_\kappa \leq b_\kappa$ for each infinite natural number κ. By Lemma 8, page 244, ${}^\circ a_\kappa \leq {}^\circ b_\kappa$. But ${}^\circ a_\kappa = L$ and ${}^\circ b_\kappa = L'$; so $L \leq L'$.

Lemma 4. Each convergent sequence is bounded.

Dem. Let (a_n) converge. Then there is a standard number L such that $a_\kappa \simeq L$ for each infinite natural number κ. Thus $|a_\kappa - L| \simeq 0$, so $|a_\kappa - L| < 1$ and it follows that $|a_\kappa| < |L| + 1$ for each infinite natural number κ. As in the proof of Theorem 1, we can show that there is a standard natural number q such that

$$\forall n(n > q \rightarrow |a_n| < |L| + 1), \qquad n \in N \tag{10}$$

is true for $\mathscr{R}$. Let

$$m = \max\{|a_1|, \ldots, |a_q|, |L| + 1\}$$

Then $|a_n| \leq m$ for each $n \in N$; i.e., each term of the sequence (a_n) is between $-m$ and m.

A sequence (a_n) is said to be a Cauchy sequence iff

$$\forall h \exists q \forall nm(n, m > q \rightarrow |a_n - a_m| < h), \qquad h > 0, \quad n, m \in N \tag{11}$$

is true for $\mathscr{R}$.

Lemma 5. (a_n) is a Cauchy sequence iff $a_\kappa \simeq a_\Omega$ for any infinite natural numbers κ and Ω.

Dem. Use the technique illustrated in the proof of Theorem 1.

Lemma 5 provides us with a nonstandard criterion for Cauchy sequences. Using this criterion, we can show that a sequence converges iff it is a Cauchy sequence.

Lemma 6. (a_n) converges iff (a_n) is a Cauchy sequence.

Dem. 1. Assume that (a_n) converges, say to L. Let κ and Ω be any infinite natural numbers. By (1), $a_\kappa \simeq L$ and $a_\Omega \simeq L$; so $a_\kappa \simeq a_\Omega$.

2. Assume that (a_n) is a Cauchy sequence. If a_κ is finite for some infinite natural number κ, then $^\circ a_\kappa$ (the standard part of a_κ) is the limit of the sequence (a_n). By Lemma 5,

$$\forall nm(n, m > \omega \rightarrow |a_n - a_m| < 1), \qquad n, m \in {}^*N$$

is true for $^*\mathscr{R}$; thus $\exists q \forall nm(n, m > q \rightarrow |a_n - a_m| < 1)$, $q, n, m \in {}^*N$, is true for $^*\mathscr{R}$. By the Transfer Theorem,

$$\exists q \forall nm(n, m > q \rightarrow |a_n - a_m| < 1), \qquad n, m, q \in N \tag{12}$$

is true for $\mathscr{R}$. Therefore, there is a finite natural number, say q, such that

$$\forall nm(n, m > q \rightarrow |a_n - a_m| < 1), \qquad n, m \in N$$

is true for $\mathscr{R}$. Thus

$$\forall n(n > q \rightarrow |a_n - a_{q+1}| < 1), \qquad n \in N$$

is true for $\mathscr{R}$. Transferring to $^*\mathscr{R}$ and choosing κ infinite, we obtain

$$|a_\kappa - a_{q+1}| < 1$$

is true for $^*\mathscr{R}$. So, $a_\kappa - a_{q+1}$ is finite; but a_{q+1} is finite since $q + 1 \in N$. Therefore a_κ is finite. Let $L = {}^\circ a_\kappa$; then (a_n) converges to L by (1). This completes our proof of Lemma 6.

Exercises

1. Use (1) to show that

$$\lim\left(\frac{n+1}{n}\right) = 1$$

2. Use (1) to show that

$$\lim\left(\frac{n+1}{n^2}\right) = 0$$

3. Use (1) to show that the sequence $((n^2 + 1)/n)$ diverges.
4. Prove Lemma 5.
5. Let (c_n) be a map of N into N such that $c_i < c_j$ if $i < j$. Then (a_{c_n}) is said to be a *subsequence* of (a_n).
   (a) Prove that $c_n \geq n$ for each $n \in N$, where (c_n) has the property given above.
   (b) Use (1) to prove that each subsequence of a convergent sequence (a_n) converges to $\lim(a_n)$.
6. Let (a_n) be an increasing sequence whose terms have a least upper bound B.
   (a) Show that $\forall h \exists q \forall n(n > q \rightarrow B - h \leq a_n \leq B)$, $h > 0$, $q, n \in N$, is true for $\mathscr{R}$.
   (b) Prove that $a_\kappa \simeq B$ for each infinite natural number κ. Use part (a) and the transfer theorem.

13.6. Continuity; Uniform Continuity

The relation symbols of $\mathscr{R}$ include symbols for each function; these symbols are also regarded as constants of $\mathscr{R}$. Let f be a relation symbol which represents a function, and let $a, b \in R$; then "fab" is in the domain of the semantical system $\mathscr{R}$, and is true for $\mathscr{R}$ iff $b = f(a)$. We shall continue to use the language of operations to represent swffs of $\mathscr{R}$.

We must give credit to the mathematicians of the 19th century, who recognized that the notion of *continuity* can be characterized by a swff of

$\mathscr{R}$. Indeed, let f be any function, and let $a \in \operatorname{dom} f$; then f is continuous at a iff

$$\forall h \exists k \forall x(|x - a| < k \rightarrow |f(x) - f(a)| < h), \qquad h, k > 0, \quad x \in \operatorname{dom} f \tag{1}$$

is true for $\mathscr{R}$.

Intuitively, f is continuous at a provided that

$$f(x) \text{ approximates } f(a) \text{ if } x \text{ approximates } a \tag{2}$$

Here, the meaning of "approximates" is vague; moreover, it is not clear which number system is involved in (2). We can make (2) precise by agreeing that "approximates" means "is infinitely close to" and recognizing that (2) refers to $^*\mathscr{R}$, not $\mathscr{R}$. In short, by (2) we mean the following:

$$f(x) \simeq f(a) \qquad \text{if} \quad x \simeq a \tag{3}$$

Here we are following the mathematicians' practice of suppressing quantifiers wherever possible; (3) is an abbreviation for the following swff of $^*\mathscr{R}$:

$$\forall x(x \simeq a \rightarrow f(x) \simeq f(a)), \qquad x \in \operatorname{dom} f \tag{4}$$

In turn, (4) can be simplified; indeed, (4) is equivalent to the following swff of $^*\mathscr{R}$:

$$\forall \varepsilon(f(a + \varepsilon) \simeq f(a)), \qquad \varepsilon \simeq 0, \quad a + \varepsilon \in \operatorname{dom} f \tag{5}$$

In (2)–(5) we have suppressed the star on $^*f$; in the transition to $^*\mathscr{R}$ the given function f grows to $^*f$, a superset of f.

Our claim is that (5) is a criterion for the continuity of f at a, $a \in R$.

Theorem 1. For any function f and any $a \in R$, (1) is true for $\mathscr{R}$ iff (5) is true for $^*\mathscr{R}$.

Dem. 1. Assume that (1) is true for $\mathscr{R}$. Choose $h > 0$, $h \in R$; by (1), there is a positive number k, $k \in R$, such that

$$\forall x(|x - a| < k \rightarrow |f(x) - f(a)| < h), \qquad x \in \operatorname{dom} f \tag{6}$$

is true for $\mathscr{R}$, so (6) is true for $^*\mathscr{R}$ when interpreted in $^*\mathscr{R}$. Let ε be an infinitesimal such that $a + \varepsilon \in \operatorname{dom} f$, and set $x = a + \varepsilon$; then $|x - \varepsilon| = |\varepsilon| < k$. Thus, by (6) interpreted in $^*\mathscr{R}$,

$$|f(a + \varepsilon) - f(a)| < h \tag{7}$$

In (7), h is any positive standard number; therefore, $f(a + \varepsilon) - f(a)$ is an infinitesimal. We conclude that $f(a + \varepsilon) \simeq f(a)$. Therefore, (5) is true for $^*\mathscr{R}$.

2. Assume that (5) is true for $^*\mathscr{R}$. Then

$$\forall x(x \simeq a \to f(x) \simeq f(a)), \qquad x \in \operatorname{dom} f \tag{8}$$

is true for $^*\mathscr{R}$. Now, $1/\omega$ is an infinitesimal; thus if $|x - a| < 1/\omega$, then $x \simeq a$. From (8),

$$\forall x(|x - a| < 1/\omega \to f(x) \simeq f(a)), \qquad x \in \operatorname{dom} f \tag{9}$$

is true for $^*\mathscr{R}$; so

$$\exists k \forall x(|x - a| < k \to f(x) \simeq f(a)), \qquad k > 0, \quad k \in {}^*R, \quad x \in \operatorname{dom} f \tag{10}$$

is true for $^*\mathscr{R}$. Let h be any positive standard number; if $f(x) \simeq f(a)$, then $|f(x) - f(a)| < h$. So, from (10),

$$\exists k \forall x(|x - a| < k \to |f(x) - f(a)| < h), \qquad k > 0, \quad k \in {}^*R, \quad x \in \operatorname{dom} f \tag{11}$$

is true for $^*\mathscr{R}$. By the transfer theorem,

$$\exists k \forall x(|x - a| < k \to |f(x) - f(a)| < h), \qquad k > 0, \quad k \in R, \quad x \in \operatorname{dom} f \tag{12}$$

is true for $\mathscr{R}$. Since h is any positive standard number, we can quantify over h, inserting a universal quantifier. So (1) is true for $\mathscr{R}$.

We now illustrate these ideas.

Example 1. We shall show that the function x^2 (the *squaring* function) is continuous at 5. Since $\operatorname{dom} x^2 = R$, it follows that $\operatorname{dom} {}^*(x^2) = {}^*R$. Let $\varepsilon \simeq 0$; by Lemma 7, page 243,

$$(5 + \varepsilon)^2 \simeq 5^2 = 25$$

So (5) is true for $^*\mathscr{R}$, where $f = x^2$ and $a = 5$. Thus, by Theorem 1, x^2 is continuous at 5.

Example 2. We shall use the nonstandard criterion (5) to prove that the sum of two functions is continuous at a if both functions are continuous at a. Let f and g be continuous at a, and let ε be any infinitesimal such that $a + \varepsilon$ is in the domain of $f + g$; then

$$\begin{aligned}
[f + g](a + \varepsilon) &= f(a + \varepsilon) + g(a + \varepsilon) && \text{(by the definition of } f + g\text{)} \\
&\simeq f(a) + g(a) && \text{(by Lemma 7, page 243)} \\
&= [f + g](a)
\end{aligned}$$

By (5), $f + g$ is continuous at a.

A function f is *continuous on* E, a subset of R, provided that f is continuous at each member of E.

Theorem 2. Let f be any function and let E be any subset of R; then f is continuous on E iff

$$\forall x\varepsilon(f(x + \varepsilon) \simeq f(x)), \qquad \varepsilon \simeq 0, \quad x + \varepsilon \in \text{dom } {}^*f, \quad x \in E \qquad (13)$$

is true for $^*\mathscr{R}$.

Dem. Apply Theorem 1.

We need some information about the set in $^*\mathscr{R}$ that corresponds to a closed interval of the form $[a, b]$, $a < b$, in $\mathscr{R}$. Now,

$$[a, b] = \{t \in R \mid a \leq t \leq b\}$$

The existence of this set can be expressed by a swff of $\mathscr{R}$; therefore, by the transfer theorem,

$$^*[a, b] = \{t \in {}^*R \mid a \leq t \leq b\}$$

Since $a, b \in R$, we see that each member of $^*[a, b]$ is finite, so has a standard part. It is important that we prove that the standard part of any member of $^*[a, b]$ is also a member of $[a, b]$.

Lemma 1. Let $x \in {}^*[a, b]$; then $^\circ x \in [a, b]$.

Dem. By assumption, $a \leq x \leq b$ (in $^*\mathscr{R}$); thus, by Lemma 8, page 244, $a \leq {}^\circ x \leq b$. So $^\circ x \in [a, b]$.

The following fact is also useful.

Lemma 2. Let f be continuous on $[a, b]$, and let $x \in {}^*[a, b]$; then $f(x)$ is finite.

Dem. By Lemma 1, $^\circ x \in [a, b]$; so f is continuous at $^\circ x$. Thus $f(x) \simeq f(^\circ x)$. Since $f(^\circ x)$ is finite, we conclude that $f(x)$ is finite.

Let us prove two important facts about continuity, using the methods of nonstandard analysis. Recall that a closed interval with end-points a and b, i.e., $\{t \in R \mid a \leq t \leq b\}$, is denoted by $[a, b]$. Note how our proofs depend upon certain tuples in $^*\mathscr{R}$ being internal (i.e., in $^*T$), and the fact that whatever is true for all tuples in T (provided it can be expressed by a swff of $\mathscr{R}$) is also true (when interpreted in $*\mathscr{R}$) for all internal tuples.

Maximum Value Theorem. Let f be continuous on $[a, b]$; then f has a maximum value on $[a, b]$, i.e., $\exists m \forall t(f(m) \geq f(t))$, $m, t \in [a, b]$.

Dem. Choose $\kappa \in *N - N$ and let $\varepsilon = (b - a)/\kappa$; then

$$(a, a + \varepsilon, a + 2\varepsilon, \ldots, a + \kappa\varepsilon) \in *T$$

So

$$(f(a), f(a + \varepsilon), \ldots, f(a + \kappa\varepsilon)) \in *T$$

This tuple has a greatest term since each tuple in T has a greatest term. Let $f(a + i\varepsilon)$ be its greatest term, where $0 \leq i \leq \kappa$, and let $m = {}^\circ(a + i\varepsilon)$. Since f is continuous at m, $f(m) \simeq f(a + i\varepsilon)$. Let $t \in [a, b]$; then $t \simeq a + j\varepsilon$ for some $j = 1, \ldots, \kappa$. Since f is continuous at t,

$$f(t) \simeq f(a + j\varepsilon) \leq f(a + i\varepsilon) \simeq f(m)$$

By Lemma 8, page 244,

$${}^\circ[f(a + j\varepsilon)] \leq {}^\circ[f(a + i\varepsilon)]$$

But ${}^\circ[f(a + j\varepsilon)] = f(t)$ and ${}^\circ[f(a + i\varepsilon)] = f(m)$; so $f(t) \leq f(m)$ for each $t \in [a, b]$.

Intermediate Value Theorem. Let f be continuous on $[a, b]$ and let K be any standard number such that $f(a) < K < f(b)$. Then there is a standard number t between a and b such that $f(t) = K$.

Dem. Choose κ and ε as in the proof of the Maximum Value Theorem, and form the same two tuples as in that proof. Since K is between the first and last terms of the tuple $(f(a), f(a + \varepsilon), \ldots, f(a + \kappa\varepsilon))$, K is between two consecutive terms of this tuple, i.e., for some $i < k$,

$$f(a + i\varepsilon) \leq K \leq f(a + i\varepsilon + \varepsilon)$$

Thus, by Lemma 8, page 244,

$${}^\circ[f(a + i\varepsilon)] \leq K \leq {}^\circ[f(a + i\varepsilon + \varepsilon)]$$

Let $t = {}^\circ(a + i\varepsilon)$; then

$$f(t) \simeq f(a + i\varepsilon) \qquad \text{and} \qquad f(t) \simeq f(a + i\varepsilon + \varepsilon)$$

since f is continuous at t. Therefore, ${}^\circ[f(a + i\varepsilon)] = {}^\circ[f(a + i\varepsilon + \varepsilon)] = f(t)$. So $f(t) \leq K \leq f(t)$; we conclude that $f(t) = K$.

We turn, now, to the concept of *uniform continuity*. The classical formulation of this concept is as follows.

Definition. A function f is *uniformly continuous* on E, a subset of R, provided that

$$\forall h \exists k \forall xy(|x - y| < k \rightarrow |f(x) - f(y)| < h), \qquad h, k > 0, \quad x, y \in E \quad (14)$$

is true for $\mathscr{R}$.

Here is Robinson's nonstandard criterion for uniform continuity.

Theorem 3. A function f is uniformly continuous on E, a subset of R, provided that

$$\forall xy(x \simeq y \rightarrow f(x) \simeq f(y)), \qquad x, y \in {}^*E \quad (15)$$

is true for ${}^*\mathscr{R}$.

Dem. Follow the method of the proof of Theorem 1, page 253.

Recall that f is continuous on E iff

$$\forall xy(x \simeq y \rightarrow f(x) \simeq f(y)), \qquad x \in E, \quad y \in {}^*E$$

is true for ${}^*\mathscr{R}$. So, the distinction between *uniform continuity* and *continuity* rests on the domain of a quantifier.

We shall now prove that each function that is continuous on a closed interval is uniformly continuous on that interval.

Theorem 4. Let f be continuous on $[a, b]$; then f is uniformly continuous on $[a, b]$.

Dem. Let $x, y \in {}^*[a, b]$, where $x \simeq y$; then both x and y are finite and ${}^\circ x = {}^\circ y$. Of course, ${}^\circ x \simeq x$ and ${}^\circ x \simeq y$. Since ${}^\circ x \in [a, b]$, f is continuous at ${}^\circ x$. Therefore, $f({}^\circ x) \simeq f(x)$ and $f({}^\circ x) \simeq f(y)$; so $f(x) \simeq f(y)$. By (15), f is uniformly continuous on $[a, b]$.

Exercises

1. Use criterion (5) to show that x^3 is continuous at a, where $a \in R$.
2. Use (5) to prove that the product of two functions is continuous at a if both functions are continuous at a.

3. Use (5) to prove that the composite function $f \circ g$ is continuous at a if (i) g is continuous at a; (ii) f is continuous at $g(a)$.

4. Prove that $(a, a + \varepsilon, a + 2\varepsilon, \ldots, a + \kappa\varepsilon) \in {}^*T$, where κ is an infinite natural number, $\varepsilon = (b - a)/\kappa$, and both a and b are standard.

5. Let $(a_1, \ldots, a_\kappa) \in {}^*T$, where κ is an infinite natural number, and let $f \in F$. Prove that $({}^*f(a_1), \ldots, {}^*f(a_\kappa)) \in {}^*T$.

6. Let $(a_1, \ldots, a_n) \in {}^*T$, where $n \in {}^*N$, and let K be a standard number such that $a_1 < K < a_n$. Prove that $a_i \le K \le a_{i+1}$, where $i \in {}^*N$ and $i < n$.

7. Prove Theorem 3.

8. Use (15) to show that $1/x$ is not uniformly continuous on the open interval $(0, 1)$.

13.7. Principles of Permanence

A principle of permanence is a statement declaring that a property possessed by each member of a set, say B, is also possessed by each member of a proper superset of B. In a sense, then, the property persists beyond B.

This is rather vague; here is an example.

First Principle of Permanence. Let S be an internal subset of ${}^*N$ such that $n \in S$ for each $n \in N$; then there is an infinite natural number κ such that $n \in S$ for each $n < \kappa$, $n \in {}^*N$.

Dem. Either S is ${}^*N$ or ${}^*N - S$ is nonempty. If the former, there is nothing to prove. Assume the latter; then ${}^*N - S$ has a smallest member (in $\mathscr{R}$, any nonempty set of natural numbers has a smallest member), say κ. But each finite natural number is in S; therefore, κ is an infinite natural number. We conclude that $n \in S$ for each $n < \kappa$, $n \in {}^*N$.

This principle of permanence casts some light on the nature of internal subsets of ${}^*N$. Each superset of N that is internal must be a superset of $\{n \in {}^*N \mid n < \kappa\}$ for some infinite natural number κ. For example, $N \cup \{\omega\}$ does not have this property, so is not an internal subset of ${}^*N$. For the same reason, $N \cup \{t \in {}^*N \mid t > \omega\}$ is not an internal subset of ${}^*N$.

Here is another example of a principle of permanence.

Second Principle of Permanence. Let S be an internal subset of ${}^*N$ such that $\kappa \in S$ for each $\kappa \in {}^*N - N$; then there is a finite natural number q such that $n \in S$ for each $n > q$, $n \in {}^*N$.

Dem. If $S = {}^*N$, there is nothing to prove. If $S \neq {}^*N$, then ${}^*N - S$ is bounded above. In $\mathscr{R}$, any set of natural numbers that is bounded above has a greatest member. Since ${}^*N - S$ is internal, we see that ${}^*N - S$ has a greatest member, say q, that is finite. Therefore, $n \in S$ for each $n > q$, $n \in {}^*N$.

It is convenient to include *sequences* as constants of the semantical system $\mathscr{R}$; the concept of a sequence is expressed by a relation symbol of $\mathscr{R}$, say "Seq." The corresponding relation of $\mathscr{R}$, which is also denoted by Seq, is the set of all sequences. Accordingly, *Seq denotes a certain relation of ${}^*\mathscr{R}$; its members are sequences in ${}^*\mathscr{R}$, i.e., maps of ${}^*N$ into ${}^*R$. Again, we say that a map of ${}^*N$ into ${}^*R$ is an *internal* sequence provided the map is in *Seq; any other map of ${}^*N$ into ${}^*R$ is said to be *external*.

For example, the sequence $(n/\omega) = (1/\omega, 2/\omega, 3/\omega, \ldots)$, which is a map of ${}^*N$ into ${}^*R$, is internal; to see this, recall that in $\mathscr{R}$ for each $a \in R$, $(an) = (a, 2a, 3a, \ldots)$ is a sequence, so in ${}^*\mathscr{R}$ for each $a \in {}^*R$, the sequence (an) is internal. Now, the internal sequence (n/ω) has the property that for each $n \in N$, $n/\omega \simeq 0$. We claim that this property persists into ${}^*N - N$; i.e., there is an infinite natural number κ such that $n/\omega \simeq 0$ for each $n < \kappa$, $n \in {}^*N$. Since the relation $\simeq$ corresponds to no relation of $\mathscr{R}$, we cannot make a direct use of the transfer theorem. Instead, we shall appeal to the following property of internal sequences.

Lemma 1. Let (s_n) be an internal sequence and let $m \in {}^*R$ be such that $|s_n| < m$ for each $n \in N$. Then there is an infinite natural number κ such that $|s_n| < m$ for each $n < \kappa$, $n \in {}^*N$.

Dem. Let $S = \{n \in {}^*N \mid |s_n| < m\}$; then S is an internal subset of ${}^*N$ such that $n \in S$ for each $n \in N$. By the First Principle of Permanence, there is an infinite natural number κ such that $n \in S$ for each $n < \kappa$, $n \in {}^*N$. This establishes Lemma 1.

Note. We point out that for any sequence (s_n) in $\mathscr{R}$ and for any $m \in R$, $\{n \in N \mid |s_n| < m\}$ is a subset of N. The existence of this subset of N can be expressed by a swff of $\mathscr{R}$; transferring to ${}^*\mathscr{R}$, we conclude that internal sequences yield internal subsets of ${}^*N$ in this way.

Here is our third principle of permanence.

Third Principle of Permanence. Let (s_n) be any internal sequence such that $s_n \simeq 0$ for each $n \in N$. Then there is an infinite natural number κ such that $s_n \simeq 0$ for each $n < \kappa$, $n \in {}^*N$.

Dem. In $\mathscr{R}$, if (a_n) is a sequence, so is (na_n). By the transfer theorem, the corresponding statement is true for $^*\mathscr{R}$. We are given that (s_n) is an internal sequence, so (ns_n) is an internal sequence. For $n \in N$, $s_n \simeq 0$, so $ns_n \simeq 0$; thus $|ns_n| < 1$ for each $n \in N$. By Lemma 1, there is an infinite natural number κ such that $|s_n| < 1/n$ for each $n < \kappa$, $n \in {}^*N$. For n infinite, $1/n$ is an infinitesimal; so $s_n \simeq 0$ for $n < \kappa$, n infinite. By assumption, $s_n \simeq 0$ for $n \in N$. We conclude that $s_n \simeq 0$ for $n < \kappa$, $n \in {}^*N$.

Next, we want to establish a principle of permanence for the case of a sequence (s_n) such that s_n is infinite for each $n \in N$. The idea is to form the sequence of reciprocals $(1/s_n)$ and to apply the Third Principle of Permanence to this sequence. First, we must ensure that $(1/s_n)$ is internal. Indeed, what if $s_n = 0$? Of course, for n finite, $s_n \neq 0$; but for some infinite n, $s_n = 0$ is possible. We shall treat this problem in $\mathscr{R}$ and use our transfer theorem to handle the problem in $^*\mathscr{R}$. Let (a_n) be any sequence in $\mathscr{R}$, and define (b_n) such that

$$b_n = \begin{cases} 1/a_n & \text{if} \quad a_n \neq 0 \\ 1 & \text{if} \quad a_n = 0 \end{cases}$$

Then (b_n) is a sequence. We can express this construction, i.e., the existence of (b_n), by a swff of $\mathscr{R}$:

$$\forall(a_i)\exists(b_i)\forall n(a_n \neq 0 \rightarrow b_n = 1/a_n \;\dot{\wedge}\; a_n = 0 \rightarrow b_n = 1)$$
$$(a_i), (b_i) \in \text{Seq}, \qquad n \in N$$

Translating to $^*\mathscr{R}$ shows that corresponding to each internal sequence (s_n) of $^*\mathscr{R}$ there is an internal sequence $(1/s_n)$, defined as above.

Here is our corollary to the Third Principle of Permanence.

Corollary 1. Let (s_n) be any internal sequence such that s_n is infinite for each $n \in N$. Then there is an infinite natural number κ such that s_n is infinite for each $n < \kappa$, $n \in {}^*N$.

Dem. The sequence $(1/s_n)$ is internal, as we have shown above; moreover, $1/s_n \simeq 0$ for each $n \in N$. By the Third Principle of Permanence, there is an infinite natural number κ such that $1/s_n \simeq 0$ for each $n < \kappa$, $n \in {}^*N$. The reciprocal of any nonzero infinitesimal is infinite; we conclude that s_n is infinite for each $n < \kappa$, $n \in {}^*N$.

Finally, we mention that principles of permanence must be formulated with some care. The following statement, which has the appearance of a principle of permanence, is clearly false. "Let (s_n) be an internal sequence such that s_n is finite for each $n \in N$. Then there is an infinite natural number κ such that s_n is finite for each $n < \kappa$, $n \in {}^*N$." To see that this statement is false, apply it to the internal sequence (n) [i.e., take $s_n = n$ for each $n \in {}^*N$].

Here is another false principle of permanence. "Let (s_n) be any internal sequence such that $s_n \simeq 0$ for each infinite n. Then there is a finite natural number q such that $s_n \simeq 0$ for each $n > q$, $n \in {}^*N$." To see that this statement is false, take $(s_n) = (1/n)$.

Exercises

1. Let S be an internal subset of $^*N$ such that

$$\forall n(n > q \rightarrow n \in S), \qquad n \in N$$

where $q \in N$. Prove that there is an infinite natural number κ such that

$$\forall n(n < \kappa \rightarrow n \in S), \qquad n \in {}^*N - N$$

2. Show that $\{n \in N \mid n > 5\} \cup \{\omega\}$ is an external subset of $^*N$.

3. Show that $\{n \in N \mid n > 10\} \cup \{n \in {}^*N \mid n \text{ is even}\}$ is an external subset of $^*N$.

4. Let S be an internal subset of $^*N$ such that

$$\forall n(n < \kappa \rightarrow n \in S), \qquad n \in {}^*N - N$$

where κ is an infinite natural number. Prove that there is a finite natural number q such that

$$\forall n(n > q \rightarrow n \in S), \qquad n \in N$$

5. Show that $\{n \in {}^*N - N \mid n < \omega\}$ is an external subset of $^*N$.

6. Show that $\{n \in {}^*N - N \mid n < \omega\} \cup \{n \in N \mid n < 10\}$ is an external subset of $^*N$.

7. Prove that each member of *Seq is a map of $^*N$ into $^*R$.

8. Exhibit an external sequence.

9. (a) Exhibit a swff of $\mathscr{R}$ that asserts the existence of $\{n \in N \mid |s_n| < m\}$, where $(s_n) \in$ Seq and $m \in R$.
   (b) Prove that $\{n \in {}^*N \mid |s_n| < m\}$ is an internal subset of $^*N$.

10. Use the proof of Lemma 1 to show directly that there is an infinite natural number κ such that

$$\forall n(n < \kappa \rightarrow |s_n| < 1), \qquad n \in {}^*N$$

provided that (i) (s_n) is an internal sequence; (ii) $\forall n(|s_n| < 1)$, $n \in N$.

11. Let (s_n) be an internal sequence such that

$$\forall n(n > q \rightarrow s_n \simeq 0), \qquad n \in N$$

where $q \in N$. Prove that there is an infinite natural number κ such that

$$\forall n(n < \kappa \rightarrow s_n \simeq 0), \qquad n \in {}^*N - N$$

12. Let (s_n) be an internal sequence such that

$$\forall n(n > q \rightarrow s_n \text{ is infinite}), \qquad n \in N$$

Prove that there is an infinite natural number κ such that

$$\forall n(n < \kappa \rightarrow s_n \text{ is infinite}), \qquad n \in {}^*N - N$$

14

Normal Semantical Systems

14.1. Equality Relations

A relation $\underline{E}$ of a semantical system $\sum$ is said to be *substitutive* (for $\sum$) provided that for each relation symbol T of $\sum$ and for each $n \in$ type T,

$$\forall x_1 \cdots x_n y_1 \cdots y_n(Tx_1 \cdots x_n \wedge Ex_1y_1 \wedge \cdots \wedge Ex_ny_n \rightarrow Ty_1 \cdots y_n) \quad (1)$$

is true for $\sum$. In this case, we also say that the relation symbol E is substitutive.

Clearly E is not substitutive if $2 \notin$ type E [since (1) is *not* a swff of $\sum$]. Also, E is substitutive if Eab is false for $\sum$ whenever a and b are distinct constants.

Example 1. Let $\sum$ be the semantical system for which

$$\text{diag} \sum = \{E01, E22, R002, R112\}$$

Here, E is substitutive. To see this, put R for T in (1) and take $n = 3$. We must show that

$$\forall x_1x_2x_3y_1y_2y_3(Rx_1x_2x_3 \wedge Ex_1y_1 \wedge Ex_2y_2 \wedge Ex_3y_3 \rightarrow Ry_1y_2y_3)$$

is true for $\sum$. Now,

$$R002 \wedge E01 \wedge E01 \wedge E22 \rightarrow R112$$

is true for $\sum$; also

$$R112 \wedge E1a \wedge E1b \wedge E22 \rightarrow Rab2$$

is true for $\sum$ for any constants a and b (since $E1a$ is false for $\sum$). We must also put E for T in (1) and take $n = 2$; i.e., we must show that

$$\forall x_1x_2y_1y_2(Ex_1x_2 \wedge Ex_1y_1 \wedge Ex_2y_2 \rightarrow Ey_1y_2) \tag{2}$$

is true for $\sum$. Each of

$$E01 \wedge E01 \wedge E10 \rightarrow E10, \qquad E01 \wedge E01 \wedge E11 \rightarrow E11$$

is true for $\sum$. Also,

$$E22 \wedge E2a \wedge E2b \rightarrow Eab$$

is true for $\sum$ for any constants a and b. It follows that E is substitutive. Of course, $\underline{R}$ is not substitutive since $2 \notin$ type R.

We are interested in a substitutive relation symbol E such that $\underline{E}$ is an equivalence relation, i.e., reflexive, symmetric, and transitive. Let us spell this out; a relation $\underline{E}$ of $\sum$ is said to be an *equivalence* relation of $\sum$ provided that type $E = \{2\}$ and each of the following swffs of $\sum$ is true for $\sum$:

$$\forall x(Exx) \tag{3}$$

$$\forall xy(Exy \rightarrow Eyx) \tag{4}$$

$$\forall xyz(Exy \wedge Eyz \rightarrow Exz) \tag{5}$$

Clearly, the *identity* relation $\{(a, a) \mid a \text{ is a constant of } \sum\}$ is a substitutive equivalence relation of $\sum$; here, we assume that $\sum$ has a relation symbol $=$ such that the swff $=ab$ is false for $\sum$ iff a and b are distinct constants of $\sum$.

Definition. $\underline{E}$ is said to be an *equality* relation of $\sum$ provided that $\underline{E}$ is a substitutive equivalence relation of $\sum$. In this case we also say that the relation symbol E is an *equality* of $\sum$.

We emphasize that type $E = \{2\}$ is a necessary condition for an equality relation $\underline{E}$.

Our criterion for an equality relation can be simplified in view of the fact that a relation is transitive if it is both substitutive and reflexive.

Lemma 1. $\underline{E}$ is an equality relation of $\sum$ iff $\underline{E}$ is substitutive, reflexive, and symmetric, and type $E = \{2\}$.

Dem. Since $\underline{E}$ is substitutive, (2) is true for $\sum$; thus

$$\forall x_1x_2y_2(Ex_1x_2 \wedge Ex_1x_1 \wedge Ex_2y_2 \rightarrow Ex_1y_2) \tag{6}$$

is true for $\sum$. But $\underline{E}$ is reflexive; so

$$\forall x_1 x_2 y_2 (Ex_1 x_2 \wedge Ex_2 y_2 \rightarrow Ex_1 y_2)$$

is true for $\sum$. This proves Lemma 1.

Here is an example of a semantical system with an equality relation.

Example 2. Let $\sum$ be the semantical system such that

$$\text{diag} \sum = \{E00, E11, E22, E01, E10, R002, R112, R012, R102\}$$

So E and R are the relation symbols of $\sum$, and 0, 1, and 2 are its constants. Here, type $E = \{2\}$ and type $R = \{3\}$. Clearly, $\underline{E}$ is reflexive and symmetric. Moreover, replacing an instance of 0 by 1 in any string in diag $\sum$ yields a string in diag $\sum$, and vice versa. So, $\underline{E}$ is substitutive. We conclude that $\underline{E}$ is an equality relation of $\sum$.

How many equalities can a semantical system have? Any number, but they are coextensive, i.e., yield the same relation.

Lemma 2. Let E_1 and E_2 be equalities of $\sum$. Then $\underline{E}_1 = \underline{E}_2$; i.e., for any constants a and b of $\sum$,

$$E_1 ab \text{ is true for } \sum \text{ iff } E_2 ab \text{ is true for } \sum \tag{7}$$

Dem. Assume that $E_1 ab$ is true for $\sum$. Since $\underline{E}_1$ and $\underline{E}_2$ are each reflexive,

$$E_2 aa \wedge E_1 aa \wedge E_1 ab$$

is true for $\sum$. From (1), since $\underline{E}_1$ is substitutive,

$$E_2 aa \wedge E_1 aa \wedge E_1 ab \rightarrow E_2 ab$$

is true for $\sum$. Therefore, $E_2 ab$ is true for $\sum$. Similarly, we see that $E_1 ab$ is true for $\sum$ if $E_2 ab$ is true for $\sum$. This proves Lemma 2.

A semantical system that does not possess an equality can, generally speaking, be extended to a semantical system with equality in several ways.

Example 3. Let $\sum$ be the semantical system such that

$$\text{diag} \sum = \{R002, R112, R012, R102\}$$

So R is the only relation symbol of $\sum$, the constants of $\sum$ are 0, 1, and 2, and type $R = \{3\}$. In Example 2 we have seen that $\sum$ can be extended to a semantical system with equality. We point out now that $\sum$ can also be extended to $\sum'$, where

$$\text{diag } \textstyle\sum' = \{E00, E11, E22, R002, R112, R012, R102\}$$

Clearly, $\underline{E}$ is an equality relation of $\sum'$. This equality relation is the *identity* relation for $\sum$.

Given a semantical system $\sum$ that does not possess an equality relation, we can always introduce the *identity* to serve as an equality relation of $\sum$. This means that we regard "=" as a relation symbol of $\sum$ whose type is $\{2\}$; moreover, for any constants a and b of $\sum$, "$a = b$" is true for $\sum$ iff a and b are the same.

Exercises

1. Let E be a relation symbol of $\sum$ such that $2 \notin \text{type } E$; show that E is not substitutive.
2. Let E be a relation symbol of $\sum$ such that $2 \in \text{type } E$ and Eab is false for $\sum$ whenever a and b are distinct constants. Show that E is substitutive.
3. Let $\sum$ be the semantical system such that

$$\text{diag } \textstyle\sum = \{E01, E00, E22, R002, R112\}$$

Is E substitutive? (No)

4. Let $\sum$ be the semantical system such that

$$\text{diag } \textstyle\sum = \{E01, E00, R002, R112\}$$

Is E substitutive? (Yes)

5. Let $\sum$ be the semantical system such that

$$\text{diag } \textstyle\sum = \{E01, R002, R112\}$$

Is E substitutive? (Yes)

6. Let $\sum$ be the semantical system such that

$$\text{diag } \textstyle\sum = \{E01, R000, R111\}$$

Is E substitutive? (Yes)

7. Let $\sum$ be the semantical system such that

$$\text{diag } \textstyle\sum = \{E01, E00, R000, R111\}$$

Is E substitutive? (No)

8. Let $\sum$ be the semantical system such that

$$\text{diag } \textstyle\sum = \{E01, E00, R00, R11\}$$

(a) Is E substitutive? (No)
(b) Is R substitutive? (Yes)

9. Let E be a relation symbol of a semantical system $\sum$ such that:

(1) type $E = \{2\}$.
(2) $\underline{E} = \{(a, a) \mid a \text{ is a constant of } \sum\}$.

Show that E is an equality relation of $\sum$.

10. Let $\sum$ be the semantical system such that

$$\text{diag } \textstyle\sum = \{E01, E00, R00, R11\}$$

(a) Is E an equality relation? (No)
(b) Is R an equality relation? (Yes)

11. Let $\sum$ be the semantical system such that

$$\text{diag } \textstyle\sum = \{E00, E11, E01, E10, R000, R111\}$$

Is E an equality relation? (No)

12. Let $\sum$ be the semantical system such that

$$\text{diag } \textstyle\sum = \{E00, E11, R000, R111\}$$

Is E an equality relation? (Yes)

13. Let $\sum$ be the semantical system such that

$$\text{diag } \textstyle\sum = \{E00, E11, E01, E10, R0, R1\}$$

Is E an equality relation? (Yes)

14. Let $\sum$ be the semantical system such that

$$\text{diag } \textstyle\sum = \{E00, E11, R0, R111\}$$

Is E an equality relation? (Yes)

15. Let $\sum$ be the semantical system such that

$$\text{diag } \textstyle\sum = \{E00, E11, E01, E10, R0, R111\}$$

Is E an equality relation? (No)

16. Let E_1 and E_2 be equality relations of a semantical system $\sum$. Prove that for any constants a and b of $\sum$,

$$(a, b) \in \underline{E}_1 \quad \text{iff} \quad (a, b) \in \underline{E}_2$$

17. Let $\sum$ be the semantical system of Example 2, and let $\sum'$ be the semantical system such that

$$\text{diag } \textstyle\sum' = \{E00, E22, R002\}$$

Let A be any swff of $\sum$ and let A' be the swff of $\sum'$ obtained from A by replacing each instance of 1 by 0. Prove that

$$A \text{ is true for } \sum \quad \text{iff} \quad A' \text{ is true for } \sum'$$

for each swff A of $\sum$. First, prove that A' is a swff of $\sum'$ provided that A is a swff of $\sum$.

14.2. Normal Semantical Systems

Let $\sum$ be a semantical system and let $\underline{E}$ be an equality relation of $\sum$. We say that $\sum$ is *normal* provided that Eab is false for $\sum$ whenever a and b are distinct constants of $\sum$. By assumption, $\underline{E}$ is reflexive; so Eaa is true for $\sum$ for each constant a of $\sum$. Thus $\underline{E}$ is $\{(a, a) \mid a \text{ is a constant of } \sum\}$, the identity relation.

In view of Lemma 2, page 265, this means that $\sum$ is normal provided that $\sum$ has an equality relation and provided that for each equality E of $\sum$, $\underline{E}$ is $\{(a, a) \mid a \text{ is a constant of } \sum\}$.

Our purpose in this section is to show that each consistent set of wffs has a normal model.

Theorem 1. Let K be consistent; then K has a normal model.

Dem. By the Strong Completeness Theorem, K has a model; let $\sum$ be a model of K under μ. Either some relation symbol of $\sum$ is an equality or no relation symbol of $\sum$ is an equality. We consider both possibilities.

1. Let E be an equality of $\sum$. Since $\underline{E}$ is an equivalence relation on C, the set of all constants of $\sum$, we can use $\underline{E}$ to partition C. This factoring operation yields the set of equivalence classes $C/\underline{E}$; here, $C/\underline{E} = \{[a] \mid a \text{ is a constant of } \sum\}$. Essentially, if $Eab \in \text{diag} \sum$, then we regard b as a name for $[a]$. If $\underline{E}$ is not the identity relation, then there is a certain redundancy in diag $\sum$; i.e., some of the strings in diag $\sum$ are identical (here we use the fact that $\underline{E}$ is substitutive). To eliminate the redundancy in C and in diag $\sum$, we shall form the semantical system $\sum'$ whose constants are the equivalence classes $[a]$, $a \in C$; moreover, R' is a relation symbol of $\sum'$ for each relation symbol R of $\sum$. The diagram of $\sum'$ is obtained from diag $\sum$ as follows:

$$R'[a_1] \cdots [a_n] \in \text{diag} \textstyle\sum' \qquad \text{iff} \qquad Ra_1 \cdots a_n \in \text{diag} \textstyle\sum \tag{1}$$

where $n \in \text{type } R$. We hasten to mention that type $R' = \text{type } R$. From (1), $E'[a][b] \in \text{diag} \sum'$ iff $Eab \in \text{diag} \sum$; we conclude that E' is the identity of

$\sum'$. So, $\sum'$ is a normal semantical system. To see that $\sum'$ is a model of K, let λ be the interpreter such that:

i. $\lambda t = [\mu t]$ for each individual t of $\prod$.
ii. $\lambda T = (\mu T)'$ for each predicate T of $\prod$.

It is easy to prove that for each wff A,

$$\lambda A \text{ is true for } \textstyle\sum' \quad \text{iff} \quad \mu A \text{ is true for } \textstyle\sum \tag{2}$$

Thus, $\sum'$ is a model of K under λ.

2. Assume that $\sum$ has no equality relation. Extend $\sum$ to another semantical system $\sum''$ by adjoining a new relation symbol whose type is $\{2\}$, say E, to the relation symbols of $\sum$; also, we adjoin each string of the form Eaa, where a is any constant of $\sum$, to diag $\sum$. Thus

$$\text{diag } \textstyle\sum'' = \text{diag } \textstyle\sum \cup \{Eaa \mid a \text{ is a constant of } \textstyle\sum\}$$

Here, E is an equality of $\sum''$; indeed,

$$\underline{E} = \{(a, a) \mid a \text{ is a constant of } \textstyle\sum\}$$

so E is the identity of $\sum''$. Therefore, $\sum''$ is normal. Of course, $\sum''$ is a model of K since $\sum$ is a model of K. This completes our proof of Theorem 1.

Exercises

1. Consider the semantical system $\sum'$ defined in the proof of Theorem 1.
   (a) What are the constants of $\sum'$?
   (b) What are the relation symbols of $\sum'$?
   (c) Let R' be any relation symbol of $\sum'$; what is the type of R'?
   (d) Characterize diag $\sum'$.
   (e) Prove that $E'[a][b] \in$ diag $\sum'$ iff $[a]$ and $[b]$ are the same constants of $\sum'$.
   (f) Let λ be the $\sum'$-interpreter defined above. Prove that for each wff A,

$$\lambda A \text{ is true for } \textstyle\sum' \quad \text{iff} \quad \mu A \text{ is true for } \textstyle\sum$$

2. Consider the semantical system $\sum''$ defined above. Prove that $\sum''$ is a model of K. What is the $\sum''$-interpreter involved here?

14.3. Löwenheim–Skolem Theorem

In this section we shall consider a surprising corollary to the Strong Completeness Theorem. This result, which nowadays is rather obvious, was first obtained by Löwenheim (1915). His proof was simplified by Skolem (1920); moreover, Skolem generalized Löwenheim's theorem.

First, we need the notion of the *cardinality* of a semantical system.

Definition. Let $\sum$ be any normal semantical system; by the *cardinality of* $\sum$ we mean the cardinal number of its set of constants.

Note that we restrict this notion to *normal* semantical systems. It would be misleading to apply this concept to a semantical system that is not equipped with the identity relation.

Now, a set is *denumerable* provided that it is equinumerous with the set of natural numbers, i.e., provided its cardinal number is $\aleph_0$. A set is *countable* if it is finite or denumerable. Thus, a normal semantical system is countable if the set of its constants is finite or denumerable.

We now present the Löwenheim–Skolem theorem.

Löwenheim–Skolem Theorem. Each countable, consistent set of wffs has a countable model.

Dem. Let K be any countable, consistent set of wffs. Each wff is a finite string of symbols; thus, the number of individuals and predicates which occur in K is countable. Let $\prod$ be a predicate calculus with denumerably many individuals including all individuals occurring in K, and whose predicates are precisely the predicates that occur in K. Recall the proof of the Strong Completeness Theorem (see pages 221–223). The idea of the proof is to extend K to a maximal-consistent, $\exists$-complete set K_ω. A model of K_ω, say $\sum$, is easily constructed from the wffs in K_ω (see page 219); indeed, the constants of $\sum$ are the individuals of the predicate calculus $\prod_\omega$. By construction, $\prod_\omega$ has denumerably many individuals; thus, $\sum$ has denumerably many constants. From our proof of Theorem 1, page 268, we see that we can construct a normal model of K_ω from $\sum$. The construction for the normal model does not increase the number of constants (since the normal model is obtained by factoring out the equivalence classes of $\sum$); so the normal model that results has at most denumerably many constants, i.e., it is finite or denumerable. We conclude that K has a countable, normal model.

This result is very significant. It shows, for example, that any axiomatization of the real number system within a countable language (a predicate calculus with countably many individuals and predicates) has a countable model, which therefore cannot be the real number system.

We conclude that it is impossible to characterize the real number system, or any other uncountable system, within a countable language.

We can compel each normal model of a set of axioms to have exactly n constants, where n is finite. For example, include in the axiom set the axioms

$$\forall x(x = a_1 \vee \cdots \vee x = a_n)$$
$$a_1 \neq a_2, \ldots, a_1 \neq a_n; \quad a_2 \neq a_3, \ldots, a_2 \neq a_n; \ldots; \quad a_{n-1} \neq a_n$$

as well as axioms that ensure that $=$ is interpreted as an equality relation. However, we cannot compel each normal model of a set of axioms to have exactly $\aleph$ constants, where $\aleph \geq \aleph_0$.

This comment relies on the following generalization of the Löwenheim–Skolem Theorem.

Theorem 1. Let K be a consistent set of wffs in a predicate calculus $\prod$ that has $\aleph$ individuals and predicates, where $\aleph \geq \aleph_0$. Then K has a normal model with at most $\aleph$ constants.

Dem. Henkin's proof of the Strong Completeness Theorem, which is formulated for the case of a predicate calculus with $\aleph_0$ individuals and predicates, carries over to the given predicate calculus $\prod$, which has $\aleph$ individuals and predicates. (The Axiom of Choice is used here.) It follows that K has a model with $\aleph$ constants, say $\sum$. By our proof of Theorem 1, page 268, $\sum$ can be modified to yield a normal model $\sum'$. By construction, the number of constants of $\sum'$ does not exceed the number of constants of $\sum$. Therefore, $\sum'$ has at most $\aleph$ constants.

Theorem 2. Let K be a countable set of wffs that possesses a normal, infinite model $\sum$. Then K has a normal model with $\aleph$ constants, for each $\aleph \geq \aleph_0$.

Dem. Let $=$ be the identity relation of $\sum$, and let the corresponding predicate be denoted by $=$ also. Since K is countable, there are at most $\aleph_0$ individuals occurring in K. Extend the given predicate calculus by including new individuals x_i, $i \in I$, where I has $\aleph$ members, $\aleph \geq \aleph_0$. The resulting predicate calculus has $\aleph$ individuals, since $\aleph + \aleph_0 = \aleph$. Let

$$\mathsf{K}' = \mathsf{K} \cup \{x_i \neq x_j \mid i, j \in I \text{ and } i \neq j\}$$

First, we shall show that K′ is consistent. By assumption, K has an infinite model $\sum$; thus, each finite subset of K′ has a model, e.g., $\sum$. By the Compactness Theorem, K′ has a model; therefore, K′ is consistent. Now, each model of K′ has at least $\aleph$ equivalence classes; so each normal model of K′ has at least $\aleph$ constants. Henkin's construction for a model of K′ (see the proof of Theorem 1), modified by factoring out equivalence classes, yields a normal model of K′ with at most $\aleph$ constants. We conclude that K′ has a normal model with exactly $\aleph$ constants.

Corollary 1. Let K be a set of wffs such that:

1. K has an infinite model.
2. m individuals and predicates occur in K.

Then for each $\aleph \geq m$, $\aleph$ infinite, there is a normal model of K with $\aleph$ constants.

Dem. Let $\aleph$ be any infinite cardinal such that $\aleph \geq m$. Extend the given predicate calculus and form K′ as in the proof of Theorem 2. Since $\aleph + m = \aleph$, the resulting predicate calculus has $\aleph$ individuals. It follows, as before, that K′ has a normal model with $\aleph$ constants.

This establishes the comment that precedes Theorem 1.

Exercises

1. Let K be a set of wffs such that:

   (i) K is consistent.
   (ii) K has m members, where m is an infinite cardinal.

   Prove that K has a normal model with at most m constants.

2. Let K be a set of wffs such that:

   (i) K has an infinite model.
   (ii) K has m members, where m is an infinite cardinal.

   Let $\aleph \geq m$, where $\aleph$ is an infinite cardinal; prove that there is a normal model of K with exactly $\aleph$ constants.

3. Exhibit a finite set of wffs K such that K has an infinite model, but K does not have a finite model.

4. Let K be a countable set of wffs such that for each $n \in N$ there is a model of K with at least n members. Prove that for each infinite cardinal $\aleph$, there is a model of K with exactly $\aleph$ constants.

14.4. Theories

In Chapter 12 we proved that for each wff A and for any set of wffs K,

$$\mathsf{K} \vdash A \qquad \text{iff} \qquad A \text{ is K-true}$$

i.e., A is deducible from K iff μA is a true swff of $\sum$ whenever $\sum$ is a model of K under μ.

This result allows us to conceptualize mathematical theories in a simple, straightforward way. Normally, a mathematical theory is regarded as possessing several parts. There are the undefined terms, i.e., the primitive concepts, of the theory; there is a postulate set for the theory (its axioms); there are the provable theorems of the theory (statements which are deducible from the given postulate set); there are the models of the theory.

From our viewpoint, the essence of a mathematical theory can be given in either of two ways: (1) the theorems of the theory; (2) the models of the theory.

These are equivalent in the sense that given (1), we can obtain (2); given (2), we can obtain (1). Of course, we can regard the set of all theorems of a theory as a postulate set for the theory. It is more usual, however, to get at the theorems of a theory by exhibiting a postulate set of minimal size, say K. As we have mentioned, a wff T is a theorem of a theory K iff K $\vdash T$. For this reason, we shall regard a mathematical theory as being constituted by a set of wffs—a postulate set. Of course, a postulate set that has no model is of little interest, so we restrict the notion of a mathematical theory to a *consistent* set of wffs.

Here is a powerful application of the fact that K $\vdash A$ iff A is K-true. Consider a mathematical theory that is characterized by an *infinite* postulate set, say K (remember, the theory *is* the postulate set). Let T be any theorem of this theory. Then K $\vdash T$; therefore, there is a *finite* subset of K, say $\{A_1, \ldots, A_n\}$, such that $\{A_1, \ldots, A_n\} \vdash T$. Thus, T is a theorem of the theory whose postulate set is $\{A_1, \ldots, A_n\}$; so, μT is a true swff of $\sum$ whenever $\sum$ is a model of $\{A_1, \ldots, A_n\}$ under μ. Here we have a genuine mathematical result.

Let us illustrate the power of this observation. The notion of a field is easily characterized by a postulate set within our language. Furthermore, it is easy to express the idea of a field that has characteristic p, where p is prime (this means that p a's added together produce the additive identity for each field element a). A field is said to have characteristic zero iff the field does not have characteristic 2, does not have characteristic 3, does not have characteristic 5, and in general, does not have characteristic p, given that p is prime. Thus, the notion of a field of characteristic zero is characterized by an infinite postulate set. Let T be a theorem of this theory; then T is deducible from a finite subset of the given infinite postulate set; this finite subset contains only a *finite* number of statements asserting that the field does not have characteristic p for various prime numbers p. Therefore, there is a largest prime number, say q, which occurs in this finite postulate set. Hence, each field with characteristic greater than q is a model of our postulate set. Since T is deducible from this finite postulate set, we see that T is true for each field of characteristic greater than q. Summarizing, if T is true for each field of characteristic zero, then T is true for each field of characteristic greater than q, where q is a prime number that depends upon the wff T.

In the same way, it is easy to see that the concept of an algebraically closed field (a field in which each polynomial possesses a zero) is characterized by an infinite postulate set. Therefore, by our result, if A is true for each algebraically closed field, then A is true for any field such that each polynomial of degree less than m possesses a zero, where m is a natural number that depends upon the wff A.

Exercises

1. Exhibit a postulate set for *groups*.

2. Exhibit a postulate set for *Abelian* groups.

3. Characterize the concept of a *field* by means of a set of wffs.

4. Characterize the concept of an *ordered* field.

5. (a) Characterize the concept of a field with characteristic two.
   (b) Characterize the concept of a field with characteristic three.

6. Characterize the concept of a *Boolean algebra.*

7. By a *loop* we mean a semantical system $\sum$ which involves just one relation symbol which represents a binary operation $+$. Truth-values are assigned so

   (i) $\forall x(x + 0 = 0 + x = x)$ is true for $\sum$.
   (ii) Each equation of the form $x + y = z$ has a unique solution given any two of x, y, and z.

   Characterize the concept of a loop within the predicate calculus.

8. Characterize the concept of a field for which each polynomial of degree two has a zero.

9. Consider an ordered field such that:

   (i) Each polynomial of odd degree has a zero.
   (ii) Each polynomial of the form $x^2 + a$ has a zero provided that $a < 0$.

   Such a field is said to be a *real-closed* ordered field. Exhibit a postulate set for this theory.

10. Characterize the concept of an Abelian group that has exactly two constants.

11. An Abelian group is said to be *torsion-free* provided that the result of adding a group element, say *a*, to itself any finite number of times is different from the additive identity—provided that *a* is not the additive identity. Characterize the notion of a torsion-free Abelian group.

12. An Abelian group is said to be *completely divisible* provided that, corresponding to each group element, say *a*, and to each natural number, say *n*, there is a group element, say *b*, such that $a = b + \cdots + b$ (n b's). Characterize this concept within the predicate calculus.

15

Axiomatic Set Theory

15.1. Introduction

At the close of the 19th century and the beginning of the 20th, the discovery by Burali-Forti and Russell of the paradoxical nature of the intuitive approach to set theory gave impetus to the axiomatization of the subject. Zermelo's postulates for set theory were published in 1908. Fraenkel (1922*a*) extended these postulates and sharpened (1922*b*) Zermelo's Axiom of Separation.

In the Zermelo–Fraenkel approach, which we shall follow, the notion of a set is restricted by the postulates. Here, the giant collections involved in a paradox are not recognized as sets; indeed, it can be proven within the system that the collection of all sets is not itself a set, and that Russell's set, the collection of all sets that are not members of themselves, is not a set (in the sense of Zermelo–Fraenkel).

Another method of eliminating the paradoxes was devised by von Neumann in 1925, and later modified by Bernays, and still later by Gödel (1940). The idea, here, is that the giant sets are harmless in themselves; a paradox arises only when such a set is regarded, at least potentially, as a *member* of some set. So arises the distinction between *sets* and *classes*. Any collection is called a class; a class that is a member of some collection is called a set as well. Some classes are not sets. The paradoxes are refuted by not allowing the collections involved to be sets.

Our aim is to write down a set of wffs K such that the constants of each model of K can be regarded as sets; K is called a postulate set for Zermelo–Fraenkel set theory. The idea is that the postulate set K compels the constants of each model of K to be *sets*. Here, we wish to deal with a semantical system $\sum$ that involves just one relation symbol $\in$; moreover, each string in the domain of $\sum$ has the form "$\in ab$," where a and b are constants of $\sum$; i.e., the type of $\in$ is $\{2\}$. For simplicity, we shall denote each string of the form "$\in ab$" by writing "$a \in b$" (read *a is a member of b*).

The basic idea is that for any constants (sets) a and b of $\sum$, we regard a as a member of b iff $\sum$ assigns "true" to the string "$a \in b$." Intuitively, each constant b represents the collection

$$\{a \mid a \text{ is a constant and } \textstyle\sum \text{ associates "true" with the string } a \in b\}$$

Remember that an interpreter is needed to make the link between the predicate calculus and the language of a semantical system. To avoid this, and in the interest of simplicity, we shall eliminate the predicate calculus from our considerations by the following well-known method. We shall consider a semantical system $\sum$ that has just one relation symbol $\in$; however, we shall not specify the constants of $\sum$. So, the domain of $\sum$ is unspecified and the truth-values that $\sum$ associates with the strings in its domain are also not specified. We do require, however, that the constants of $\sum$ and the assignment of truth-values to the strings in dom $\sum$ be such that the Zermelo–Fraenkel axioms are true for $\sum$. If so, then $\sum$ is called a model of Zermelo–Fraenkel set theory.

Eliminating the predicate calculus requires that we utilize not only the swffs of $\sum$ but also the *quasi-swffs*. Recall that a quasi-swff is an expression obtained from a swff by replacing some of its constants by placeholders. For example, "$x \in y$" is a quasi-swff since "$a \in b$" is a swff (here, a and b are constants of $\sum$). For example, we can use a quasi-swff to characterize a specific collection of constants of $\sum$; e.g., the *intersection* of sets a and b is the collection of all constants, say x, such that $x \in a \wedge x \in b$ is true for $\sum$. Thus "$a \cap b$" denotes the collection

$$\{x \mid x \in a \wedge x \in b \text{ is true for } \textstyle\sum\} \tag{1}$$

Here, "$x \in a \wedge x \in b$" is a quasi-swff. We point out that (1) specifies a set provided that the swff

$$\exists y \forall x (x \in y \leftrightarrow x \in a \wedge x \in b\} \tag{2}$$

is true for $\sum$. So, $a \cap b$ is a *set* provided that there is a constant of $\sum$, say c, such that

$$x \in c \qquad \text{iff} \qquad x \in a \wedge x \in b$$

for each constant x of $\sum$. If so, then we regard "$a \cap b$" as a name for c and we regard (1) as a name for c.

Hereafter, we shall freely drop the phrase "is true for $\sum$" (more accurately, we shall *suppress* this phrase); remember that we regard swffs as informal statements. Indeed, the language of semantical systems is the usual language of mathematics. We shall freely use the language of mathematics to express swffs of $\sum$. For example, we shall regard (1) as a name for the set whose existence is asserted by (2).

Here are the Zermelo–Fraenkel axioms for set theory; they consist of six axioms and one collection of axioms. In the following, $\varnothing$ is a name for $\{x \mid x \in x \wedge \neg(x \in x)\}$, the empty set; the expressions "$x = y$" and "$x \subset y$" are abbreviations for certain quasi-swffs of $\sum$:

"$x = y$" is an abbreviation for $\forall t(t \in x \leftrightarrow t \in y)$
"$x \subset y$" is an abbreviation for $\forall t(t \in x \rightarrow t \in y)$

The reference to *functional in x* will be explained in Section 15.3. We shall discuss each axiom, or axiom scheme, later.

Zermelo–Fraenkel Axioms

1. *Axiom of Extensionality.*

$$\forall xy(x = y \rightarrow \forall z(x \in z \rightarrow y \in z))$$

2. *Axiom Scheme of Replacement.* For each set s, the collection

$$\{y \mid \exists x(x \in s \wedge F(x, y))\}$$

is a set provided that $F(x, y)$ is functional in x.

3. *Axiom of Power Set.* For each set s, the collection

$$\{x \mid x \subset s\}$$

is a set.

4. *Axiom of Sum Set.* For each set s, the collection

$$\{x \mid \exists y(y \in s \wedge x \in y)\}$$

is a set.

5. *Axiom of Infinity.* There is a set W such that $\varnothing \in W$ and

$$\forall x(x \in W \rightarrow x \cup \{x\} \in W)$$

Here $a \cup b$ (the *union* of sets a and b) is the collection $\{x \mid x \in a \vee x \in b\}$, which is a set by the Axiom of Sum Set.

6. *Axiom of Regularity.* Each nonempty set s has a member x such that $x \cap s = \varnothing$.

7. *Axiom of Choice.* Let s be a disjointed set such that $\varnothing \notin s$; then there is a set t such that

$$\forall z \exists x(z \in s \rightarrow t \cap z = \{x\})$$

Note. s is disjointed if no two members of s have a common member. Each set t is called a *selection set* of s; notice that t selects x from z for each $z \in s$.

As we have mentioned, the idea is to choose the domain of $\sum$ (so its constants) and to assign truth-values to each string in dom $\sum$ so that the Zermelo–Fraenkel axioms are true for $\sum$; in this case, $\sum$ is called a model of Zermelo–Fraenkel set theory. Some insight into this procedure is provided by the following example, in which we choose dom $\sum$ so that some of the Zermelo–Fraenkel axioms are true for $\sum$, but some are not true for $\sum$.

Example 1. Let $\sum$ be the semantical system with constants a, b, and c such that

$$\text{diag} \textstyle\sum = \{a \in b, \quad a \in c, \quad b \in c, \quad c \in c\}$$

Then $a = \varnothing$, the empty set, $b = \{a\}$, and $c = \{a, b, c\}$. This semantical system is *not* acceptable as a model of Zermelo–Fraenkel set theory for several reasons:

1. The Axiom of Power Set is not true for $\sum$. The Zermelo–Fraenkel set theory requires that the collection of all subsets of any set is also a set. The set b has two subsets within the system, namely a and b; but the collection $\{a, b\}$ is not recognized as a set within this system.

2. The Axiom Scheme of Replacement is not true for $\sum$. The Theorem of Separation, which follows from these axioms, requires that the collection

$$\{y \mid y \in c \wedge \neg(y \in y)\} = \{a, b\}$$

be a set of the system. Similarly, the Theorem of Separation requires that the collection

$$\{y \mid y \in c \wedge y \in y\} = \{c\}$$

be a set of the system.

3. The Axiom of Infinity is not true for $\sum$. Try $W = b$; then the axiom requires that $a \cup \{a\} \in W$. Now, $a \cup \{a\} = a \cup b = b$, which is not a member of b. Try $W = c$; we require that $b \cup \{b\} = \{a, b\} \in c$. But $\{a, b\}$ is not a member of c; indeed, neither $\{b\}$ not $\{a, b\}$ is a set within the system.

We point out that the remaining axioms of Zermelo–Fraenkel set theory are each true for $\sum$. The axioms are discussed in the remainder of this chapter.

The Axiom of Choice is often given a preferred position in Zermelo–Fraenkel (ZF) set theory. It is customary to write "ZFC" or "ZF + AC" to emphasize that the Axiom of Choice is included.

Some authors develop set theory without the Axiom of Regularity, but include the Axiom of Choice. The resulting set theory, which is based on six Zermelo–Fraenkel axioms (Extensionality, Replacement, Power Set, Sum Set, Infinity, and Choice) is also called ZF set theory.

The famous logician Skolem also made a substantial contribution to axiomatic set theory; so ZF set theory is sometimes called Zermelo–Fraenkel–Skolem set theory.

Although it has not been proven that ZF set theory is consistent, it is widely believed that it is. In 1938, Gödel demonstrated that ZFC is consistent if ZF (without Choice) is consistent. Paul Cohen, in 1963, proved that the Axiom of Choice is independent of the other Zermelo–Fraenkel axioms.

Exercises

1. Let $\sum$ be the semantical system with one relation symbol $\in$ of type $\{2\}$ and constants a and b such that

$$\text{diag} \textstyle\sum = \{a \in b\}$$

Which of the Zermelo–Fraenkel axioms are true for $\sum$; which axioms are false for $\sum$?

2. Let $\sum$ be the semantical system with one relation symbol $\in$ of type $\{2\}$ and constants a and b such that

$$\text{diag} \textstyle\sum = \{a \in b, \quad b \in b\}$$

Which of the Zermelo–Fraenkel axioms are true for $\sum$; which axioms are false for $\sum$?

3. Let $\sum$ be the semantical system with one relation symbol $\in$ of type $\{2\}$ and constants a, b, and c such that

$$\text{diag} \textstyle\sum = \{a \in a, \quad a \in b, \quad a \in c, \quad b \in c, \quad c \in c\}$$

Which of the Zermelo–Fraenkel axioms are true for $\sum$; which axioms are false for $\sum$?

4. Let $\sum$ be the semantical system with one relation symbol $\in$ of type $\{2\}$ and constants a, b, and c such that

$$\text{diag} \textstyle\sum = \{c \in b, \quad a \in c, \quad b \in c\}$$

Which of the Zermelo–Fraenkel axioms are true for $\sum$; which axioms are false for $\sum$?

15.2. Axiom of Extensionality

In Section 14.1 we discussed the general notion of an *equality* relation. Recall that an equality relation is a substitutive equivalence relation. Now, a semantical system $\sum$ that has just one relation symbol $\in$ can be equipped with an equality relation E provided that E is an equivalence relation such that

$$\forall xyuv(x \in y \wedge Exu \wedge Eyv \rightarrow u \in v) \tag{1}$$

$$\forall xyuv(Exy \wedge Exu \wedge Eyv \rightarrow Euv) \tag{2}$$

are both true for the resulting semantical system.

Lemma 1. (2) is true if E is an equivalence relation.

Dem. Let x, y, u, and v be constants of $\sum$ such that

$$Exy \wedge Exu \wedge Eyv$$

Then $Eux \wedge Exy$; so Euy (since E is transitive). Thus

$$Euy \wedge Eyv$$

So Euv.

The idea is to adjoin the relation symbol E to $\sum$, and to adjoin all strings of the form "Exy" to dom $\sum$, where x and y are any constants of $\sum$. Call the resulting semantical system $\sum'$; here, truth-values are assigned in accordance with the following definition.

$$Exy \text{ is true for } \textstyle\sum' \text{ iff } \forall t(t \in x \leftrightarrow t \in y) \text{ is true for } \textstyle\sum \tag{3}$$

Theorem 1. E is substitutive provided that

$$\forall xy(Exy \rightarrow \forall z(x \in z \rightarrow y \in z)) \tag{4}$$

is true for $\sum'$.

Dem. From (3) it follows that E is an equivalence relation; so, by Lemma 1, (2) is true for $\sum'$. We must show that (1) is true for $\sum'$ provided that (4) is true for $\sum'$. Accordingly, assume (4). Let x, y, u, and v be constants of $\sum'$ such that $x \in y \wedge Exu \wedge Eyv$. We shall show that $u \in v$. Now, $Exu \wedge x \in y$; by (4), $u \in y$. But Eyv; so, by (3), $u \in v$. This proves that (1) is true for $\sum'$. We conclude that E is substitutive.

The significance of this result is that E is an *equality* relation of $\sum'$ if (4) is true for $\sum'$. To this purpose, we take (4) as an axiom, called the *Axiom of Extensionality*.

Axiom of Extensionality. $\forall xy(Exy \rightarrow \forall z(x \in z \rightarrow y \in z))$.

This axiom guarantees that the relation $\underline{E}$ defined by (3) is an equality relation. As we suggested on page 268, we can regard $\underline{E}$ as the *identity* relation of $\sum'$ by the device of selecting one constant, say *a*, from each equivalence class under $\underline{E}$, and taking each constant in an equivalence class to be a name for the chosen representative of that equivalence class. Under this viewpoint $\sum'$ is essentially a *normal* semantical system.

For simplicity, we shall automatically adjoin the relation $\underline{E}$, as defined by (3), to each semantical system $\sum$; this will eliminate the need to mention $\sum'$ explicitly.

Since we regard E as the identity, we plan in due course on replacing "E" by "=." However, we shall continue writing "E" in order to maintain the distinction between the equality relation of $\sum$ (more accurately, $\sum'$) and the use of "=" as an abbreviation for "is a name for"; e.g., in definitions.

For any model $\sum$ of Zermelo–Fraenkel set theory, constants x and y are equal (i.e., Exy is true for $\sum$) provided that x and y have the same members. Therefore, we can get at any constant x of $\sum$, up to equality, provided that we know the members of x, i.e., the constants y for which "$y \in x$" is true for $\sum$. This observation is the basis of the following well-known convention for naming sets: "$\{a, b, c, \ldots\}$" is a name for any constant x such that

$$a \in x, \quad b \in x, \quad c \in x, \ldots$$

are each true for $\sum$, whereas "$y \in x$" is false for $\sum$ in case

$$\neg Eya, \quad \neg Eyb, \quad \neg Eyc, \ldots$$

are each true for $\sum$. (Of course, it is possible that there is no such constant x.)

For example, "$\{a\}$" denotes any constant of $\sum$, say x, such that:

(i) $a \in x$.
(ii) $\forall y(\neg Eya \rightarrow \neg(y \in x))$.

There may be another constant of $\sum$, say z, that satisfies (i) and (ii); however, under (3), we see that Exz and Ezx; i.e., x and z are equal.

Notice that if Eab, then $\{a\} = \{b\}$; i.e., "$\{a\}$" and "$\{b\}$" are names for the same constants of $\sum$. Stretching our convention slightly, we shall also accept "$\{a, b\}$" as a name for a constant of $\sum$; in this case, $\{a, b\} = \{a\}$ (here, we are assuming that Eab).

We mentioned earlier that each *empty* set (if any) of $\sum$ is denoted by "$\varnothing$"; i.e., $x = \varnothing$ provided that $\forall y(\neg(y \in x))$.

Example 1. Let $\sum$ be the semantical system with constants a, b, c, d, e such that

$$\text{diag} \sum = \{a \in b, \quad b \in c, \quad c \in c, \quad d \in c, \quad a \in d\}$$

and $x \in y$ is a swff of $\sum$ for any constants x and y. Then $a = \varnothing$, $b = \{a\}$, $c = \{b, c, d\}$, $d = \{a\}$, and $e = \varnothing$. Introducing E as in (3), we find that

$$Eae, \quad Eea, \quad Ebd, \quad Edb$$

i.e., a and e are equal, and b and d are equal. Thus $c = \{b, c\}$. Is the Axiom of Extensionality true for $\sum$? Since Eae, we must verify that $\forall z(a \in z \rightarrow e \in z)$ is true for $\sum$. But $a \in b \rightarrow e \in b$ is false for $\sum$. We conclude that the Axiom of Extensionality is false for $\sum$.

A semantical system for which any two constants are distinct under the binary relation defined by (3) automatically satisfies the Axiom of Extensionality. We now exhibit a semantical system, constructed from the system of Example 1, which satisfies this axiom nontrivially.

Example 2. Let $\sum$ be the semantical system with constants a, b, c, d, e such that

$$\text{diag} \sum = \{a \in b, \quad b \in c, \quad c \in c, \quad d \in c, \quad a \in d, \quad e \in b, \quad e \in d\}$$

and $x \in y$ is a swff of $\sum$ for any constants x and y. Then

$$a = \varnothing, \quad b = \{a, e\}, \quad c = \{b, c, d\}, \quad d = \{a, e\}, \quad e = \varnothing \tag{5}$$

Introducing E as in (3), we see that

$$Eae, \quad Eea, \quad Ebd, \quad Edb$$

since $a = \varnothing = e$ and $b = \{a, e\} = d$. Now, the Axiom of Extensionality is true for $\sum$. To see this we must show that

$$\forall z(a \in z \leftrightarrow e \in z) \qquad \text{and} \qquad \forall z(b \in z \leftrightarrow d \in z)$$

From (5), it is clear that these swffs are true for $\sum$. We conclude that the Axiom of Extensionality is true for $\sum$. Notice that the diagram of $\sum$ does not distinguish between a and e, and does not distinguish between b and d. Accordingly, $b = \{a, e\} = \{a\}$. Later (see page 290), we shall demonstrate that some of the other Zermelo–Fraenkel axioms are not true for $\sum$.

We emphasize that a collection of constants of a semantical system $\sum$ is not necessarily a *set* within the system. Let us pin down our use of the term *set*.

Definition. A collection of constants of $\sum$, say $\{a, b, c, \ldots\}$, is said to be a *set of* $\sum$ provided that there is a constant of $\sum$, say x, such that for each constant y,

$$y \in x \quad \text{iff} \quad Eya \vee Eyb \vee Eyc \vee \cdots$$

i.e., $y \in x$ iff y equals some member of the collection.

Exercises

1. Let E be the binary relation defined by (3). Prove that E is an equivalence relation.
2. Show that the binary relation defined by (3) is not necessarily an *equality* relation.
3. Let $\sum$ be the semantical system with constants a, b, c, d, e, f such that

$$\text{diag} \sum = \{b \in a, \quad c \in a, \quad a \in b, \quad a \in c, \quad d \in d\}$$

and $x \in y$ is a swff of $\sum$ for any constants x and y. Let E be the binary relation defined by (3).

(a) Is the Axiom of Extensionality true for $\sum$?
(b) Simplify $\sum$ by eliminating redundancies.

15.3. Axiom Scheme of Replacement

Any placeholder t that occurs in a swff is either preceded by a quantifier or lies within the scope of "$\forall t$." On the other hand, this is not necessarily true for placeholders that occur in a quasi-swff (see page 118). An occurrence of a placeholder t is said to be *bound* in a quasi-swff if it is directly preceded by a quantifier, or lies within the scope of "$\forall t$"; an occurrence of a placeholder t is *free* in a quasi-swff if that occurrence of t is not bound.

A quasi-swff in which exactly one placeholder has a free occurrence is said to be a *unary predicate*. A quasi-swff in which exactly two placeholders have free occurrences is said to be a *binary predicate*. For example, each of the following quasi-swffs is a unary predicate:

$$a \in x, \quad y \in b, \quad a \in z \vee z \in a, \quad \forall x(x \in y)$$

Here, a and b are constants; x, y, and z are placeholders. Each of the following quasi-swffs is a binary predicate:

$$Exy, \quad \neg(x \in y), \quad y \in x, \quad \forall z(z \in x \rightarrow Ezy)$$

We shall denote a unary predicate whose free placeholder is t by writing $F(t)$; we shall denote a binary predicate whose free placeholders are s and t by writing $F(s, t)$. We may use any upper-case letter in place of F.

We need the notion of a binary predicate $F(x, y)$ that is *functional* in x. Here, we are working with a specific semantical system $\sum$ which possesses a binary relation E defined by (3), page 282, and which satisfies the Axiom of Extensionality. Therefore, E is an equality relation for $\sum$. In this context, we say that $F(x, y)$ is functional in x provided that corresponding to each constant, say a, of $\sum$ there is at most one constant, say b, such that the swff $F(a, b)$ is true for $\sum$. Thus, $F(x, y)$ is functional in x provided that the swff

$$\forall xyz(F(x, y) \wedge F(x, z) \rightarrow Eyz) \tag{1}$$

is true for $\sum$.

If $F(x, y)$ is functional in x, then

$$\{(a, b) \mid F(a, b) \text{ is true for } \textstyle\sum\}$$

is a function in the normal sense. The domain of this function is

$$\{a \mid \exists y(F(a, y) \text{ is true for } \textstyle\sum)\}$$

and its range is

$$\{b \mid \exists x(F(x, b) \text{ is true for } \textstyle\sum)\}$$

Example 1. Let $\sum$ be the semantical system with constants a, b, c, d, e, f such that

$$\operatorname{diag} \textstyle\sum = \{b \in a, \quad c \in a, \quad a \in b, \quad a \in c, \quad d \in d\}$$

Introducing E by (3), page 282, we see that Ebc and Eef. The Axiom of Extensionality is true for $\sum$; so E is an equality of $\sum$. It is easy to verify that the binary predicate

$$x \in y \wedge (y \in b \vee y \in d)$$

is functional in x. The corresponding function is

$$\{(b, a), (c, a), (d, d)\}$$

which simplifies to $\{(b, a), (d, d)\}$, since b and c are equal.

For a different semantical system $\sum$, the binary predicate of the above example may fail to be functional in x.

For any semantical system for which the relation E defined by (3), page 282, satisfies the Axiom of Extensionality, each of the following binary predicates is functional in x:

$$Exy, \quad y \in x \wedge \forall w(w \in x \to Ewy), \quad Exy \wedge a \in y$$

For each of the following binary predicates, there is a semantical system (of the sort described) in which the predicate is not functional in x:

$$x \in y, \quad \neg(x \in y), \quad \neg Exy, \quad Exy \vee x \in y$$

The following fact is easy to prove. (We continue to assume that $\sum$ satisfies Extensionality.)

Lemma 1. Let $P(y)$ be any unary predicate; then the binary predicate $Exy \wedge P(y)$ is functional in x.

Dem. This is left as an exercise.

We now present a family of postulates for Zermelo–Fraenkel set theory.

Axiom Scheme of Replacement. For each set s,

$$\{y \mid \exists x(x \in s \wedge F(x, y))\} \text{ is a set}$$

provided that $F(x, y)$ is functional in x.

Remember that we are practicing the convention that the mere writing of a swff announces that it is true for $\sum$. Also, we use the term "set" as a synonym for "constant."

This axiom scheme contains a postulate for each constant of $\sum$ and for each binary predicate. Having fixed s and $F(x, y)$ in this way, the resulting postulate asserts that if $F(x, y)$ is functional in x, then there is a constant of $\sum$, say a, such that for each constant y of $\sum$,

$$y \in a \quad \text{iff} \quad \exists x(x \in s \wedge F(x, y))$$

Again, we have omitted two instances of the phrase "is true for $\sum$."

Lemma 2. If $\sum$ is a model of Zermelo–Fraenkel set theory, then the members of the singleton sets in any set s of $\sum$ constitute a set of $\sum$.

Dem. Let

$$F(x, y) = y \in x \wedge \forall w(w \in x \to Ewy) \tag{2}$$

which is functional in x. Then, for any constants x and y of $\sum$,

$$F(x, y) \quad \text{iff} \quad x = \{y\} \tag{3}$$

Thus, by the Axiom Scheme of Replacement,

$$\{y \mid \exists x(x \in s \wedge x = \{y\})\}$$

is a set of $\sum$, provided that s is a set of $\sum$. This proves Lemma 2.

Example 2. Let $\sum$ be the semantical system with constants a, b, c, d such that

$$\text{diag} \sum = \{b \in a, \quad a \in b, \quad a \in d, \quad b \in d, \quad c \in d, \quad d \in d\}$$

Then

$$a = \{b\}, \quad b = \{a\}, \quad c = \varnothing, \quad d = \{a, b, c, d\}$$

Here, $\underline{E}$ is the identity relation; so the Axiom of Extensionality is automatically true for $\sum$. As we have observed in our proof of Lemma 2, the Axiom Scheme of Replacement requires that

$$\{y \mid \exists x(x \in d \wedge x = \{y\})\}$$

is a set of $\sum$. The members of d that are singleton sets are

$$\{b\} \quad \text{and} \quad \{a\}$$

So, this axiom scheme requires that $\{a, b\}$ is a set of $\sum$. Clearly, $\{a, b\}$ is *not* a set of $\sum$. We conclude that the Axiom Scheme of Replacement is not satisfied by $\sum$.

We emphasize that the Axiom Scheme of Replacement serves to guarantee the existence of many sets, provided that certain other sets exist. To this purpose, we must exhibit a suitable binary predicate, functional in one of its placeholders. However, by Lemma 1, each unary predicate gives rise to a binary predicate that is functional in one of its placeholders. This fact yields the following useful theorem.

Theorem of Separation. Let $P(y)$ be any unary predicate and let s be any set of $\sum$; then

$$\{y \mid y \in s \wedge P(y)\}$$

is a set of $\sum$.

Dem. By Lemma 1, the binary predicate $Exy \wedge P(y)$ is functional in x. Therefore,

$$\{y \mid \exists x(x \in s \wedge Exy \wedge P(y))\}$$

is a set of $\sum$, say a. For each constant y of $\sum$,

$$\exists x(x \in s \wedge Exy \wedge P(y)) \quad \text{iff} \quad y \in s \wedge P(y)$$

Thus, $y \in a$ iff $y \in s \wedge P(y)$. We conclude that $a = \{y \mid y \in s \wedge P(y)\}$; so $\{y \mid y \in s \wedge P(y)\}$ is a set of $\sum$.

The significance of the Theorem of Separation is that each collection consisting of the members of a set s that satisfy a unary predicate is also a set. Let us pin down the *subset* concept.

Definition. A collection of constants $\{a, b, c, \ldots\}$ is said to be a *subset* of a set s provided that $a \in s$, $b \in s$, $c \in s, \ldots$ and provided that the collection is a set. We write $\{a, b, c, \ldots\} \subset s$.

The point is that a collection $\{a, b, c, \ldots\}$ is not necessarily a set; i.e., there may not be a constant, say y, such that for each constant z,

$$z \in y \quad \text{iff} \quad Eza \vee Eab \vee Ezc \vee \cdots$$

However, the Theorem of Separation guarantees that the collection obtained by separating out the members of s that satisfy a unary predicate $P(y)$ is a set.

We introduced $a \cap b$, the intersection of sets a and b, on page 278. There we pointed out that the collection $a \cap b$ is a set of $\sum$ provided that there is a constant, say c, such that

$$x \in c \quad \text{iff} \quad x \in a \wedge x \in b$$

for each constant x of $\sum$. The Theorem of Separation allows us to prove that $a \cap b$ is a set of $\sum$ if a and b are sets of $\sum$.

Theorem 1. Let a and b be any sets of $\sum$; then $a \cap b$ is a set of $\sum$.

Dem. We point out that "$y \in b$" is a unary predicate. Thus, by the Theorem of Separation,

$$\{y \mid y \in a \wedge y \in b\}$$

is a set of $\sum$; i.e., the collection $a \cap b$ is a set of $\sum$.

We shall now prove that if $\sum$ is a model of Zermelo–Fraenkel set theory, then the empty set is a set of $\sum$. Indeed, in Lemma 3 which follows, we shall assume only that $\sum$ satisfies the Axiom of Extensionality and the Axiom Scheme of Replacement (actually, the Theorem of Separation).

Lemma 3. The empty set is a set of $\sum$.

Dem. Each semantical system has at least one constant; thus $\sum$ has a constant, say s. By the Theorem of Separation,

$$\{y \mid y \in s \wedge \neg Eyy\}$$

is a set, since $\neg Eyy$ is a unary predicate. Therefore, there is a constant of $\sum$, say a, such that for each constant y,

$$y \in a \quad \text{iff} \quad y \in s \wedge \neg Eyy \tag{4}$$

Since E is an equality relation, no constant of $\sum$ satisfies the condition on the RHS of (4). Therefore, a has no members; i.e., a is the empty set. This proves that the empty set is a set of $\sum$.

A set is said to be *finite* if it has exactly n members, where $n = 0, 1, 2, \ldots$. A set is said to be *infinite* if it is not finite. Let us prove the following result:

Theorem 2. Each finite subcollection of a set is a set.

Dem. Let s be any set of $\sum$, and let $a_1, \ldots, a_n$ be members of s, $n \in N$; so $\{a_1, \ldots, a_n\}$ is a finite collection of members of s. Let $F(y)$ be the unary predicate

$$Eya_1 \vee \cdots \vee Eya_n$$

By the Theorem of Separation,

$$\{y \mid y \in s \wedge F(y)\}$$

is a set of $\sum$; i.e., $\{a_1, \ldots, a_n\}$ is a set of $\sum$. This completes our proof.

Using this result, we can now demonstrate that the semantical system $\sum$ of Example 2, page 284, is *not* a model of Zermelo–Fraenkel set theory. Recall that $\{b, c, d\}$ is a set of $\sum$. Certainly, the collection $\{b\}$ is a subset of $\{b, c, d\}$; but $\{b\}$ is not a set of $\sum$. This contradicts Theorem 2. We conclude that $\sum$ does not satisfy the Axiom Scheme of Replacement.

Throughout the remaining sections of this chapter we shall relax our formal approach to the language of $\sum$. For example, we shall freely use "$=$" to represent an equality of $\sum$, writing "$x = y$" for "Exy." Moreover, we shall represent the predicates "$\neg(x \in y)$" and "$\neg(x = y)$" by "$x \notin y$" and "$x \neq y$," respectively.

Exercises

1. Let $\sum$ be the semantical system of Example 1. Show that the binary predicate

$$x \in y \wedge a \in x$$

is functional in x.

2. Exhibit a semantical system $\sum$ so that the binary predicate

$$x \in y \wedge a \in x$$

is *not* functional in x.

3. (a) Exhibit a $\sum$ so that the binary predicate

$$x \in y \wedge \neg x \in a$$

is not functional in x. Here, a is a constant.

   (b) Exhibit $\sum$ so that the binary predicate of part (a) is functional in x.

4. Prove that the binary predicate

$$y \in x \wedge \forall w(w \in x \rightarrow Ewy)$$

is always functional in x.

5. (a) Prove that the binary predicate $Exy \wedge a \in y$ is always functional in x.
   (b) Is the binary predicate $Exy \wedge a \in y$ always functional in y?

6. Prove Lemma 1.

7. (a) Let a be a constant of $\sum$; prove that $\varnothing \subset a$.
   (b) Let a be a constant of $\sum$; prove that $a \subset a$.
   (c) Let a and b be constants of $\sum$; prove that $a \cap b \subset a$.

8. Let $\sum$ satisfy the Axiom of Extensionality and the Axiom Scheme of Replacement.

   (a) Prove that $\{x \mid x \in s \wedge x \in x\}$ is a set of $\sum$, provided that s is a set of $\sum$.
   (b) Prove that $\{x \mid x \in s \wedge \neg(x \in x)\}$ is a set of $\sum$, provided that s is a set of $\sum$.
   (c) Prove that $\{x \mid x \in s \wedge \neg Exa\}$ is a set of $\sum$, provided that s is a set of $\sum$.
   (d) Prove that $\{x \mid x \in a \wedge \neg(x \in b)\}$ is a set of $\sum$, provided that a and b are sets of $\sum$.

9. Let $\sum$ be a semantical system that satisfies the Axiom of Extensionality, and let $F(x, y)$ be any binary predicate that is functional in x. Prove that $\{y \mid \exists x(x \in \varnothing \wedge F(x, y))\} = \varnothing$.

10. Let $\sum$ be the semantical system with one constant a and one relation symbol $\in$, such that "$a \in a$" is false for $\sum$.

    (a) Show that the Axiom of Extensionality is true for $\sum$.
    (b) Prove that the Axiom Scheme of Replacement is satisfied by $\sum$. *Hint:* See Exercise 9.
    (c) Is $\varnothing$ a set of $\sum$?
    (d) Is $\{\varnothing\}$ a set of $\sum$?

11. (a) Lemma 3 asserts that $\varnothing$ is a set of $\sum$ if $\sum$ satisfies the Axiom of Extensionality and the Axiom Scheme of Replacement. Under the same assumptions, is $\{\varnothing\}$ necessarily a set of $\sum$? Justify your answer.
    (b) Is the statement "$\{\varnothing\}$ is a set of $\sum$" deducible from the Axiom of Extensionality and the Axiom Scheme of Replacement? Justify your answer.

15.4. Axiom of Power Set

The Axiom of Power Set asserts that the collection of all subsets of a set is a set. Remember that a collection x is a subset of a set s iff x is a set and $\forall t(t \in x \rightarrow t \in s)$. The point is that a collection of members of a set s is not necessarily a set.

We shall now adopt the following axiom; i.e., we shall consider only semantical systems that satisfy the Axiom of Extensionality, the Axiom Scheme of Replacement, and the following axiom.

Axiom of Power Set. For each set s, $\{x \mid x \subset s\}$ is a set.

Given that s is a set, the set $\{x \mid x \subset s\}$ is called the *power set* of s, and is denoted by "$\mathscr{P}s$."

We pointed out previously (see Exercise 11 above) that $\{\varnothing\}$ is not necessarily a set of $\sum$, where $\sum$ satisfies the Axiom of Extensionality and the Axiom Scheme of Replacement. Adding the Axiom of Power Set compels the collection $\{\varnothing\}$ to be a set.

We now assume that $\sum$ is a semantical system that satisfies the Axiom of Power Set as well as the axioms introduced in the two preceding sections.

Lemma 1. $\{\varnothing\}$ is a set of $\sum$.

Dem. By Lemma 3, page 290, $\varnothing$ is a set of $\sum$. By the Axiom of Power Set, $\{x \mid x \subset \varnothing\}$ is a set of $\sum$. The empty set is the only subset of $\varnothing$; thus

$$\{x \mid x \subset \varnothing\} = \{\varnothing\}$$

This proves that $\{\varnothing\}$ is a set of $\sum$.

More generally, we can prove that $\{s\}$ is a set if s is a set.

Theorem 1. $\{s\}$ is a set of $\sum$ provided that s is a set of $\sum$.

Dem. Since s is a set, so is $\mathscr{P}s$. By the Theorem of Separation,

$$\{x \mid x \in \mathscr{P}s \wedge Exs\}$$

is a set of $\sum$. Thus $\{s\}$ is a set of $\sum$.

This result is very significant; it discloses the fact that $\sum$ has infinitely many distinct constants.

Theorem 2. $\sum$ has infinitely many sets.

Dem. By Theorem 1, each of the following sets is a set of $\sum$:

$$\varnothing, \quad \{\varnothing\}, \quad \{\{\varnothing\}\}, \ldots \tag{1}$$

Here, each set s in the list is followed by the singleton set $\{s\}$. Thus, each set in (1), except $\varnothing$, is a singleton set. Assume that two sets in the list (1), say $\underbrace{\{\cdots}_{m}\{\varnothing\}\cdots\}$ and $\underbrace{\{\cdots}_{n}\{\varnothing\}\cdots\}$, are equal. Now, singleton sets $\{a\}$ and $\{b\}$ are equal iff a and b are equal. Let $m < n$ and apply this observation m times to the given sets. It follows that the given sets are equal iff $\varnothing$ and $\underbrace{\{\cdots}_{n-m}\{\varnothing\}\cdots\}$ are equal. These sets are equal iff $n = m$ (since no singleton set equals the empty set). This proves that no two sets in (1) are equal. Thus, $\sum$ has infinitely many sets.

Example 1. Let $\sum$ be the semantical system whose sets are given by (1). So, each set of $\sum$ is a singleton set, except $\varnothing$. Clearly, the Axiom of Extensionality and the Axiom Scheme of Replacement are satisfied by $\sum$. Take $s = \{\varnothing\}$; then $\mathscr{P}s = \{\varnothing, \{\varnothing\}\}$, which has two members. Therefore, this collection is *not* a set of $\sum$. This proves that the Axiom of Power Set is not true for $\sum$.

Before presenting an example of a semantical system for which the Axiom of Power Set is true, we need a little more information.

Lemma 2. Let s be a set with n members, $n \in N$; then $\mathscr{P}s$ has 2^n members.

Dem. By Theorem 2, page 290, each collection whose members are in s is a set, and so is a subset of s. Let $s = \{a_1, \ldots, a_n\}$; to form a subset of s we either accept a_i in the subset or we reject a_i, $i = 1, \ldots, n$. Therefore, there are 2^n distinct ways of forming subsets of s; thus s has 2^n subsets.

Example 2. Let $\sum$ be the semantical system whose sets are each finite, such that each finite set is a set of $\sum$. The Axiom of Extensionality and the Axiom Scheme of Replacement are satisfied by $\sum$. By Lemma 2, $\mathscr{P}s$ is finite if s is finite; therefore, $\mathscr{P}s$ is a set of $\sum$ if s is a set of $\sum$. So, the Axiom of Power Set is true for $\sum$.

We emphasize that each set of the semantical system of Example 2 is finite. Therefore, we cannot deduce the existence of an infinite set from the Axiom of Power Set. To this purpose we require the Axiom of Infinity, which we shall consider later.

Again, let $\sum$ be any semantical system that satisfies the Axiom of Power Set and the axioms of the preceding sections. Let us show that for each $n \in N$, there is a set of $\sum$ with at least n members.

Lemma 3. For each $n \in N$, there is a set of $\sum$ with at least n members.

Dem. Choose $n \in N$; we shall apply Lemma 2 repeatedly. Now, $\{\varnothing\}$ is a set of $\sum$ with one member. By Lemma 2,

$$\begin{array}{llll} s_1 = \mathscr{P}\{\varnothing\} & \text{has} & 2 & \text{members} \\ s_2 = \mathscr{P}s_1 & \text{has} & 2^2 = 4 & \text{members} \\ s_3 = \mathscr{P}s_2 & \text{has} & 2^4 = 16 & \text{members} \end{array}$$

In general, if s_m has k members, then $s_{m+1} = \mathscr{P}s_m$ has 2^k members. For each natural number j, $j \leq 2^j$. It follows that s_n has at least n members. This proves Lemma 3.

Next, we want to prove that for each $n \in N$, there is a set of $\sum$ with exactly n members. Here, we need the following fact.

Lemma 4. For each $n \in N$, there is a set of $\sum$ that contains each of $\{\varnothing\}, \{\{\varnothing\}\}, \ldots, \underbrace{\{\cdots}_{n}\{\varnothing\}\cdots\}$.

Dem. Let $b_1 = \{\varnothing\}$, $b_2 = \{b_1\}, \ldots, b_n = \{b_{n-1}\}$, and let $\mathscr{P}(\mathscr{P}s) = \mathscr{P}^2 s$, $\mathscr{P}(\mathscr{P}^2 s) = \mathscr{P}^3 s$, and, in general, $\mathscr{P}(\mathscr{P}^{k-1} s) = \mathscr{P}^k s$. Then

$$\begin{aligned} \mathscr{P}\varnothing &= \{\varnothing\} = b_1 \\ \mathscr{P}^2\varnothing &= \{\varnothing, \{\varnothing\}\} = \{\varnothing, b_1\} \\ \mathscr{P}^3\varnothing &= \{\varnothing, \{\varnothing\}, \{b_1\}, \ldots\} = \{\varnothing, b_1, b_2, \ldots\} \\ \mathscr{P}^4\varnothing &= \{\varnothing, \{\varnothing\}, \{b_1\}, \{b_2\}, \ldots\} = \{\varnothing, b_1, b_2, b_3, \ldots\} \end{aligned}$$

Continuing, we see that

$$\mathscr{P}^{n+1}\varnothing = \{\varnothing, b_1, b_2, \ldots, b_n, \ldots\}$$

Thus, each of the given sets $b_1, \ldots, b_n$ is a member of $\mathscr{P}^{n+1}\varnothing$. This proves Lemma 4.

We shall now prove that for each $n \in N$, there is a set of $\sum$ with exactly n members. Actually, we shall prove a little more.

Theorem 3. Let $a_1, \ldots, a_n$ be any n sets of $\sum$, where $n \in N$. Then $\{a_1, \ldots, a_n\}$ is a set of $\sum$.

Dem. We have already shown [see (1)] that

$$\{\varnothing\},\quad \{\{\varnothing\}\}, \ldots, \quad \{\cdots\{\varnothing\}\cdots\}$$

are distinct sets of $\sum$. We can express

$$(x = \{\varnothing\} \wedge y = a_1) \vee \cdots \vee (x = \underbrace{\{\cdots\{}_{n}\varnothing\}\cdots\} \wedge y = a_n) \qquad (2)$$

by a binary predicate functional in x, which we call $F(x, y)$. For example, the first disjunct of (2) is represented in $F(x, y)$ by

$$Exb \wedge \forall t(t \in b \leftrightarrow \forall z(z \notin t)) \wedge Eya_1$$

By Lemma 4, there is a set of $\sum$, say s, that contains each set listed in (1). Thus, by the Axiom Scheme of Replacement,

$$\{y \mid \exists x(x \in s \wedge F(x, y))\}$$

is a set of $\sum$; i.e., $\{a_1, \ldots, a_n\}$ is a set of $\sum$.

Taking $n = 2$ in Theorem 3 yields the following result:

Theorem of Pairs. $\{a, b\}$ is a set of $\sum$ if a and b are sets of $\sum$.

We can now prove the following:

Theorem on Existence of Ordered Pairs. $\{\{a\}, \{a, b\}\}$ is a set of $\sum$ if a and b are sets of $\sum$.

Dem. By Theorem 1, page 292, $\{a\}$ is a set of $\sum$. By the Theorem of Pairs, $\{a, b\}$ is a set of $\sum$. Thus, by the Theorem of Pairs, $\{\{a\}, \{a, b\}\}$ is a set of $\sum$.

We take "(a, b)" to be a name for the set $\{\{a\}, \{a, b\}\}$, which is called an *ordered pair*.

Theorem on Decomposition of Ordered Pairs. Let (a, b) and (c, d) be any ordered pairs of $\sum$; then $(a, b) = (c, d)$ iff $a = c \wedge b = d$.

Dem. 1. Assume that $a = c \wedge b = d$. Then $\{a\} = \{c\}$ and $\{a, b\} = \{c, d\}$; so $\{\{a\}, \{a, b\}\} = \{\{c\}, \{c, d\}\}$, i.e., $(a, b) = (c, d)$.

2. Assume that $(a, b) = (c, d)$. Thus $\{\{a\}, \{a, b\}\} = \{\{c\}, \{c, d\}\}$. Consequently either $\{a\} = \{c\}$ or $\{a\} = \{c, d\}$. Similarly either $\{a, b\} = \{c\}$ or $\{a, b\} = \{c, d\}$. It is left as an exercise to show that in each case we can conclude that $a = c$ and $b = d$.

Lemma 5. $(a, b) \neq (b, a)$ if $a \neq b$.

Dem. By the preceding theorem, $(a, b) = (b, a)$ iff $a = b \wedge b = a$.

Lemma 6. For each set a, $(a, a) = \{\{a\}\}$.

Dem. Consider the definition of (a, a).

We now consider a famous result.

Theorem 4. Russell's Paradox. $\{x \mid x \notin x\}$ is *not* a set of $\sum$.

Dem. Assume that $\{x \mid x \notin x\}$ is a set of $\sum$. Then there is a constant a such that for each constant x

$$x \in a \leftrightarrow x \notin x \tag{3}$$

In particular, putting a for x in (3) yields

$$a \in a \leftrightarrow a \notin a \tag{4}$$

which is false for $\sum$. Therefore, the assumption of this argument is false. We conclude that $\{x \mid x \notin x\}$ is not a set of $\sum$.

It follows from Russell's Paradox that the collection of all sets of $\sum$ is not itself a set.

Corollary 1. $\{x \mid x = x\}$ is *not* a set of $\sum$.

Dem. Assume that the collection $\{x \mid x = x\}$ is a set of $\sum$, say s. Then, by the Theorem of Separation,

$$\{y \mid y \in s \wedge y \notin y\}$$

is a set of $\sum$. Thus $\{y \mid y \notin y\}$ is a set of $\sum$. This contradicts Theorem 4. We conclude that the collection $\{x \mid x = x\}$ is not a set of $\sum$.

Exercises

1. Let $\sum$ be the semantical system with constants a, b, and c such that

$$\operatorname{diag} \textstyle\sum = \{b \in b, \quad a \in c\}$$

   (a) Is the Axiom of Extensionality true for $\sum$?
   (b) Is the Axiom Scheme of Replacement true for $\sum$?
   (c) Is the Axiom of Power Set true for $\sum$?

2. Let $s = \{s\}$; is "$s \in s$" true or false?

3. Let $\sum$ be the semantical system with constants a, b, and c such that

$$\operatorname{diag} \textstyle\sum = \{a \in a, \quad b \in b, \quad c \in c\}$$

   (a) Is the Axiom of Extensionality true for $\sum$?
   (b) Is the Axiom Scheme of Replacement true for $\sum$?
   (c) Is the Axiom of Power Set true for $\sum$?

4. Prove that $\{a\} \neq \{b\}$ iff $a \neq b$.

5. Express "$x = \{\{\varnothing\}\}$" by a unary predicate.

6. Complete the proof of the Theorem on Decomposition of Ordered Pairs.

15.5. Axiom of Sum Set

The Theorem of Separation allows us to prove that the intersection of any sets a and b of $\sum$ is a set of $\sum$. It is important that the *union* of any sets a and b be a set of $\sum$. Of course, Theorem 3, page 295, ensures that $a \cup b$ is a set if both a and b are finite. However, if a or b is infinite, we have no way, as yet, to guarantee that $a \cup b$ is a set. To this purpose we need an axiom.

Suppose, for the moment, that we postulate that $a \cup b$ is a set whenever a and b are sets; then $a_1 \cup \cdots \cup a_n$ is a set if $a_1, \ldots, a_n$ are sets. However, this axiom does not allow us to conclude that the union of infinitely many sets is a set. With this in mind, we formulate our axiom as follows.

Axiom of Sum Set. For each set s

$$\{x \mid \exists y(y \in s \wedge x \in y)\}$$

is a set.

We now assume that $\sum$ is a semantical system that satisfies this axiom as well as the axioms of the preceding sections.

The set $\{x \mid \exists y(y \in s \wedge x \in y)\}$ consists of all the members of the members of s, and is denoted by "$\bigcup s$," called the *union* of s. We also write

$$\text{“}\bigcup s = \bigcup_{y \in s} y\text{”}$$

since $s = \{y \mid y \in s\}$

Lemma 1. $\bigcup \varnothing = \varnothing$.

Dem. Apply the definition.

Lemma 2. $\bigcup \{s\} = s$.

Dem. Apply the definition.

Lemma 3. $\bigcup \{a, b\} = a \cup b$ for any sets a and b.

Dem. Recall that $a \cup b = \{x \mid x \in a \vee x \in b\}$. Now,

$$\begin{aligned}\bigcup \{a, b\} &= \{x \mid \exists y(y \in \{a, b\} \wedge x \in y)\} \\ &= \{x \mid x \in a \vee x \in b\}\end{aligned}$$

Thus $\bigcup \{a, b\} = a \cup b$.

Similarly, we can extend the notion of the *intersection* of sets a and b to the intersection of any *set* of sets.

Definition. For each set s, $\bigcap s$ denotes the collection

$$\{x \mid x \in \bigcup s \wedge \forall y(y \in s \rightarrow x \in y)\}$$

This is the collection of all constants that are members of *each* member of s.

First, we must show that $\bigcap s$ is a set if s is a set.

Theorem 1. $\bigcap s$ is a set if s is a set.

Dem. Now, $\forall y(y \in s \rightarrow x \in y)$ is a unary predicate (here, s is a constant); moreover, $\bigcup s$ is a set. Thus, by the Theorem of Separation,

$$\{x \mid x \in \bigcup s \wedge \forall y(y \in s \rightarrow x \in y)\}$$

is a set.

Lemma 4. $\bigcap\varnothing = \varnothing$.

Dem.

$$\begin{aligned}\bigcap\varnothing &= \{x \mid x \in \bigcup\varnothing \wedge \forall y(y \in s \rightarrow x \in y)\}\\ &= \{x \mid x \in \varnothing \wedge \forall y(y \in s \rightarrow x \in y)\}\\ &= \varnothing\end{aligned}$$

since "$x \in \varnothing$" is false for each constant x of $\sum$.

Lemma 5. For each set s, $\bigcap s \subset \bigcup s$.

Dem. We must show two things:

(i) $\bigcap s$ is a set.
(ii) If $x \in \bigcap s$, then $x \in \bigcup s$, for each constant x.

Now, (i) is true for $\sum$ by Theorem 1. And (ii) follows immediately from the definition of $\bigcap s$.

Lemma 6. $\bigcap\{a, b\} = a \cap b$ for any sets a and b.

Dem. Recall that $a \cap b = \{x \mid x \in a \wedge x \in b\}$. Now,

$$\begin{aligned}\bigcap\{a, b\} &= \{x \mid x \in \bigcup\{a, b\} \wedge \forall y(y \in \{a, b\} \rightarrow x \in y)\}\\ &= \{x \mid x \in a \cup b \wedge (x \in a \wedge x \in b)\}\\ &= \{x \mid x \in a \wedge x \in b\}\end{aligned}$$

Thus $\bigcap\{a, b\} = a \cap b$.

We now present an important fact about $\bigcap s$.

Theorem 2. For each $y \in s$, $\bigcap s \subset y$.

Dem. Let $y \in s$; we must show that for each constant x,

$$x \in \bigcap s \rightarrow x \in y \tag{1}$$

Assume that $x \in \bigcap s$; then, by definition,

$$x \in \bigcup s \wedge \forall y(y \in s \rightarrow x \in y)$$

But $y \in s$; thus, $x \in y$. This establishes (1) and completes our proof.

Exercises

1. Is the Axiom of Sum Set true for the semantical system of Example 1, page 293?
2. Is the Axiom of Sum Set true for the semantical system of Example 2, page 293?

3. Can the existence of an infinite set be deduced from the Axiom of Sum Set and the axioms of the preceding sections? Justify your answer.

4. Prove Lemma 1.

5. Prove Lemma 2.

6. Prove that

$$\bigcup\{\{x\} \mid x \in s\} = s$$

for any set s of Σ.

7. Let a, b, and c be any sets of Σ. Prove that:
(a) $a \cap (b \cup c) = (a \cap b) \cup (a \cap c)$.
(b) $a \cup (b \cap c) = (a \cup b) \cap (a \cup c)$.
(c) $a \cap (b \cap c) = (a \cap b) \cap c$.
(d) $a \cup (b \cup c) = (a \cup b) \cup c$.

8. Let s be any set; prove that for each $y \in s$, $y \subset \bigcup s$.

15.6. Axiom of Infinity

The existence of an infinite set cannot be deduced from the axioms of the preceding sections. However, to express mathematical concepts in terms of set theory requires that certain infinite collections be sets. Accordingly, we now introduce an appropriate axiom.

Axiom of Infinity. There is a set W such that:

(i) $\varnothing \in W$.
(ii) $\forall x(x \in W \rightarrow x \cup \{x\} \in W)$.

Since $\varnothing \in W$, (ii) allows us to churn out infinitely many members of W. First, let us agree to define the symbols $c_0, c_1, c_2, \ldots$ as follows.

Definition. $c_0 = \varnothing$ and $c_{n+1} = c_n \cup \{c_n\}$ for each $n = 0, 1, 2, \ldots$.

Thus

$$\begin{aligned}
c_1 &= c_0 \cup \{c_0\} = \varnothing \cup \{\varnothing\} = \{\varnothing\} = \{c_0\} \\
c_2 &= c_1 \cup \{c_1\} = \{c_0\} \cup \{c_1\} = \{c_0, c_1\} \\
c_3 &= c_2 \cup \{c_2\} = \{c_0, c_1\} \cup \{c_2\} = \{c_0, c_1, c_2\} \\
c_4 &= c_3 \cup \{c_3\} = \{c_0, c_1, c_2\} \cup \{c_3\} = \{c_0, c_1, c_2, c_3\}
\end{aligned}$$

Since $\varnothing \in W$, we see from (ii) that each of the c's is in W. The c's are all distinct; indeed,

$$c_n = \{c_0, c_1, \ldots, c_{n-1}\}$$

has exactly n members. Therefore, W has infinitely many members, the c's. We shall verify this observation in a moment.

We now assume that $\sum$ satisfies the Axiom of Infinity as well as the axioms of the preceding sections. Therefore, there is a set of $\sum$, say W, such that (i) and (ii) are true. Each of the c's is in W; suppose that there is a member of W, say a_0, which is different from each of the c's. By (ii), $a_0 \cup \{a_0\} = a_1$ is in W, $a_1 \cup \{a_1\} = a_2$ is in W; continuing, we obtain a sequence of a's in W, where $a_{n+1} = a_n \cup \{a_n\}$ for $n = 0, 1, 2, \ldots$.

Our first goal is to demonstrate the existence of a "smallest" infinite set that contains each of the c's. This set is to satisfy the Axiom of Infinity, so must possess the property described by (i) and (ii). A set that has the property described by (i) and (ii) is called a *successor set*.

Definition. We say that a set s is a successor set provided that $\varnothing \in s$ and $\forall x(x \in s \rightarrow x \cup \{x\} \in s)$.

By the Axiom of Infinity there is at least one set of $\sum$ that is a successor set. The following theorem is basic to the work of this section.

Uniqueness Theorem. There is a unique set of $\sum$, say M, such that:

1. M is a successor set.
2. M is a subset of each successor set.

Dem. By the Axiom of Infinity there is a successor set W. By the Axiom of Power Set, $\mathscr{P}W$ is a set; moreover, the collection of all subsets of W that are successor sets is a set. Indeed, by the Theorem of Separation,

$$H = \{h \mid h \in \mathscr{P}W \wedge \varnothing \in h \wedge \forall x(x \in h \rightarrow x \cup \{x\} \in h)\}$$

is a set. Thus, by Theorem 1, page 298, $\bigcap H$ is a set, which we shall call M. By construction, $\varnothing \in h$ for each $h \in H$; so $\varnothing \in M$. We shall now show that $\forall x(x \in M \rightarrow x \cup \{x\} \in M)$. Let $x \in M$; by Theorem 2, page 299, $M \subset h$ for each $h \in H$. So, $x \in h$ for each $h \in H$. Since each $h \in H$ is a successor set, it follows that $x \cup \{x\} \in h$; so $x \cup \{x\} \in \bigcap H$, i.e., $x \cup \{x\} \in M$. This proves that M is a successor set. To establish condition 2, let V be any successor set. It follows that $M \cap V$ is a successor set. Now, $M \cap V \in \mathscr{P}W$ since $M \cap V \subset M \subset W$; thus $M \cap V \in H$. But M is a subset of each member of H; so $M \subset M \cap V$. Thus $M \subset V$. Finally, to prove uniqueness, let s be any successor set that is a subset of each successor set. Then $M \subset s$ and $s \subset M$. So $M = s$. This completes our proof.

In our proof of the Uniqueness Theorem, notice that $M = \bigcap H$, where H is the set of all subsets of W that are successor sets; thus M is a subset of each member of H. We have proved that M is a subset of *each* successor set. In this sense, M is the *smallest* successor set.

Let x be any set; then $x \cup \{x\}$ is a set. Let us denote the latter set by x' (read "the successor of x").

Definition. For each constant x of $\sum$, $x' = x \cup \{x\}$.

We shall now set out to obtain some information about the set M. Our technique is based on the following corollary to the Uniqueness Theorem.

Principle of Mathematical Induction. Let s be a subset of M that is a successor set; then $s = M$.

Dem. By the Uniqueness Theorem, $M \subset s$. We are given that $s \subset M$; thus $s = M$.

Our proof of the following lemma illustrates how this principle serves to establish facts about M.

Lemma 1. For each $x \in M$, $\forall y(y \in x \rightarrow y \subset x)$.

Dem. Let $s = \{x \mid x \in M \land \forall y(y \in x \rightarrow y \subset x)\}$; so s is a subset of M. We shall show that s is a successor set.

1. Show that $\varnothing \in s$. Clearly, $\varnothing \in M \land \forall y(y \in \varnothing \rightarrow y \subset \varnothing)$.
2. Assume that $x \in s$; we shall show that $x' \in s$. By assumption,

$$x \in M \land \forall y(y \in x \rightarrow y \subset x)$$

Now, M is a successor set and $x \in M$; so $x' \in M$. It remains to prove that $\forall y(y \in x' \rightarrow y \subset x')$. Let $y \in x'$; then $y \in x$ or $y = x$. If $y \in x$, then $y \subset x$ (since $x \in s$), so $y \subset x'$. If $y = x$, then $y \subset x \cup \{x\}$, since $x \subset x \cup \{x\}$. Thus $x' \in s$. This proves that s is a successor set. Thus, by the Principle of Mathematical Induction, $s = M$. Therefore, for each $x \in M$, $\forall y(y \in x \rightarrow y \subset x)$. This completes our proof of Lemma 1.

We need Lemma 1 to establish the following basic property of M.

Lemma 2. $\forall x(x \in M \rightarrow x \notin x)$.

Dem. Let $s = \{x \mid x \in M \wedge x \notin x\}$; then s is a subset of M. We shall show that s is a successor set.

1. Show that $\varnothing \in s$. Clearly, $\varnothing \in M \wedge \varnothing \notin \varnothing$.
2. Assume that $x \in s$; we shall show that $x' \in s$. By assumption, $x \in M \wedge x \notin x$; thus $x' \in M$ (since M is a successor set). Now, $x' = x \cup \{x\}$; so, for each constant y, $y \in x'$ iff $y \in x \vee y = x$. Therefore,

$$x' \in x' \qquad \text{iff} \qquad x' \in x \vee x' = x \tag{1}$$

If $x' \in x$, then $x' \subset x$ by Lemma 1, and it follows that $x \in x$ (since $x \in x'$). This contradiction proves that $x' \notin x$. If $x' = x$, then $x \in x$ (since $x \in x'$). Therefore, $x' \neq x$. We conclude from (1) that $x' \notin x'$. Thus $x' \in s$. This proves that s is a successor set. Thus, by the Principle of Mathematical Induction, $s = M$. Therefore, for each $x \in M$, $x \notin x$; i.e., $\forall x(x \in M \rightarrow x \notin x)$.

We point out that each of the c's defined on page 300 is a member of M (since each of the c's is a member of each successor set). Using Lemma 2, we can prove that the c's are distinct sets of $\sum$. If not, then $c_m = c_n$, where $m \neq n$. Take $m < n$ (i.e., the step-by-step construction of the c's yields c_m before c_n). But each c obtained before c_n is a member of c_n; thus $c_m \in c_n$, so $c_n \in c_n$. This contradicts Lemma 2. We conclude that $c_m \neq c_n$ if $m \neq n$; i.e., the sets yielded by the step-by-step construction are distinct. Moreover, $c_n = \{c_0, c_1, \ldots, c_{n-1}\}$ has exactly n members.

It is tempting to speculate that $M = \{c_0, c_1, c_2, \ldots\}$, since this collection, if it is a set, is a successor set and is a subset of each successor set. However, $\{c_0, c_1, c_2, \ldots\}$ is not necessarily a set of $\sum$. If not, then there is a member of M, say a, that is different from each c. Some insight into this situation is provided by the following facts about M.

Lemma 3. For each $x \in M$, $\varnothing \in x \vee x = \varnothing$.

Dem. Let $s = \{x \mid x \in M \wedge \varnothing \in x'\}$; then s is a subset of M. We shall show that s is a successor set.

1. Show that $\varnothing \in s$. Clearly, $\varnothing \in M \wedge \varnothing \in \{\varnothing\}$.
2. Let $x \in s$; we shall show that $x' \in s$. By assumption, $x \in M \wedge \varnothing \in x'$; thus $x' \in M$. Now $x' = x \cup \{x\}$; so $\varnothing \in x$ or $\varnothing = x$. If $\varnothing \in x$, then $\varnothing \in x'$. If $x = \varnothing$, then $x' = x \cup \{x\} = \varnothing \cup \{\varnothing\} = \{\varnothing\}$; so $\varnothing \in x'$. Therefore $x' \in s$. This proves that s is a successor set. By the Principle of Mathematical Induction, $s = M$. Therefore, for each $x \in M$, $\varnothing \in x \vee x = \varnothing$.

Lemma 4. For each $x \in M$, $\forall y(y \in x \rightarrow y' \in x \vee y' = x)$.

Dem. Let $s = \{x \mid x \in M \wedge \forall y(y \in x \rightarrow y' \in x')\}$; then s is a subset of M. We shall show that s is a successor set.

1. Show that $\varnothing \in s$. Clearly, $\varnothing \in M \wedge \forall y(y \in \varnothing \rightarrow y' \in \varnothing')$.
2. Let $x \in s$; we shall show that $x' \in s$. By assumption,

$$x \in M \wedge \forall y(y \in x \rightarrow y' \in x')$$

Thus, $x' \in M$. We must show that $\forall y(y \in x' \rightarrow y' \in x' \vee y' = x')$. Let $y \in x'$; then $y \in x \vee y = x$. If $y \in x$, then $y' \in x'$ since $x \in s$. If $y = x$, then $y' = y \cup \{y\} = x \cup \{x\} = x'$. This proves that $x' \in s$. We conclude that s is a successor set. By the Principle of Mathematical Induction, $s = M$. This establishes Lemma 4.

Here is an important fact about M.

Lemma 5. No member of M is a successor set.

Dem. Assume that some member of M, say x, is a successor set. By the Uniqueness Theorem, $M \subset x$; thus $x \in x$ (since $x \in M$). This contradicts Lemma 2. We conclude that no member of M is a successor set.

Corollary 1. For each $x \in M$, $x \neq \varnothing$, $\exists y(y \in x \wedge y' \notin x \wedge y' = x)$.

Dem. Lemmas 4 and 5.

We want to show that each member of M can be regarded as a *chain*. To begin with, we need the fact that each member of a member of M is a member of M.

Lemma 6. For each $x \in M$, $\forall y(y \in x \rightarrow y \in M)$.

Dem. This is left as an exercise; use the Principle of Mathematical Induction.

Lemma 7. For each $x, y \in M$, $x \neq y \rightarrow x \in y \vee y \in x$.

Dem. Let $s = \{x \mid x \in M \wedge \forall y(x \neq y \wedge y \in M \rightarrow x \in y \vee y \in x)\}$; so s is a subset of M. We shall show that s is a successor set.

1. Show that $\varnothing \in s$. Apply Lemma 3.
2. Let $x \in s$; we shall show that $x' \in s$. By assumption,

$$\forall y(y \neq x \wedge y \in M \rightarrow x \in y \vee y \in x) \tag{2}$$

We must show that

$$\forall y(y \neq x' \wedge y \in M \rightarrow x' \in y \vee y \in x') \tag{3}$$

Take $y \neq x'$ and $y \in M$. If $y = x$, then $y \in x'$. So, assume that $y \neq x$. From (2), either $x \in y$ or $y \in x$. We now consider both possibilities.

Case 1: $x \in y$. By Lemma 4, $x' \in y \vee x' = y$. But $y \neq x'$, so $x' \in y$.

Case 2: $y \in x$. $x' = x \cup \{x\}$, so $y \in x'$.

This establishes (3), so $x' \in s$. Thus s is a successor set; we conclude that $s = M$. This completes our proof.

Corollary 2. $\forall xy(x \subset y \vee y \subset x)$, $x, y \in M$.

Dem. If $x = y$, then $x \subset y$. If $x \neq y$, then $x \in y$ or $y \in x$ by Lemma 7. Thus, by Lemma 1, $x \subset y \vee y \subset x$.

Corollary 3. Trichotomy Law. $\forall xy(x \neq y \rightarrow x \in y$ or else $y \in x)$, $x, y \in M$.

Dem. Let x and y be distinct members of M. By Lemma 7, $x \in y$ or $y \in x$. We want to show that $\neg(x \in y$ and $y \in x)$. Assume that $x \in y$ and $y \in x$. By Lemma 1, $x \subset y$ and $y \subset x$; so $x = y$. This contradicts our assumption that $x \neq y$; we conclude that $\neg(x \in y$ and $y \in x)$, so $x \in y$ or else $y \in x$.

Corollary 4. For each $x, y \in M$, x is a proper subset of y iff $x \in y$.

Dem. The details are left as an exercise.

We are interested in the possibility that $M \neq \{c_0, c_1, c_2, \ldots\}$; the following theorem sheds light on this situation.

Theorem 1. Let $a \in M$, where a is different from each c_n; then each $c_n \in a$.

Dem. Now $c_0 = \varnothing$; so $a \neq \varnothing$. Thus, by Lemma 3, $\varnothing \in a$; i.e., $c_0 \in a$. By Lemma 4, since $a \neq c_1$, $c_1 \in a$. Continuing, Lemma 4 allows us to show that each $c_n \in a$.

This theorem asserts that each c_n is a member of *each* member of M that is not equal to some c_n.

Corollary 5. Let $a \in M$, where a is different from each c_n; then a is infinite.

Dem. By Theorem 1, each $c_n \in a$. There are infinitely many c's, since no two c's are equal.

Moreover, if $a \in M$ and is different from each c, then a has infinitely many members other than the c's. To see this, suppose that a has only finitely many members different from the c's, say $d_1, \ldots, d_n$. Now, "$x \neq d_1 \wedge \cdots \wedge x \neq d_n$" is a unary predicate; so, by the Theorem of Separation,

$$\{x \mid x \in a \wedge x \neq d_1 \wedge \cdots \wedge x \neq d_n\}$$

is a set, namely $\{c_0, c_1, c_2, \ldots\}$. This contradicts our assumption that the collection $\{c_0, c_1, c_2, \ldots\}$ is not a set of $\sum$. We have proven the following:

Theorem 2. If $\{c_0, c_1, c_2, \ldots\}$ is not a set, then there is a member of M, say a, such that a has infinitely many members in addition to the c's.

Indeed, each member of M that is different from each of the c's has infinitely many members in addition to all the c's.

The assumption that M has a member different from each of the c's allows us to conclude that M has infinitely many members different from each of the c's. This is due to the fact that each member of a member of M is also a member of M (see Lemma 6).

Corollary 6. If $\{c_0, c_1, c_2, \ldots\}$ is not a set, then M has infinitely many members different from each of the c's.

We point out that if $\{c_0, c_1, c_2, \ldots\}$ is not a set of $\sum$, then the collection that consists of all members of M that are different from each of the c's also is *not* a set of $\sum$.

Turning to the question of explaining basic notions of mathematics in terms of set theory, we point out that the set M constructed in our proof of the Uniqueness Theorem can be identified with N, the set of all natural numbers; indeed, we define N to be M. Under this approach, x is a natural number iff $x \in M$. In the exercises we point out that the relational system $\langle M, ' \rangle$ is a *Peano System*, i.e., has the following properties.

1. $'$ is a unary operation on M.
2. $c_0 \in M$.
3. $\forall x(x' \neq c_0)$, $x \in M$.
4. $\forall xy(x \neq y \rightarrow x' \neq y')$, $x, y \in M$.
5. $S = M$ if S is both a successor set and a subset of M.

This depends on interpreting "subset" in the sense of the semantical system $\sum$. Consequently, Peano's induction postulate 5 applies only to a collection of natural numbers S that is a set of $\sum$. If the collection $\{c_0, c_1, c_2, \ldots\}$ is a set of $\sum$, then we obtain the classical Peano System; otherwise, we obtain a nonstandard Peano System. In the former case, our Peano System yields the number systems of classical mathematics, via a well-known construction. In the latter case, our Peano System yields the number systems of nonstandard analysis, which we considered in Chapter 12.

We mention that any two Peano Systems are isomorphic, provided that each system refers to the same set theory $\sum$. Of course, different set theories may yield nonisomorphic Peano Systems.

It is easy to introduce an *order* relation on the Peano System $\langle M, ' \rangle$.

Definition. Let $x, y \in M$; then $x < y$ (read *x is less than y*) if $x \in y$.

An order relation is transitive and obeys the trichotomy law. In view of Corollary 3, we see that $<$ satisfies the trichotomy law; the transitivity of $<$ follows from Lemma 1. Thus, $<$ is an order relation on M.

Exercises

1. Prove that each c is a set.
2. Prove that each finite collection of c's is a set.
3. Prove that $a \cap b$ is a successor set if both a and b are successor sets.
4. Prove Lemma 6.
5. Prove Corollary 4.
6. Prove that $\forall x(x' \neq c_0)$, $x \in M$.
7. Prove that $\forall xy(x \neq y \rightarrow x' \neq y')$, $x, y \in M$.
8. Prove that $\forall x(x' \in M)$, $x \in M$.
9. Show that the relational system $\langle M, ' \rangle$ is a Peano System.
10. Prove that $\forall yz(y \in x \wedge z \in y \rightarrow z \in x)$, for each $x \in M$.

11. Prove that $\forall xy(x \subset y \rightarrow y \notin x)$, $x, y \in M$.

12. Let $x, y \in M$, where $y' = x$; then y is said to be an *immediate predecessor* of x. Prove that:

    (a) c_0 has no immediate predecessor.
    (b) For each $x \in M$, $x \neq c_0$, x has a *unique* immediate predecessor.

13. Prove that $\forall x \exists y(x = c_0 \vee x = y')$, $x, y \in M$.

14. Prove that $\forall x(x \neq x')$, $x \in M$.

15. Prove that $\forall xy(x' = y' \rightarrow x = y)$, $x, y \in M$.

16. A subset of M, say s, is said to be a *chain* if

$$\forall xy(x \in s \wedge x = y' \rightarrow y \in s), \qquad x, y \in M$$

    (a) Show that each member of M is a member of some chain.
    (b) Exhibit a chain that contains each member of M.
    (c) We say that s is the *smallest* chain that contains z if (i) $z \in s$ and s is a chain; (ii) s is a subset of each chain that contains z. Prove that for each $z \in M$, there is a smallest chain that contains z.
    (d) Exhibit the smallest chain that contains c_0.
    (e) Prove that $z' \notin s$ if s is the smallest chain that contains z.
    (f) Let $z \in M$; prove that $z = c_0$ if the smallest chain that contains z is the smallest chain that contains c_0.
    (g) Let $y, z \in M$; prove that $z = y$ if the smallest chain that contains z is the smallest chain that contains y.

15.7. Nonstandard Set Theory

It is usually assumed that the collection $\{c_0, c_1, c_2, \ldots\}$ introduced on page 300 is a set. In Section 15.6 we have presented some basic properties of the set M and have indicated something of its nature in case $M \neq \{c_0, c_1, c_2, \ldots\}$.

In support of that investigation we shall now prove that there is a semantical system that satisfies the axioms considered so far, and for which $M \neq \{c_0, c_1, c_2, \ldots\}$. To this purpose, we shall use Robinson's method for proving the existence of $*\mathscr{R}$.

Let $\sum$ be any model of Zermelo–Fraenkel set theory (e.g., a semantical system for which $M = \{c_0, c_1, c_2, \ldots\}$ and which satisfies all of the Zermelo–Fraenkel axioms). Let K be the collection of all swffs of $\sum$ that are true for $\sum$. Let "ω" be a symbol that is not in the vocabulary of $\sum$, and form the postulates

$$\omega \in M, \quad c_0 \in \omega, \quad c_1 \in \omega, \quad c_2 \in \omega, \ldots \tag{1}$$

where M is the unique successor set of $\sum$ that is a subset of each successor set of $\sum$. Let K′ be the postulate set obtained by adjoining the postulates of (1) to K. Each finite subset of K′ has a model; indeed, $\sum$ is a model of each finite collection of postulates in K′. Therefore, by the Compactness Theorem, K′ has a model, say $^*\sum$, under an interpreter *.

Of course, each constant of $\sum$ occurs in some swff of K, and therefore must be interpreted by the interpreter * in $^*\sum$; i.e., for each constant x of $\sum$, there is a unique constant, say $^*x$, of $^*\sum$ which * regards as the interpretation of x in $^*\sum$. Similarly, * interprets the relation symbol $\in$ of $\sum$, in $^*\sum$; we shall denote the corresponding relation symbol of $^*\sum$ by "$^*\in$."

In particular, * associates some constant of $^*\sum$ with M; under our agreement, this constant of $^*\sum$ is denoted by "$^*M$." It is easy to see that $^*M$ is the unique successor set of $^*\sum$ that satisfies the Uniqueness Theorem, page 301.

We want to prove that $^*\sum$ is a model of Zermelo–Fraenkel set theory. To simplify our notation we shall assume that for each constant x of $\sum$, $^*x$ is x; i.e., we shall replace each instance of $^*x$ in any string in dom $^*\sum$ by x. In the same way, we can assume that $^*\omega$ is ω. We shall suppress the * which appears in the relation symbol $^*\in$ of $^*\sum$. The symbols for equality and the other defined predicates of $^*\sum$ are the same for all semantical systems of this sort; there are no *'s to suppress here. In the passage to $^*\sum$, a constant x of $\sum$ may acquire a new member y, where y is *not* a constant of $\sum$; or x may become a member of a set z, where z is *not* a constant of $\sum$. Bear in mind that if $a \notin b$ for $\sum$, then $\neg(a \in b)$ is in K; so $a \notin b$ is true for $^*\sum$.

In particular, $^*\sum$ is a model of K under the interpreter *; thus, $^*\sum$ is an elementary extension of $\sum$, i.e., we have the following result:

Transfer Theorem. For each swff A of $\sum$, A is true for $\sum$ iff A is true for $^*\sum$.

We are now in a position to prove that $^*\sum$ is a model of Zermelo–Fraenkel set theory. We point out that each ZF axiom can be expressed by a swff of $\sum$. For example, consider the axiom in the Axiom Scheme of Replacement, which involves $F(x, y)$, a specific binary predicate functional in x; this axiom is expressed by the swff

$$\forall uvw(F(u, v) \wedge F(u, w) \rightarrow v = w) \rightarrow \forall s \exists t \forall y(y \in t \leftrightarrow \exists x(x \in s \wedge F(x, y))) \tag{2}$$

which is true for $\sum$ by assumption. By the Transfer Theorem, (2) is true for $^*\sum$. Of necessity, the expression "$F(x, y)$" is in the vocabulary of ZF set theory; i.e., it involves only the relation symbol $\in$, placeholders, and logical connectives [we use defined symbols such as predicates and names for certain constants to assist us in writing down $F(x, y)$].

Similarly, we see that the other ZF axioms are true for $*\sum$. We conclude that $*\sum$ is a model of ZF set theory.

Although "M" is a generic name for the unique successor set of a semantical system that is a subset of each successor set of the system, we shall continue to denote this set of $*\sum$ by "$*M$"; we do this to avoid confusing $*M$ with the corresponding set of $\sum$.

Using Lemma 2, page 302, we obtain from (1) that ω is different from each of the c's; moreover $\omega \in *M$, so

$$*M \neq \{c_0, c_1, c_2, \ldots\}$$

We conclude from the Principle of Mathematical Induction that the collection $\{c_0, c_1, c_2, \ldots\}$ is *not* a set of $*\sum$.

Definition. Each model of the ZF axioms for which $\{c_0, c_1, c_2, \ldots\}$ is not a set is called a *nonstandard* model of ZF set theory.

We have established the following:

Theorem 1. $*\sum$ is a nonstandard model of ZF set theory.

Now $\omega \in *M$ and ω is different from each of the c's. By Theorem 2, page 306, ω has infinitely many members different from each of the c's, say $b_1, b_2, b_3, \ldots$. By Corollary 1, page 304, there is a member of ω, say b_1, such that $b_1' = \omega$; of course, $b_1' \notin \omega$. We shall represent the set ω as follows:

$$\omega = \{c_0, c_1, c_2, \ldots; \ldots, b_3, b_2, b_1\}$$

Notice our use of a semicolon to separate the c's from the other members of ω; the three dots to the right of the semicolon indicate infinitely many b's. The successor of each member of ω (if it has a successor in ω) is displayed immediately to its right; of course, the successor of each member of ω, except b_1, is a member of ω. Moreover, $\cdots \in b_3 \in b_2 \in b_1$; i.e., $b_{n+1} \in b_n$ for $n = 1, 2, 3, \ldots$. Of course, each c is a member of each b.

Following a similar procedure, we represent the set $*M$ as follows:

$$*M = \{c_0, c_1, c_2, \ldots; \ldots, \omega, \omega', \omega'', \ldots\}$$

The three dots to the left of the first ω represent the b's; i.e.,

$$*M = \{c_0, c_1, c_2, \ldots; \ldots, b_3, b_2, b_1, \omega, \omega', \omega'', \ldots\}$$

The three dots to the right of ω'' should not be read as "and so on"; rather, the three dots merely indicate that $*M$ has infinitely more members. Of course, $^*M$ is a successor set, so the successor of each member of $^*M$ is also a member of $^*M$. The point is that we cannot assume that the collection $\{\omega, \omega', \omega'', \ldots\}$, with one member for each standard natural number, is a set of $^*\sum$.

We now draw attention to an important property of the set ω, the fact that it is *groundless*. We must define this notion.

Definition. A set x is said to be *groundless* if there is an infinite sequence of sets $a_1, a_2, a_3, \ldots$, not necessarily all distinct, such that

$$\cdots \in a_3 \in a_2 \in a_1 \in x$$

A set x is *grounded* if x is not groundless.

We pause to clarify the term "infinite sequence" which appears in this definition. Here, we mean that: (i) a_1 is a specific set; (ii) for each standard natural number n, $n > 1$, we can determine a_n, possibly in terms of the preceding a's.

For example, both $^*M$ and ω are groundless. Moreover, if y is a set such that $y \in y$, then y is groundless. To see this, notice that $y, y, y, \ldots$ is an infinite sequence such that $\cdots \in y \in y \in y \in y$. Furthermore, if x is groundless and $a_1, a_2, a_3, \ldots$ is an infinite sequence of sets such that $\cdots \in a_3 \in a_2 \in x$, then a_n is groundless for each standard natural number n.

Here is the well-known paradox of grounded sets. Let G be the set of all grounded sets; i.e., $x \in G$ iff x is grounded. Either G is grounded or G is groundless. If the former, then $G \in G$; thus G is groundless. If the latter, then there is an infinite sequence of sets $a_1, a_2, a_3, \ldots$ such that $\cdots \in a_3 \in a_2 \in a_1 \in G$. Clearly a_1 is groundless; thus $a_1 \notin G$.

This paradox is easy to resolve. It is based on the assumption that G is a set. Admitting that G is merely a collection, which may or may not be a set, we see that the force of the paradox is to prove that G is *not* a set.

15.8. Axiom of Regularity

We now assume that the following axiom is true for $\sum$.

Axiom of Regularity. Each nonempty set s has a member x such that $x \cap s = \varnothing$.

If s is a set of $\sum$ and $s \neq \varnothing$, this axiom requires that some member of s be such that none of its members is a member of s. This axiom is also known as the *Axiom of Foundation* and the *Axiom of Groundedness.*

Theorem 1. $\forall s(s \notin s)$.

Dem. Let s be any set of $\sum$. Then $\{s\}$ is a singleton set of $\sum$. By the Axiom of Regularity, $s \cap \{s\} = \varnothing$; therefore, $s \notin s$.

Lemma 1. $\forall xy(x \in y \rightarrow y \notin x)$.

Dem. Let x and y be sets of $\sum$ such that $x \in y$. Now, $\{x, y\}$ is a set. Thus, by the Axiom of Regularity, $x \cap \{x, y\} = \varnothing$ or $y \cap \{x, y\} = \varnothing$. By assumption, x is a member of the set $y \cap \{x, y\}$; we conclude that $x \cap \{x, y\} = \varnothing$, so $y \notin x$.

Lemma 2. Let n be a standard natural number and let $a_1, \ldots, a_n$ be any n sets such that $a_1 \in a_2 \in \cdots \in a_{n-1} \in a_n$. Then $a_n \notin a_1$.

Dem. By Theorem 3, page 295, $s = \{a_1, \ldots, a_n\}$ is a set. By assumption, $a_1 \in a_2 \cap s, a_2 \in a_3 \cap s, \ldots, a_{n-1} \in a_n \cap s$; so $a_i \cap s \neq \varnothing$ if $i \neq 1$. Therefore, by the Axiom of Regularity, $a_1 \cap s = \varnothing$; so $a_n \notin a_1$.

This result can be extended as follows.

Theorem 2. Let $a_1, a_2, a_3, \ldots$ be an infinite sequence of sets such that the collection $\{a_1, a_2, a_3, \ldots\}$ is a set. Then there is a standard natural number n such that $a_{n+1} \notin a_n$.

Dem. Assume that $a_{n+1} \in a_n$ for each standard natural number n. Then $a_{n+1} \in a_n \cap \{a_1, a_2, a_3, \ldots\}$ for each standard natural number n. Thus $a_n \cap \{a_1, a_2, a_3, \ldots\} \neq \varnothing$ for each standard natural number n. This contradicts the Axiom of Regularity.

We caution that Theorem 2 does not imply that each set of $\sum$ is grounded. There may be an infinite sequence of sets $a_1, a_2, a_3, \ldots$ such that the collection $\{a_1, a_2, a_3, \ldots\}$ is *not* a set of $\sum$. Indeed, for the model $^*\sum$ (see page 309), the collection $\{c_0, c_1, c_2, \ldots\}$ is not a set. It is a misnomer to call the Axiom of Regularity the "Axiom of Groundedness."

15.9. Axiom of Choice

Let s be any nonempty, finite set, say $s = \{a_1, \ldots, a_n\}$; moreover, assume that $\varnothing \notin s$ and that any two members of s are disjoint, i.e., $a_i \cap a_j = \varnothing$ if $i \neq j$. Then there is a set of $\sum$, say t, such that $t \cap a_i$ is a singleton set for each $i = 1, \ldots, n$. The set t is said to be a *choice* set for s, since it allows us to choose a member from each member of s (form $t \cap a_i$). Indeed, t can be regarded as a set formed by choosing exactly one member from each of the a's.

To show that a suitable collection t is a set of $\sum$, choose $b_1 \in a_1, \ldots, b_n \in a_n$, and form $t = \{b_1, \ldots, b_n\}$. By Theorem 3, page 295, the collection is a set.

The task of proving theorems of a mathematical nature is eased by assuming that any infinite, disjointed set that does not contain the empty set also possesses a choice set. Zermelo observed that this assumption about sets was widely used (even *unconsciously* used), so he explicitly introduced it as an axiom in 1904.

Axiom of Choice. Let s be a disjointed set such that $\varnothing \notin s$; then there is a set t such that

$$\forall z \exists x (z \in s \rightarrow t \cap z = \{x\}) \tag{1}$$

In case $s = \varnothing$, take $t = \varnothing$; in this case, any set t satisfies (1).

16

Complete Theories

16.1. Vaught's Test

A wff A is *defined* in a set of wffs K provided that each predicate of A occurs in a wff of K, and provided that each individual that is free in A is also free in some wff of K.

One of the important ideas of model theory is the notion of a *complete* theory. Recall that any set of wffs is called a *theory*.

Definition. We say that a set of wffs K is *complete* if $K \vdash A$ or $K \vdash \neg A$ for each wff A that is defined in K.

For example, the empty set is complete and each contradictory set is complete. In view of Corollary 1, page 216, each maximal-consistent set of wffs is complete.

Lemma 1. Each consistent set of wffs has a complete superset.

Dem. By Theorem 1, page 216, each consistent set has a maximal-consistent superset.

Many theories are not complete. A postulate set for groups is not complete, since some groups are Abelian, while others are not. Therefore, a wff that expresses the commutativity of the group operation is not deducible from the group postulates; neither is its negation deducible from the group postulates. We conclude that this theory is not complete.

Similarly, we can prove that a postulate set for algebraically closed fields is not complete. Here we can use the fact that there is an algebraically closed field with characteristic p whenever p is prime or zero. Therefore, a wff that asserts that a field has characteristic 2 is not deducible from the postulate set; neither is its negation deducible from the postulate set. We conclude that this theory is not complete.

We pause to point out the value of this concept. Let K be a complete theory, and suppose that we can prove, by a purely mathematical argument, that a certain wff A is true for a particular model of K when interpreted in that semantical system. Then $\neg A$ is not deducible from K. Since K is complete, either $K \vdash A$ or $K \vdash \neg A$. Therefore $K \vdash A$. We conclude from Theorem 1, page 224, that A is K-true; i.e., μA is true for $\sum$ whenever $\sum$ is a model of K under μ.

Here is a test for completeness due to Vaught (1954).

Vaught's Test. K is complete if:

1. Each model of K is infinite.
2. There is a transfinite cardinal $\aleph$, which is not less than the number of wffs in K, such that any two similar normal models of K with cardinal $\aleph$ are isomorphic.
3. If $\sum$ is a model of K under both μ and λ, then μF is λF for each predicate F in K and $\mu t = \lambda t$ for each individual t that is free in K.

Dem. Let K be a set of wffs that satisfies the three conditions of Vaught's test. Suppose that K is *not* complete. Then there is a wff A, which is defined in K, such that neither A nor $\neg A$ is deducible from K. So, both $K \cup \{\neg A\}$ and $K \cup \{A\}$ are consistent; thus, each of these sets has a model. Let $\sum_1$ be a model of $K \cup \{\neg A\}$, and let $\sum_2$ be a model of $K \cup \{A\}$ that is similar to $\sum_1$; by (1), both $\sum_1$ and $\sum_2$ are semantical systems with transfinite cardinals. By Exercise 2, page 272, with $\aleph$ as in (2), there are normal semantical systems $\sum_3$ and $\sum_4$, each with cardinal $\aleph$, such that $\sum_3$ is a model of $K \cup \{\neg A\}$, and $\sum_4$ is a model of $K \cup \{A\}$. We can assume that $\sum_3$ and $\sum_4$ are similar. By (2), $\sum_3$ and $\sum_4$ are isomorphic, i.e., $\sum_3$ and $\sum_4$ are essentially the same. Thus, $\sum_3$ is a model of both $K \cup \{\neg A\}$ and $K \cup \{A\}$. From (3), there is essentially just one interpreter, say μ, such that $\sum_3$ is a model of K under μ. Therefore, A and $\neg A$ are both true for $\sum_3$ under μ. This contradiction proves that our assumption that K is not complete is false. This establishes Vaught's test.

Now that we have established this important test, it is time to look at it from a general point of view. Vaught's test asserts that a theory is complete provided that certain algebraic conditions are satisfied. We say that the conditions of the test are *algebraic* because they refer directly to the models of the theory. Thus, by demonstrating that a given theory has the algebraic properties mentioned in Vaught's test, one in fact has established the completeness of the theory. Here, then, is a striking example of the powerful methods that result from the fusion of two disciplines.

To illustrate the application of Vaught's test, we shall now prove that the theory of algebraically closed fields with fixed characteristic is complete. Let K be the usual postulate set for this theory; we shall show that the three conditions of Vaught's test are fulfilled. It is easy to see that no algebraically closed field is finite; in fact, a field in which each polynomial of degree two possesses a zero is of necessity infinite. Referring to (2), let $\aleph$ be any uncountable transfinite cardinal; then $\aleph$ is not less than the number of wffs in K (which is $\aleph_0$). But Steinitz (1910) has shown that two algebraically closed fields are isomorphic if they have the same characteristic and the same degree of transcendence over their prime field. Since the degree of transcendence of an uncountable field over its prime field is the cardinal number of the field, it follows that any two algebraically closed fields that possess the same characteristic and have cardinal $\aleph$ also have the same degree of transcendence over their prime field and so are isomorphic. This establishes (2). Finally, we point out that the postulate set for the concept of a field has property (3). The conditions of Vaught's test have been satisfied, so K is complete.

Notice that by purely algebraic observations we have established the important logical result: The postulate set characterizing the theory of algebraically closed fields with fixed characteristic is complete.

16.2. Diagrammatic Sets

The two concepts *maximal consistent* and $\exists$-*complete* are important because of the fact that we can construct a model for any set of wffs that is both maximal consistent and $\exists$-complete (see page 219). Consider the related problem: Characterize syntactically the set of all swffs of $\sum$, a given semantical system, that are true for $\sum$. Here we need the notion of the *full diagram* of a semantical system, which is due to Abraham Robinson.

Let $\prod$ be the predicate calculus obtained from $\sum$ as follows: The predicates of $\prod$ are the relation symbols of $\sum$ and the individuals of $\prod$ are the constants and placeholders of $\sum$. This allows us to drop the distinction between wffs and swffs; in particular, we may speak of a wff being true for $\sum$.

Here is Robinson's notion of the *full diagram* of a semantical system.

Definition. A set of wffs D is said to be the *full diagram* of a semantical system $\sum$ provided that $A \in \mathsf{D}$ iff A is true for $\sum$, and either A is atomic or else A is $\neg B$, where B is atomic.

Our problem is to characterize syntactically the set of all wffs of $\prod$ that are true for $\sum$. Let W be the set of all wffs of $\prod$.

Theorem 1. $\{A \in \mathsf{W} \mid A \text{ is true for } \sum\}$ is maximal consistent and $\exists$-complete.

Dem. Denote the given set of wffs by K. Certainly K is consistent, since $\sum$ is a model of K under the identity interpreter. To show that K is maximal consistent, let B be any wff of $\prod$ such that $B \notin \mathsf{K}$. Then $\neg B \in \mathsf{K}$, so $\mathsf{K} \cup \{B\}$ is contradictory. This proves that K is maximal consistent. To show that K is $\exists$-complete, let $\exists tC \in \mathsf{K}$. Since $\exists tC$ is true for $\sum$, there is a constant a of $\sum$ such that $\mathsf{S}_t^a[C]$ is true for $\sum$. Therefore $\mathsf{S}_t^a[C] \in \mathsf{K}$; we conclude that K is $\exists$-complete.

We now display the connection between the full diagram of $\sum$ and the set of all wffs of $\prod$ that are true for $\sum$.

Theorem 2. Let K be any superset of D, the full diagram of $\sum$, such that K is both maximal consistent and $\exists$-complete. Then $\mathsf{K} = \{A \in \mathsf{W} \mid A \text{ is true for } \sum\}$.

Dem. We point out that $\sum$ is the semantical system obtained from K by applying the construction on page 219. By Lemma 1, page 219, A is true for $\sum$ iff $A \in \mathsf{K}$, for each $A \in \mathsf{W}$. This establishes Theorem 2.

Corollary 1. Within $\prod$, there is exactly one superset of D that is both maximal consistent and $\exists$-complete, namely $\{A \in \mathsf{W} \mid A \text{ is true for } \sum\}$.

Of course, each wff that is deducible from D, the full diagram of $\sum$, is true for $\sum$. The converse is not necessarily true. However, we can show that each wff that is true for $\sum$ and has a special form is deducible from D.

Theorem 3. $\mathsf{D} \vdash A$ provided that A is quantifier-free and is true for $\sum$.

Dem. Assuming that there is a quantifier-free wff that is true for $\sum$ and is not deducible from D, let A be the shortest such wff (counting connectives). There are just three possibilities about A.

1. Suppose that A is atomic. Then $A \in \mathsf{D}$, so $\mathsf{D} \vdash A$. This contradiction shows that A is not atomic.

2. Suppose that $\neg$ is the main connective of A; then $A = \neg B$ for some wff B. If B is atomic, then $\neg B \in \mathsf{D}$, so $\mathsf{D} \vdash A$; therefore, B is not atomic. Assume that $B = \neg C$, so $A = \neg(\neg C)$; then C is true for $\sum$ and is shorter than A. Thus $\mathsf{D} \vdash C$; so $\mathsf{D} \vdash A$. Assume that $B = E \vee F$, so $A = \neg(E \vee F)$. Then $E \vee F$ is not true for $\sum$; thus $\neg E$ is true for $\sum$ and $\neg F$ is true for $\sum$. It follows that $\mathsf{D} \vdash \neg E$ and $\mathsf{D} \vdash \neg F$; thus $\mathsf{D} \vdash \neg E \wedge \neg F$, so $\mathsf{D} \vdash A$. This contradiction proves that $\neg$ is not the main connective of A.

3. Suppose that $\vee$ is the main connective of A; then $A = B \vee C$ for some wffs B and C. Then B is true for $\sum$ or C is true for $\sum$ (or both). If the former, then $\mathsf{D} \vdash B$; if the latter, $\mathsf{D} \vdash C$. In either case, $\mathsf{D} \vdash B \vee C$, i.e., $\mathsf{D} \vdash A$. This contradiction establishes Theorem 3.

Theorem 4. $\mathsf{D} \vdash A$ provided that A is in prenex normal form, "$\forall$" does not appear in the prefix of A, and A is true for $\sum$.

Dem. If the theorem is false, then there is a "shortest" wff (i.e., the length of its prefix is minimum) that meets the conditions of the theorem but is not deducible from D. This wff has the form $\exists tB$. By assumption, there is an individual a such that $\mathsf{S}_t^a[B]$ is true for $\sum$. Either $\mathsf{S}_t^a[B]$ is quantifier-free, or $\mathsf{S}_t^a[B]$ is in prenex normal form. In the former case, $\mathsf{D} \vdash \mathsf{S}_t^a[B]$ by Theorem 3. In the latter case, "$\forall$" does not appear in the prefix of $\mathsf{S}_t^a[B]$, and the prefix of $\mathsf{S}_t^a[B]$ is shorter than the prefix of $\exists tB$; therefore, $\mathsf{D} \vdash \mathsf{S}_t^a[B]$ by assumption. Hence, in either case, $\mathsf{D} \vdash \mathsf{S}_t^a[B]$. But $\vdash \mathsf{S}_t^a[B] \rightarrow \exists tB$; thus $\mathsf{D} \vdash \exists tB$. This completes our proof.

It is vital that we extend the scope of Theorems 3 and 4. We shall achieve this by constructing the semantical system involved from the symbols of a given predicate calculus, rather than constructing a predicate calculus from a given semantical system. Consider a predicate calculus that has infinitely many individuals; our purpose is to select a set of wffs of the given predicate calculus, which we shall use to construct our semantical system $\sum$. Consider the following definition.

Definition. We shall say that a set of wffs, say $\mathscr{D}$, is *diagrammatic* provided that:

(a) $\mathscr{D}$ is nonempty.
(b) If $A \in \mathscr{D}$, then either A is atomic or else $A = \neg B$, where B is atomic.

(c) Let P be a predicate that occurs in $\mathscr{D}$ and let α be any P-string whose terms occur in $\mathscr{D}$; then either $P\alpha \in \mathscr{D}$ or else $\neg P\alpha \in \mathscr{D}$.

For example, let type $F = \{1\}$ and let type $G = \{2\}$; then each of the following sets is diagrammatic:

$\{Fx\}$, $\{Fy, \neg Fz\}$, $\{Gxx, Gxy, \neg Gyx, \neg Gyy\}$,
$\{Fx, \neg Fy, \neg Gxx, \neg Gxy, Gyx, Gyy\}$
$\{Gst \mid s \text{ and } t \text{ are individuals}\}$
$\{Gst \mid s \text{ and } t \text{ are distinct individuals}\} \cup \{\neg Gss \mid s \text{ is an individual}\}$

Notice that a diagrammatic set $\mathscr{D}$ involves a set of individuals, say S, and a set of predicates such that $A \in \mathscr{D}$ iff $A = P\alpha$ or else $A = \neg P\alpha$, where P is a predicate in the given set and the terms of α are members of S.

Given a diagrammatic set $\mathscr{D}$, it is easy to construct a semantical system $\sum$ from the symbols appearing in $\mathscr{D}$ by following the procedure on page 215. This means that the constants of $\sum$ are the individuals that occur in $\mathscr{D}$, and the relation symbols of $\sum$ are the predicates that occur in $\mathscr{D}$. Furthermore, $P\alpha \in \text{diag} \sum$ iff $P\alpha \in \mathscr{D}$. Thus each diagrammatic set describes a semantical system in a natural way. We mention that for each predicate P that occurs in $\mathscr{D}$, $\{\alpha \mid P\alpha \in \mathscr{D}\}$ is a relation of $\sum$.

Now that we have constructed $\sum$ from the given diagrammatic set $\mathscr{D}$, let us consider the full diagram of $\sum$. First, we must construct a predicate calculus from the symbols of $\sum$. The resulting predicate calculus is contained within the given predicate calculus; furthermore, the full diagram of $\sum$ is $\mathscr{D}$, the given diagrammatic set. This comment allows us to generalize Theorems 3 and 4 as follows.

Theorem 5. Let $\mathscr{D}$ be any diagrammatic set and let $\sum$ be the semantical system constructed from $\mathscr{D}$ as above. Then $\mathscr{D} \vdash A$ provided that A is quantifier-free and is true for $\sum$.

Theorem 6. Let $\mathscr{D}$ be any diagrammatic set and let $\sum$ be the semantical system constructed from $\mathscr{D}$ as above. Then $\mathscr{D} \vdash A$ provided that A is in prenex normal form, $\forall$ does not appear in the prefix of A, and A is true for $\sum$.

Here is a fact that we shall soon utilize.

Theorem 7. Let $\mathscr{D}$ be any diagrammatic set and let $\sum$ be the semantical system constructed from $\mathscr{D}$ as above. Let A be a swff of $\sum$ such that neither A nor $\neg A$ is deducible from $\mathscr{D}$. Then there is a wff C

in prenex normal form and an individual t free in C, such that C is a swff of $\sum$, neither $\exists tC$ nor $\neg(\exists tC)$ is deducible from $\mathscr{D}$, and $\exists tC$ is not true for $\sum$.

Dem. Consider the given wff A; either A is true for $\sum$ or else $\neg A$ is true for $\sum$. We may as well assume that A is true for $\sum$. If A is quantifier-free, then $\mathscr{D} \vdash A$ by Theorem 5; so A is not quantifier-free. We may assume, then, that A is in prenex normal form (see the Fundamental Theorem about Prenex Normal Form, page 199). By Theorem 6, $\forall$ appears in the prefix of A. It follows that there is a swff of $\sum$ that has the form $\exists tC$ and is such that neither $\exists tC$ nor $\neg(\exists tC)$ is deducible from $\mathscr{D}$. We may also assume that no swff of $\sum$ that has these properties has a shorter prefix than has $\exists tC$. We want to show that the swff $\exists tC$ is not true for $\sum$. Assume, for the moment, that $\exists tC$ is true for $\sum$; then there is a constant a such that $\mathsf{S}_t^a[C]$ is true for $\sum$. Moreover, by Theorem 6, $\forall$ appears in the prefix of C; thus, $\mathsf{S}_t^a[C]$ is in prenex normal form, has a shorter prefix than $\exists tC$, and is a swff of $\sum$. Therefore, $\mathscr{D} \vdash \mathsf{S}_t^a[C]$ or else $\mathscr{D} \vdash \neg \mathsf{S}_t^a[C]$. Since $\mathsf{S}_t^a[C]$ is true for $\sum$, a model of $\mathscr{D}$ under the identity interpreter, it follows that $\mathscr{D} \vdash \mathsf{S}_t^a[C]$. But $\vdash \mathsf{S}_t^a[C] \rightarrow \exists tC$. We conclude that $\mathscr{D} \vdash \exists tC$, which is a contradiction. Therefore $\exists tC$ is not true for $\sum$. This establishes Theorem 7.

16.3. Simplifying the Concept of a Model

Our notion of a diagrammatic set enables us to express the important concept of a *model* of a consistent set of wffs within the framework of the predicate calculus involved.

In Section 12.6 we presented Henkin's proof of the Strong Completeness Theorem. There, given a consistent set of wffs K, we constructed a semantical system $\sum$ such that $\sum$ is a model of K under the identity interpreter; moreover, the full diagram of $\sum$ is a set of wffs from some extension (see page 217) of the given predicate calculus. For this reason, we may restrict the notion of a *model* of a nonempty set of wffs by restricting ourselves to the identity interpreter. But each semantical system obtained via the identity interpreter is characterized by its full diagram. Therefore, the notion of a model of a nonempty set of wffs can be characterized within the framework of the predicate calculus involved.

Definition. We shall say that $\mathscr{D}$ is a *model of* K, a nonempty set of wffs, provided that:

1. There is an extension of the given predicate calculus in which $\mathscr{D}$ is diagrammatic.

2. The semantical system constructed from $\mathscr{D}$ is a model of K under the identity interpreter.

For example, $\{Fy\}$ is a model of the set $\{\forall yFy, \exists zFz\}$. Here, the extension of the given predicate calculus is itself; clearly, $\{Fy\}$ is diagrammatic, and the semantical system constructed from $\{Fy\}$ has one constant y and one relation symbol F; its full diagram is $\{Fy\}$.

Here is another example. Let I be the set of all individuals of a predicate calculus, and let a be a symbol that does not occur in the given predicate calculus. Then the diagrammatic set

$$\{Ft \mid t \in I\} \cup \{\neg Fa\}$$

is a model of the set of wffs $\{Ft \mid t \in I\} \cup \{\exists y(\neg Fy)\}$. Here, the extension of the given predicate calculus is constructed by adjoining "a" to the set of individuals I of the given predicate calculus. The semantical system obtained from the above diagrammatic set has just one relation symbol, namely F; its constants are the members of $I \cup \{a\}$. Note that type $F = \{1\}$.

In view of these observations we can simplify the Strong Completeness Theorem as follows.

Theorem 1. Strong Completeness Theorem. A set of wffs K is consistent iff K has a model (in the sense of the preceding definition).

This is an important result, which we shall use in Section 16.4.

The following facts are easy to verify. Throughout, $\mathscr{D}$ is any diagrammatic set, and A and B are wffs defined in $\mathscr{D}$.

Lemma 1. $\mathscr{D}$ is a model of $\mathscr{D}$.

Lemma 2. $\mathscr{D}$ is a model of $\{\neg A\}$ iff $\mathscr{D}$ is not a model of $\{A\}$.

Lemma 3. $\mathscr{D}$ is a model of $\{A \vee B\}$ iff $\mathscr{D}$ is a model of $\{A\}$, or $\mathscr{D}$ is a model of $\{B\}$.

Lemma 4. $\mathscr{D}$ is a model of $\{\forall tA\}$ iff $\mathscr{D}$ is a model of $\{\mathrm{S}_t^s[A] \mid s$ is an individual that occurs in $\mathscr{D}\}$.

Since we are expressing the algebraic notion of a semantical system by means of a diagrammatic set of wffs, a notion of the predicate calculus, we may as well characterize the algebraic notion of *isomorphic* systems within our predicate calculus.

Definition. Let $\mathscr{D}$ and $\mathscr{D}^*$ be any diagrammatic sets; we shall say that $\mathscr{D}$ and $\mathscr{D}^*$ are *isomorphic* provided there is a one–one mapping, say ν, of the individuals of $\mathscr{D}$ onto the individuals of $\mathscr{D}^*$, such that $\nu\mathscr{D} = \mathscr{D}^*$.

Note. "$\nu\mathscr{D}$" denotes the set of wffs obtained from $\mathscr{D}$ by replacing each individual occurring in a wff of $\mathscr{D}$ by its image under ν.

For example, let

$$\mathscr{D} = \{Gxy, Gyy, Gxx, Gyx\} \qquad \text{and} \qquad \mathscr{D}^* = \{Gyt, Gtt, Gyy, Gty\}$$

Then $\nu\mathscr{D} = \mathscr{D}^*$, where $\nu x = y$ and $\nu y = t$. Note that $\mathscr{D}$ is a model of $\{Gxy, Gyy\}$, while $\mathscr{D}^*$ is not a model of $\{Gxy, Gyy\}$. However, the semantical system obtained from $\mathscr{D}^*$, whose full diagram is $\mathscr{D}^*$, is a model of $\{Gxy, Gyy\}$ under the interpreter yielded by ν.

The following theorems are important.

Theorem 2. Let $\mathscr{D}$ and $\mathscr{D}^*$ be any isomorphic diagrammatic sets, and let ν be any one–one mapping of the individuals of $\mathscr{D}$ onto the individuals of $\mathscr{D}^*$, such that $\nu\mathscr{D} = \mathscr{D}^*$. Let A be any wff; then $\mathscr{D}$ is a model of $\{A\}$ iff $\mathscr{D}^*$ is a model of $\{\nu A\}$.

Dem. Apply the Strong Fundamental Theorem about Wffs, page 148.

Corollary 1. Let $\mathscr{D}$ and $\mathscr{D}^*$ be isomorphic diagrammatic sets such that $\nu\mathscr{D} = \mathscr{D}^*$; let $\sum^*$ be the semantical system constructed from $\mathscr{D}^*$. Then $\sum^*$ is a model of K under the interpreter yielded by ν if $\mathscr{D}$ is a model of K.

Dem. Let K be a set of wffs such that $\mathscr{D}$ is a model of K. By Theorem 2, $\mathscr{D}^*$ is a model of $\nu\mathsf{K}$; therefore, $\sum^*$ is a model of K under the interpreter yielded by ν.

Theorem 3. $\mathsf{K} \vdash A$ iff each model of K in which A is defined is also a model of $\{A\}$.

Dem. 1. Suppose that $\mathsf{K} \vdash A$; then, by statement II, page 213, $\mathscr{D}$ is a model of $\{A\}$ whenever $\mathscr{D}$ is a model of K in which A is defined.

2. Suppose that $\mathscr{D}$ is a model of $\{A\}$ whenever $\mathscr{D}$ is a model of K in which A is defined. If A is not deducible from K, then $\mathsf{K} \cup \{\neg A\}$ is consistent, so has a model $\mathscr{D}'$ by Theorem 1. Thus $\mathscr{D}'$ is a model of K,

and A is defined in $\mathscr{D}'$. Therefore, $\mathscr{D}'$ is a model of $\{A\}$. But $\mathscr{D}'$ is a model of $\{\neg A\}$. This contradiction establishes our theorem.

Definition. Let K be any nonempty set of wffs, let B and B' be any sets of individuals, and let ν be any one–one mapping of B onto B'. We shall say that ν is a K-*mapping* provided that $\nu t = t$ for each individual t that is free in a wff of K.

For example, $\nu = \{(x, s), (y, t), (z, z)\}$ is a K-mapping, where $\mathsf{K} = \{Fz, \forall yGyy\}$. Here, $B = \{z, y, x\}$ and $B' = \{t, z, s\}$.

Definition. We shall say that $\mathscr{D}$ is a *prime* model of K provided that:

1. $\mathscr{D}$ is a model of K.
2. Given any model of K, say $\mathscr{D}'$, there is a K-mapping ν such that $\nu\mathscr{D} \subset \mathscr{D}'$.

We can interpret this notion algebraically as follows. A system $\mathscr{P}$ is a prime model of a theory K iff each model of K (in the algebraic sense) contains a subsystem isomorphic to $\mathscr{P}$. In other words, $\mathscr{P}$ is a prime model of a theory K iff each model of K is essentially an extension of $\mathscr{P}$.

Theorem 4. $\mathsf{K} \vdash A$ provided that $\mathsf{K} \cup \mathscr{D} \vdash A$, $\mathscr{D}$ is a prime model of K, and A is defined in K.

Dem. Since $\mathsf{K} \cup \mathscr{D} \vdash A$, there is a finite subset of $\mathscr{D}$, say $\{Z_1, \ldots, Z_m\}$, such that $\mathsf{K} \vdash Z_1 \wedge \cdots \wedge Z_m \rightarrow A$. Let $x_1, \ldots, x_t$ be the individuals that are free in the wff $Z_1 \wedge \cdots \wedge Z_m$ and are not free in K. Let

$$Y = \exists x_1 \cdots \exists x_t(Z_1 \wedge \cdots \wedge Z_m)$$

then $\mathsf{K} \vdash Y \rightarrow A$ and $\mathscr{D}$ is a model of $\{Y\}$. Next, let $\mathscr{D}'$ be any model of K. Since $\mathscr{D}$ is a prime model of K, there is a K-mapping ν such that $\nu\mathscr{D} \subset \mathscr{D}'$. Hence, $\nu Y = Y$; but by Theorem 2, $\nu\mathscr{D}$ is a model of $\{\nu Y\}$, so $\nu\mathscr{D}$ is a model of $\{Y\}$. Therefore, by Theorem 6, page 320, $\nu\mathscr{D} \vdash Y$. But $\mathscr{D}'$ is a model of $\nu\mathscr{D}$, since $\nu\mathscr{D} \subset \mathscr{D}'$; therefore, $\mathscr{D}'$ is a model of $\{Y\}$. We have established that $\mathscr{D}'$ is a model of $\{Y\}$ whenever $\mathscr{D}'$ is a model of K. Hence, by Theorem 3, page 323, $\mathsf{K} \vdash Y$. Since $\mathsf{K} \vdash Y \rightarrow A$, it follows that $\mathsf{K} \vdash A$.

The ideas of this section are due to Robinson (1956); in Section 16.4 we shall see how he uses these notions to develop a valuable test for the completeness of a theory.

16.4. Robinson's Test

A highly ingenious and valuable test for the completeness of a theory has been developed by Abraham Robinson. We shall devote this section to the study of Robinson's method. First, we need the notion of a *model-complete* theory.

Definition. A nonempty set K is said to be *model complete* provided $K \cup \mathscr{D}$ is complete whenever $\mathscr{D}$ is a model of K.

Our first goal is to characterize the notion of a model-complete theory. To this purpose, we require the notion of a *primitive* wff.

Definition. A wff, say Y, is said to be *primitive* provided that Y is in prenex normal form, $\forall$ does not appear in the prefix of Y, and the matrix of Y has the form "$A_1 \wedge \cdots \wedge A_m$," where each A_i is atomic, or $A_i = \neg B_i$ and B_i is atomic.

Theorem 1. If K is model complete, then $\mathscr{D} \vdash Y$ whenever $\mathscr{D}$ and $\mathscr{D}'$ are models of K such that:

1. $\mathscr{D} \subset \mathscr{D}'$.
2. Y is primitive and Y is defined in $\mathscr{D}$.
3. $\mathscr{D}' \vdash Y$.

Dem. Since $\mathscr{D}$ is a model of K, $K \cup \mathscr{D}$ is complete. But Y is defined in $K \cup \mathscr{D}$; therefore, $K \cup \mathscr{D} \vdash Y$ or else $K \cup \mathscr{D} \vdash \neg Y$. If the latter, then $\mathscr{D}'$ is a model of $\{\neg Y\}$, since $\mathscr{D}'$ is a model of $K \cup \mathscr{D}$. But $\mathscr{D}' \vdash Y$; thus $\mathscr{D}'$ is a model of $\{Y\}$, a contradiction. This proves that $K \cup \mathscr{D} \vdash Y$. But $\mathscr{D}$ is a model of $K \cup \mathscr{D}$; so $\mathscr{D}$ is a model of $\{Y\}$. Applying Theorem 6, page 320, we see that $\mathscr{D} \vdash Y$. This proves Theorem 1.

Theorem 2. If K is not model complete, there is a primitive wff Y and models of K, say $\mathscr{D}$ and $\mathscr{D}'$, such that:

1. $\mathscr{D} \subset \mathscr{D}'$.
2. $\mathscr{D}' \vdash Y$.
3. Y is defined in $\mathscr{D}$.
4. Y is not deducible from $\mathscr{D}$.

Dem. We shall construct Y, $\mathscr{D}$, and $\mathscr{D}'$, which possess the above properties. Since K is not model complete, there is a model of K, say $\mathscr{D}_1$, and a wff A defined in $\mathscr{D}_1$, such that neither A nor $\neg A$ is deducible from

$\mathsf{K} \cup \mathscr{D}_1$. By the proof of Theorem 7, page 320, there is a wff in prenex normal form with shortest prefix, say $\exists tC$, such that neither $\exists tC$ nor $\neg\exists tC$ is deducible from $\mathsf{K} \cup \mathscr{D}_1$. Furthermore, $\mathscr{D}_1$ is not a model of $\{\exists tC\}$. Next, consider *all* pairs $(\mathscr{D}, B)$ such that $\mathscr{D}$ is a model of K, and B is a wff defined in $\mathscr{D}$ and in prenex normal form, such that neither B nor $\neg B$ is deducible from $\mathsf{K} \cup \mathscr{D}$. Choose a pair from this collection of pairs such that the wff involved has a prefix of minimum length; for example, the pair $(\mathscr{D}, \exists sE)$. This means that B_2 or $\neg B_2$ is deducible from $\mathsf{K} \cup \mathscr{D}_2$ if $\mathscr{D}_2$ is a model of K, B_2 is defined in $\mathscr{D}_2$ and is in prenex normal form, and if the prefix of B_2 is shorter than the prefix of $\exists sE$.

Since $\neg\exists sE$ is not deducible from $\mathsf{K} \cup \mathscr{D}$, $\mathsf{K} \cup \mathscr{D} \cup \{\exists sE\}$ is consistent; hence, it has a model $\mathscr{D}'$ (by Theorem 1, page 322). Thus $\mathscr{D} \subset \mathscr{D}'$. Since $\mathscr{D}'$ is a model of $\{\exists sE\}$, there is an individual occurring in $\mathscr{D}'$, say a, such that $\mathscr{D}'$ is a model of $\{\mathsf{S}^a_s[E]\}$. Consider the pair $(\mathscr{D}', \mathsf{S}^a_s[E])$; $\mathscr{D}'$ is a model of K, and $\mathsf{S}^a_s[E]$ is defined in $\mathscr{D}'$ and its prefix is shorter than that of $\exists sE$ (if $\mathsf{S}^a_s[E]$ is quantifier-free, apply Theorem 5, page 320). Therefore, $\mathsf{K} \cup \mathscr{D}' \vdash \mathsf{S}^a_s[E]$ or $\mathsf{K} \cup \mathscr{D}' \vdash \neg\mathsf{S}^a_s[E]$; since $\mathscr{D}'$ is a model of $\mathsf{K} \cup \mathscr{D}' \cup \{\mathsf{S}^a_s[E]\}$, we conclude that $\mathsf{K} \cup \mathscr{D}' \vdash \mathsf{S}^a_s[E]$. But $\vdash \mathsf{S}^a_s[E] \to \exists sE$; thus $\mathsf{K} \cup \mathscr{D}' \vdash \exists sE$.

Continuing our proof, it follows that there is a finite subset of $\mathscr{D}'$, say $\{Z_1, \ldots, Z_m\}$, such that

$$\mathsf{K} \vdash Z_1 \wedge \cdots \wedge Z_m \to \exists sE$$

Let $x_1, \ldots, x_t$ be the individuals that are free in $Z_1 \wedge \cdots \wedge Z_m$ and do not occur in $\mathscr{D}$. Let

$$Y = \exists x_1 \cdots \exists x_t(Z_1 \wedge \cdots \wedge Z_m)$$

Then $\mathsf{K} \vdash Y \to \exists sE$. Since $\mathscr{D}$ is a model of K, we see that $\mathscr{D}$ is a model of $\{Y \to \exists sE\}$. Clearly, Y is primitive, $\mathscr{D}' \vdash Y$ (by Theorem 6, page 320, since $\mathscr{D}'$ is a model of $\{Y\}$), and Y is defined in $\mathscr{D}$. If $\mathscr{D} \vdash Y$, then $\mathscr{D}$ is a model of $\{Y\}$; hence, $\mathscr{D}$ is a model of $\{\exists sE\}$. However, by Theorem 7, page 320, $\mathscr{D}$ is ***not*** a model of $\{\exists sE\}$! This contradiction proves that Y is not deducible from $\mathscr{D}$, and completes our proof of Theorem 2.

Corollary 1. K is model complete iff $\mathscr{D} \vdash Y$ whenever $\mathscr{D}$ and $\mathscr{D}'$ are models of K such that:

1. $\mathscr{D} \subset \mathscr{D}'$.
2. Y is primitive and is defined in $\mathscr{D}$.
3. $\mathscr{D}' \vdash Y$.

Note. $\mathscr{D} \vdash Y$ iff $\mathscr{D}$ is a model of $\{Y\}$ (by Theorem 6, page 320).

Robinson uses the criterion expressed in this corollary to demonstrate that various theories are model complete.

Finally, we present the connection between model-completeness and completeness.

Robinson's Test. K is complete if K is model complete and has a prime model.

Dem. Let A be any wff defined in K, and let $\mathscr{D}$ be a prime model of K. Since $K \cup \mathscr{D}$ is complete, either $K \cup \mathscr{D} \vdash A$ or else $K \cup \mathscr{D} \vdash \neg A$. If the former, then $K \vdash A$ by Theorem 4, page 324; if the latter, then $K \vdash \neg A$. Thus, K is complete. This establishes Robinson's test.

One point needs to be emphasized. The two conditions of Robinson's test are *algebraic* conditions. This is because our notion of a *model* of a set of wffs (see page 321) corresponds to the usual algebraic notion of a model, or realization, of a theory. Furthermore, the condition that K has a *prime* model corresponds to the algebraic notion of a theory with the property that each model of the theory is essentially (up to isomorphism) an extension of one model of the theory.

Examining the criterion for model completeness from a purely algebraic viewpoint, we see that a theory K is model complete iff each primitive wff Y that is meaningful in $\mathscr{M}$, where $\mathscr{M}$ is a given model of K (in the algebraic sense), is true for $\mathscr{M}$ if Y is true in some extension of $\mathscr{M}$ that is also a model of K (again, in the algebraic sense).

To illustrate, let K be the concept of an algebraically closed field. Then a primitive wff Y of this theory, which is meaningful in $\mathscr{M}$, a given model of K, asserts the existence of a solution of a given set of equations and inequalities; these equations and inequalities involve certain unknowns and possibly certain field elements of the algebraically closed field $\mathscr{M}$. Robinson has shown that by a simple trick the inequalities can be replaced by equations; hence, Y asserts the existence of a solution of a finite system of polynomial equations. But any system of polynomial equations with coefficients in a field $\mathscr{M}$ has a solution in the algebraic closure of $\mathscr{M}$, provided that the system has a solution in some extension of $\mathscr{M}$. Since we are assuming that Y has a solution in $\mathscr{M}'$, an extension of $\mathscr{M}$ that is also a model of K, we conclude that Y has a solution in $\mathscr{M}$—since $\mathscr{M}$ is algebraically closed. Thus, the concept of an algebraically closed field is model complete.

It is evident that the concept of an algebraically closed field is not complete, since there exist algebraically closed fields with differing characteristics. Let K be a set of wffs that characterizes the concept of an algebraically closed field; then, by Robinson's Test, K does not have a prime model.

Moreover, it is well known that the concept of an algebraically closed field of characteristic p possesses a prime model; clearly, this concept is model complete, since the concept of an algebraically closed field is model complete. Applying Robinson's test, we conclude that the concept of an algebraically closed field of characteristic p is complete.

Robinson has applied his test to demonstrate the completeness of many other important theories [see Robinson (1956)].

Bibliography

ABBOTT, J. C.
1969 *Sets, Lattices, and Boolean Algebras*, Allyn and Bacon, Boston, Massachusetts, 282 pp.

ABIAN, A.
1965 *The Theory of Sets and Transfinite Arithmetic*, W. B. Saunders, Philadelphia, Pennsylvania, 406 pp.

ADDISON, J. W., HENKIN, L., and TARSKI, A.
1965 (editors) *The Theory of Models*, North-Holland, Amsterdam, 494 pp.

BELL, J. L., and SLOMSON, A. B.
1971 *Models and Ultraproducts: An Introduction*, North-Holland, Amsterdam, 322 pp.

BENACERRAF, P., and PUTNAM, H.
1964 (editors) *Philosophy of Mathematics*, Prentice-Hall, Englewood Cliffs, New Jersey, 536 pp.

BERRY, G. D. W.
1953 On the Ontological Significance of the Löwenheim–Skolem Theorem. *Academic Freedom, Logic and Religion*, Am. Phil. Assn., Eastern Division, 2, Univ. of Pennsylvania Press, Philadelphia, Pennsylvania, pp. 39–55.

BETH, E. W.
1959 *The Foundations of Mathematics*, North-Holland, Amsterdam, 741 pp.

COHEN, P. J.
1966 *Set Theory and the Continuum Hypothesis*, W. A. Benjamin, New York, 154 pp.

CROSSLEY, J. N., ASH, C. J., BRICKHILL, C. J., STILLWELL, J. C., and WILLIAMS, N. H.
1972 *What is Mathematical Logic?*, Oxford University Press, London, 82 pp.

DAVIS, M., and HERSH, R.
1972 Nonstandard Analysis. *Sci. Am.* (June), pp. 78–86.

EISENBERG, M.
1971 *Axiomatic Theory of Sets and Classes*, Holt, Rinehart and Winston, New York, 366 pp.

Exner, R. M., and Rosskopf, M. F.
1959 *Logic in Elementary Mathematics*, McGraw-Hill, New York, 274 pp.
1970 Proof. Chapter 8 of the Thirty-Third Yearbook, National Council of Teachers of Mathematics, pp. 196–240.

Fraenkel, A. A.
1922*a* Zu den Grundlagen der Cantor–Zermeloschen Mengenlehre. *Math. Annalen* **86**, 230–237.
1922*b* Der Begriff "Definit" und die Unabhängigkeit des Auswahlaxioms. *Sitzungsber. Preuss. Akad. Wiss., Phys.-Math. K.*, pp. 253–257.
1925 Untersuchungen über die Grundlagen der Mengenlehre. *Math. Z.* **22**, 250–273.
1966 *Set Theory and Logic*, Addison-Wesley, Reading, Massachusetts, 95 pp.

Fraenkel, A. A., Bar-Hillel, Y., and Levy, A.
1973 *Foundations of Set Theory*, North-Holland, Amsterdam, 404 pp.

Gödel, K.
1930 Die Vollständigkeit der Axiome des logischen Funktionenkalküls. *Monatsh. Math. Phys.* **37**, 349–360.
1940 *The Consistency of the Axiom of Choice and of the Generalized Continuum-Hypothesis with the Axioms of Set Theory*, Annals of Mathematics Studies No. 3, Princeton University Press, Princeton, New Jersey, 69 pp.

Götlind, E.
1947 Ett Axiomsystem för Utsagokalkylen. *Norsk matematisk tidsskrift* **29**, 1–4.

Hatcher, W. S.
1968 *Foundations of Mathematics*, W. B. Saunders, Philadelphia, Pennsylvania, 327 pp.

Henkin, L.
1949 The Completeness of the First-Order Functional Calculus. *J. Symbolic Logic* **14**, 159–166.

Hermes, H.
1973 *Introduction to Mathematical Logic*, Springer-Verlag, New York, 242 pp.

Hilbert, D., and Ackermann, W.
1950 *Principles of Mathematical Logic*, Chelsea Publishing Co., New York, 172 pp

Hintikka, J.
1969 (editor) *The Philosophy of Mathematics*, Oxford University Press, London, 186 pp.

Kleene, S. C.
1952 *Introduction to Metamathematics*, D. van Nostrand Co., New York, 550 pp.

Lakatos, I.
1967 (editor) *Problems in the Philosophy of Mathematics*, North-Holland, Amsterdam, 241 pp.

Levitz, H.
1974 Non-Standard Analysis: An Exposition. *L'Enseignement Mathématique*, IIe Serie, Vol. 20, pp. 9–32.

LIGHTSTONE, A. H.

1964 *The Axiomatic Method: An Introduction to Mathematical Logic*, Prentice-Hall, Englewood Cliffs, New Jersey, 246 pp.

1965 *Symbolic Logic and the Real Number System*, Harper & Row, New York, 225 pp.

1968 Group Theory and the Principle of Duality. *Can. Math. Bull.* **11**, 43–50.

1969 The Notion of "Consequence" in the Predicate Calculus. *Math. Magazine* **42**, 57–60.

1972 Infinitesimals. *Am. Math. Monthly* **79**, 242–251.

1973 Infinitesimals and Integration. *Math. Magazine* **46**, 20–30.

LIGHTSTONE, A. H., and ROBINSON, A.

1957 Syntactical Transforms. *Trans. Am. Math. Soc.* **86**, 220–245.

1975 *Nonarchimedean Fields and Asymptotic Expansions*, North-Holland, Amsterdam, 203 pp.

LÖWENHEIM, L.

1915 Über Möglichkeiten im Relativkalkül. *Math. Annalen* **76**, 447–470.

LUXEMBURG, W. A. J.

1962 *Non-Standard Analysis. Lectures on A. Robinson's Theory of Infinitesimals and Infinitely Large Numbers*, Pasadena, California (mimeographed notes), 150 pp.

1969 (editor) *Applications of Model Theory to Algebra, Analysis, and Probability*, Holt, Rinehart and Winston, New York, 307 pp.

1973 What is Nonstandard Analysis? Papers in the Foundations of Mathematics. *Am. Math. Monthly* **80** (Part II), 38–67.

MENDELSON, E.

1964 *Introduction to Mathematical Logic*, D. van Nostrand, Princeton, New Jersey, 300 pp.

MONK, J. D.

1969 *Introduction to Set Theory*, McGraw-Hill, New York, 193 pp.

MORLEY, M. D.

1973 (editor) *Studies in Model Theory*, Mathematical Assn. of America, Studies in Mathematics, Vol. 8, 197 pp.

MOSTOWSKI, A.

1966 *Thirty Years of Foundational Studies*, Basil Blackwell, Oxford, 180 pp.

MYHILL, J. R.

1953 On the Ontological Significance of the Löwenheim–Skolem Theorem. *Academic Freedom, Logic and Religion*, Am. Phil. Assn., Eastern Division, 2, Univ. of Pennsylvania Press, Philadelphia, Pennsylvania, pp. 57–70.

RASIOWA, H.

1949 Sur un Certain Système d'Axiomes du Calcul des Propositions. *Norsk matematisk tidsskrift* **31**, 1–3.

ROBINSON, A.

1951 *On the Metamathematics of Algebra*, North-Holland, Amsterdam, 195 pp.

1956 *Complete Theories*, North-Holland, Amsterdam, 129 pp.

1961 Non-Standard Analysis. *Proc. Roy. Acad. Sci., Amst. A* **64**, 432–440.

1963 *Introduction to Model Theory and to the Metamathematics of Algebra*, North-Holland, Amsterdam, 284 pp.

1965 Topics in Non-Archimedean Mathematics. *The Theory of Models*, edited by J. W. Addison, L. Henkin, and A. Tarski, North-Holland, Amsterdam, pp. 285–298.

1966 *Non-Standard Analysis*, North-Holland, Amsterdam, 293 pp.

1967*a* The Metaphysics of the Calculus. *Problems in the Philosophy of Mathematics*, edited by I. Lakatos, North-Holland, Amsterdam, pp. 28–40.

1967*b* Nonstandard Arithmetic. *Bull. Am. Math. Soc.* **73**, 818–843.

1973*a* Function Theory on Some Nonarchimedean Fields. Papers in the Foundations of Mathematics. *Am. Math. Monthly* **80** (Part II), 87–109.

1973*b* Model Theory as a Framework for Algebra. *Studies in Model Theory*, edited by M. D. Morley, Mathematical Assn. of America, pp. 134–157.

ROSSER, J. B.

1953 *Logic for Mathematicians*. McGraw-Hill, New York, 530 pp.

1969 *Simplified Independence Proofs*. Academic Press, New York, 217 pp.

SKOLEM, TH.

1920 Logisch-Kombinatorische Untersuchungen über die Erfüllbarkeit oder Beweisbarkeit Mathematischer Sätze nebst einem Theoreme über Dichte Mengen. *Skrifter utgit av Videnskapsselskapet i Kristiania*, I, No. 4, 36 pp.

1930 Einige Bemerkungen zu der Abhandlung von E. Zermelo: "Über die Definitheit in der Axiomatik." *Fund. Math.* **15**, 337–341.

1934 Über die Nicht-Charakterisierbarkeit der Zahlenreihe mittels Endlich oder Abzählbar Unendlich Vieler Aussagen mit Ausschliesslich Zahlenvariablen, *Fund. Math.* **23**, 150–161.

1970 *Selected Works in Logic*, edited by J. E. Fenstad, Universitetsforlaget, Oslo, 732 pp.

STEEN, L. A.

1971 New Models of the Real-Number Line. *Sci. Am.* (August), pp. 92–99.

STEINITZ, E.

1910 Algebraische Theorie der Körper. *J. Reine Angew. Math.* **137**, 167–309.

STOLL, R. R.

1961 *Set Theory and Logic*, W. H. Freeman, San Francisco, California, 474 pp.

VAN HEIJENOORT, J.

1967 (editor) *From Frege to Gödel—A Source Book in Mathematical Logic, 1879–1931*, Harvard University Press, Cambridge, Massachusetts, 660 pp.

VAUGHT, R. L.

1954 Applications of the Löwenheim–Skolem–Tarski Theorem to Problems of Completeness and Decidability. *Indagationes Math.* **16**, 467–472.

VOROS, A.

1973 Introduction to Nonstandard Analysis. *J. Math. Phys.* **14**, 292–296.

WANG, H.

1963 *A Survey of Mathematical Logic*, Science Press, Peking, and North-Holland, Amsterdam, 651 pp.

WHITEHEAD, A. N., and RUSSELL, B.

1910 *Principia Mathematica*, Vol. 1, Cambridge University Press, Cambridge, 666 pp.

ZAKON, E.
1969 Remarks on the Nonstandard Real Axis. *Applications of Model Theory to Algebra, Analysis, and Probability*, edited by W. A. J. Luxemburg, Holt, Rinehart and Winston, New York, pp. 195–227.

ZERMELO, E.
1904 Beweis, Dass Jede Menge Wohlgeordnet Werden Kann. *Math. Annalen* **59**, 514–516.
1908 Untersuchungen über die Grundlagen der Mengenlehre I. *Math. Annalen* **65**, 261–281.

ZULAUF, A.
1969 *The Logical and Set-Theoretical Foundations of Mathematics.* Oliver and Boyd, Edinburgh, 259 pp.

Symbol Index

Subject Index